D.W. Lawlor

Photosynthesis:
Molecular, Physiological
and Environmental Processes

Second e

Longman Scientific & Technical
Longman Group UK Limited,
Longman House, Burnt Mill, Harlow
Essex, CM20 2JE, England
and Associated Companies throughout the world

First published 1987
Reprinted 1990
Second edition 1993

British Library Cataloguing in Publication Data

A catalogue record for this book is available from the British Library

ISBN 0-582-086574

Library of Congress Cataloging-in-Publication Data
Lawlor, D. W. (David W.). 1941–
    Photosynthesis : molecular, physiological, and environmental
processes / D.W. Lawlor. — 2nd ed.
        p.   cm.
    Includes bibliographical references and index.
    1. Photosynthesis. I. Title.
QK882.L37   1993
581.1′3342--dc20

Set by 4 in 10/12 pt Plantin
Printed in Hong Kong
WC/01

Contents

Preface to second edition

Understanding of photosynthesis has developed rapidly since preparation of the material for the first edition. Several developments have advanced the subject significantly. Crystallization of the photosynthetic reaction centres of a purple bacterium and high resolution X-ray analysis of the structure has greatly increased understanding of the primary events in photosynthesis; for this work Deisenhofer, Michel and Huber and their co-workers from the Max Planck Institute for Biophysics, Germany were awarded the Nobel Prize for Chemistry in 1988. Also, another molecular structure, that of ribulose bisphosphate carboxylase-oxygenase was analysed in detail by X-ray spectrography, providing detailed molecular understanding of the mechanisms of the enzyme functions. Yet another development has been a description of the genetic map of chloroplast DNA and the location of genes for chloroplast constituents. This has greatly increased appreciation of how the complex activities of the chloroplast are organized.

At the other end of the size and time scale is the problem of global climate change, associated with accumulation of CO_2 in the atmosphere. Increased public awareness of this potentially enormous problem demands of plant scientists both objective appraisal of the magnitudes of the effects on vegetation and also suggestions for its control; it will be a test of our ability to use photosynthesis in all its aspects, to reduce the burden of CO_2 in the atmosphere. Perhaps, photosynthetic energy sources will become, eventually, the way societies will fuel their energy requirements.

Changes in this edition reflect these developments and are also an attempt to rectify omissions as well as errors and misplaced emphases. I thank the many people who gave valuable comments and criticism, including Professors Govindjee, Sharkey, Schnarrenberger and D.-P. Häder; the latter contributed much to the ideas in the updated German edition (Thieme Verlag). Also reviewers who, generally, had positive comments encouraged the development of this edition. My wife deserves special thanks for her practical effort throughout as well as essential encouragement. I also thank the staff of Longman for their encouragement and tolerance during preparation of this edition.

D.W. Lawlor
Rothamsted Experimental Station
March 1992

Preface to first edition

This book provides a simplified description of the partial processes of photosynthesis at the molecular, organelle, cell and organ levels of organization in plants, which contribute to the complete process. It considers how photophysics and biochemistry determine the physiological characteristics of plants and production of plant dry matter. The text links the fundamentals of light capture by pigment molecules to the generation of high energy organic molecules and their consumption in carbon dioxide, nitrate and sulphate reduction. The mechanisms are related to the structure and function of the leaf and to control of energy and material fluxes. Photosynthesis in leaves is analysed as the resultant of light activation, biochemical demand, and the supply of CO_2. Photosynthetic processes are related to plant environment, sun and shade leaves for example, and C4 and CAM photosynthesis are analysed as ecophysiological variants of a basic process. Leaf photosynthesis is put into context as one, albeit the major, determinant of productivity of vegetation. Plant productivity is the result of the interaction of many sub-systems, all driven by the primary energy capture. Photosynthesis by the whole plant is to be seen as a series of balanced but dynamic interactions between individual molecular and physiological mechanisms.

The text is intended for undergraduate and graduate study in plant biology courses and for non-specialists in other disciplines who wish to understand photosynthetic mechanisms and control. The approach is qualitative, but there is some emphasis on quantitative aspects to encourage progress towards more rigorous analysis of the photosynthetic system, for example by modelling and systems analysis. References are mainly from the secondary or review literature, from which detailed arguments and the primary literature may be obtained. Several recent textbooks have considered more specialized aspects of photosynthesis and there are many excellent reviews; this book is intended to point the reader to them. I hope that this text will enable readers to appreciate the hierarchy of organization that enables plants to produce dry matter within the confines of the environment. With current exciting possibilities for genetically engineering plants for specific purposes, to give greater dry matter production or improved growth in particular environments for example, it is essential that the role of molecular events in the functioning of the whole system in relation to environment should be well understood.

I extend my thanks to all those who have helped in producing this book; to Professor Terry Mansfield who suggested that I should write it and Professor Dennis Baker who gave much valued editorial advice, suggestions and support throughout. Longman I thank for the opportunity to attempt this project and for their often tried patience and support. Dr Alfred Key's suggestions and constructive criticisms on a large part of the text were very valuable. Also Dr Stephen Gutteridge, Professor Peter Lea, Dr Keith Parkinson and Dr Roger Wallsgrove contributed ideas, information and comments on parts of the text. I have gained from their efforts but the omissions, distortions and factual mistakes are mine. Many people have given me much appreciated help, support and encouragement to explore the complexities of photosynthesis and plant functions, my parents and amongst other Professors A.J. Rutter, P.J. Kramer, C.P. Whittingham, H. Fock, C.B. Osmond and G. Farquhar.

In production of the manuscript Mrs Anita Webb and Janet Why and my wife gave excellent technical help with infinite patience over a long period; I am most grateful. The unstinting support of Gudrun, Kirsten and Kurt is beyond thanks; they deserve better reward.

D.W. Lawlor
Rothamsted Experimental Station
June 1985

CHAPTER *1*

Introduction to the photosynthetic process

'Life is woven out of air by light' — I. Moleschott

Photosynthesis is the process by which living organisms convert the energy of light into the chemical energy of organic molecules. This process exploits solar energy (inexhaustible even on the scale of evolutionary processes) to provide the energy for the complex physico-chemical reactions of living organisms. Photosynthesis provides the energy for the whole of the living world. In a simplified form the role of photosynthesis may be written:

$$\text{low energy inorganic chemical state} + \text{light energy} \xrightarrow[\text{organism}]{\text{photosynthetic}} \text{high energy chemical state}$$

$$[\mathbf{1.1}]$$

Sunlight is the ultimate energy source for all biological processes on earth. Other sources of energy, such as inorganic molecules, e.g. hydrogen sulphide (H_2S), may be broken down and the energy exploited; these sources were important — perhaps dominant — in earth's early history. However, they are finite and, although renewable by geochemical processes, of limited long-term importance. For much of biological time the exploitation of light to change matter from a lower to a higher energy state has been essential for life. Energy is needed to rearrange electrons in molecules and to synthesize chemical bonds, but a complex process may not take place spontaneously and a mechanism is required; this book examines the mechanisms and how they function to capture energy and transduce it to form complex biochemical products from simple inorganic molecules.

According to the law of thermodynamics, biological processes will tend to go from a high energy to a low energy condition losing energy in the process until equilibrium is achieved — this is the state of death and decay — unless energy is available to drive and maintain the reaction in the reverse direction. Living organisms are in an unstable thermodynamic state and require energy to keep chemical constituents in a highly ordered condition and to do work against the thermochemical energy gradient, in accumulating matter, such as ions or gases from the environment, or to grow, move, etc., all of which characterize the living state in contrast to the world of inanimate matter.

The movement of matter, chemical interconversion or changes in energy state cannot proceed with absolute efficiency, and involve the loss of some of the energy, usually as heat, at a temperature very close to that of the environment, so that it cannot be used to do other work, and the energy is finally radiated to the universe. Once a biological system has accumulated free energy it can convert it to different chemical forms or into physical energy or exchange it between organisms, etc., but with time the useful energy will be lost and thermodynamic equilibrium (i.e. death) will be attained.

Without continuous supply of high energy 'food' living organisms cannot survive. All non-photosynthetic organisms, such as animals, fungi and bacteria are dependent upon preformed materials. Some organisms, principally bacteria, can utilize the energy of bonds in inorganic molecules as an energy source. However only those organisms able to use the supply of energy from the sun can increase the total free energy of living material and are independent of the limitations imposed by other energy sources. Photosynthesis by some bacteria, blue–green algae (also called cyano-bacteria), algae and higher plants is achieved by a mechanism able to capture the fleeting energy of a light particle and make it available to biochemistry. Given an abundant supply of sunlight they can survive, grow and multiply using only inorganic forms of matter readily available in their environment.

Sunlight is the only form of energy which adds to the total energy supply of the earth and drives not only the weather and geochemical events, but also the biological cycles. Solar energy dominates the earth although geochemical processes also contribute to the energy balance. The earth is bathed in a sea of energy in the form of electromagnetic waves, differing in wavelength and energy, derived from the thermonuclear reactions in the sun. Short wavelength radiation, such as X-rays, is highly energetic and may destroy complex molecules by ionization. Ultraviolet light, of wavelengths greater than X-rays but shorter than visible light, breaks bonds within organic molecules and destroys many biological tissues. Infra-red radiation, of longer wavelength than visible light and of lower energy, causes chemical bonds to stretch and vibrate but is not very active in biological processes; however, it is important in the energy (heat) balance of the biosphere. The energy of visible light is sufficient to cause changes in the energy states of the valency electrons of many molecules and can be used by living organisms to effect the transition from a low to a high energy state. Molecules which absorb visible light and are relatively stable, function to transduce physical energy to chemical form, allowing the evolution of complex organic molecular 'living' systems, using light as their ultimate energy source.

Light was exploited early in evolution as a source of energy to drive biological processes. Considering the need for a continuous supply of energy, it is perhaps not surprising that sunlight provides the energy used in the biosphere and is the basis of life. Light also serves to control many biological processes, e.g. day length regulates development of many plants, but this does not add significantly to the global energy and does not drive the major energy fluxes in the biosphere. The physical characteristics of light and of the molecules with which it reacts, are crucial to the process of capturing energy and will be considered in Chapter 2.

Concepts of photosynthesis

Photosynthesis may be generalized as the capture of the energy of a photon of light by a pigment molecule, the formation of an electronic excited state, the use of this 'excited electron' to reduce a chemical substance and to form 'energy-rich' molecules such as adenosine triphosphate. These reduced substances and energy-rich molecules are used to form other, complex organic molecules. The electron lost from the excited pigment is replaced by an electron from another source in the environment.

Pigment (P) + light → excited pigment (P*)

$$P* + \text{acceptor molecule (A)} \rightarrow P \text{ minus electron (P}^+) +$$
$$A \text{ plus electron (A}^-) \rightarrow P^- \text{ addition of electron} \rightarrow P \qquad [1.2]$$

In photosynthetic bacteria many compounds may donate electrons to the oxidized pigment P^+, e.g. the sulphur bacteria use H_2S and liberate elemental sulphur (S):

$$H_2S + \text{light} + \text{bacteriochlorophyll} \rightarrow S + 2H^+ + 2e^- \qquad [1.3]$$

Cyanobacteria (blue–green algae), algae, bryophytes, ferns, gymnosperms and angiosperms use water as the source of electrons and oxygen (O_2) is released:

$$2H_2O + \text{light} + \text{chlorophyll} \rightarrow O_2 + 4H^+ + 4e^- \qquad [1.4]$$

In both types of photosynthetic system H^+ accumulates in the photosynthetic mechanism forming a proton concentration gradient, the energy of which is used to produce ATP. The energized electron may be used to reduce, ultimately, a range of inorganic substances. One of the most important is carbon dioxide (CO_2), leading to the synthesis of sugars and other organic molecules based on carbon; the basic process is $A^- + H^+ + CO_2 \rightarrow (CH_2O)$ where the H^+ 'follows' the electronically negative state to balance the charges and comes from the environment.

In plants using water to supply electrons, the reaction is:

$$CO_2 + H_2O + \frac{\text{light}}{\text{energy}} \xrightarrow[\text{containing plants}]{\text{chlorophyll-}} (CH_2O) + O_2 + \frac{\text{chemical}}{\text{energy}} \qquad [1.5]$$

With the formation of glucose ($C_6H_{12}O_6$) the energy required is

2879 kJ mol^{-1}. This reaction requires energy in the form of 'high energy bonds' of the phosphorylated compound adenosine triphosphate (ATP) and reducing power as the reduced pyridine nucleotide, nicotinamide adenine dinucleotide phosphate (NADPH); these are synthesized by complex biochemical processes driven by light energy. The transformations of energy and material require many individual chemical steps (perhaps thousands), if the processes required to form the whole organism are counted, as they must be in a complex system.

ATP and reductant are also used to assimilate other inorganic compounds. Nitrate ions (NO_3^-) are reduced to ammonia, which is then consumed in the synthesis of amino acids:

$$NO_3^- + 9\,H^+ + 8\,e^- \xrightarrow[\substack{\text{enzymes (nitrate and} \\ \text{nitrite reductase)}}]{\text{light, chlorophyll,}} NH_3 + 3\,H_2O \qquad [1.6]$$

Photosynthetic bacteria and blue−green algae, but not higher algae or plants, assimilate atmospheric nitrogen to form ammonia:

$$N_2 + 6\,H^+ + 6\,e^- \xrightarrow[\text{enzyme (nitrogenase)}]{\text{light, chlorophyll,}} 2\,NH_3 \qquad [1.7]$$

Sulphate ions are also reduced before entering metabolism:

$$SO_4^{2-} + 9\,H^+ + 8\,e^- \xrightarrow[\text{enzymes}]{\text{light, chlorophyll,}} HS^- + 4\,H_2O \qquad [1.8]$$

Many algae, in the absence of oxygen, are able to produce hydrogen gas from water using light energy captured by chlorophyll; the enzyme hydrogenase catalyses the reaction.

$$H_2O \xrightarrow[\text{hydrogenase}]{\text{light, chlorophyll,}} H_2 + \tfrac{1}{2}O_2 \qquad [1.9]$$

These examples (considered in more detail later) show that photosynthesis is more than the assimilation of CO_2 with the production of oxygen but is a process with many possible products and capable of being used biologically in many ways.

Photosynthesis as an oxidation−reduction process

All the photosynthetic processes summarized in eqns 1.2−1.9 involve oxidation and reduction. Reduction is the transfer of an electron (e^-) or electron plus proton (H^+) from a donor (D) molecule to an acceptor (A); the donor is oxidized and the acceptor reduced. An electrically neutral compound becomes negatively charged and may accept a proton from water to restore electrical neutrality:

Donor D + acceptor A → D$^+$ + A$^-$;   A$^-$ + H$^+$ → AH;

D$^+$ + e$^-$ → D   [1.10]

Van Niel established that photosynthesis in all organisms conforms to donor + acceptor reacting under the influence of light in the presence of pigment and the necessary mechanism. The oxidized donor molecule may accumulate in the environment, e.g. oxygenic photosynthesis is responsible for oxygen in the atmosphere. Biological reduction−oxidation (redox) reactions are usually catalysed by enzymes. Examples in plant metabolism are the reduction of oxaloacetic acid to malic acid by malate dehydrogenase.

Reduction and oxidation reactions are of fundamental importance to our understanding of the mechanisms of photosynthesis. The primary reaction of photosynthesis, linking the physical energy of chlorophyll molecules excited by light with biochemical processes, is the transfer of electrons from a special form of chlorophyll to an acceptor molecule driven by the energy captured. The acceptor is reduced and the special form of chlorophyll oxidized. Electrons are then donated from different sources to the oxidized chlorophyll, reducing it and allowing the process to be repeated (Fig. 1.1). The generation of the reduced acceptor and energy-rich compounds, NADPH and ATP, respectively, requires light; these reactions are therefore called 'light reactions'. NADPH and ATP are consumed to reduce other inorganic substances and to convert them into organic molecules. Given a supply of these compounds the chemical reactions of photosynthesis may proceed without light and are thus called 'dark reactions', but they can

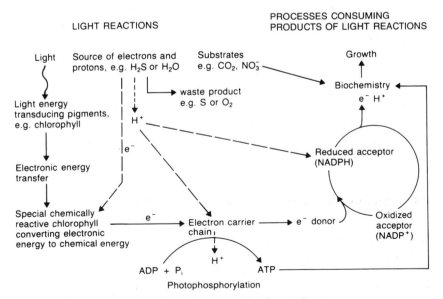

FIG. 1.1 Essential features of photosynthesis in all organisms.

and do proceed in both light and darkness. The term 'non-light requiring reactions' expresses this important point.

Energy and electron transport

In photosynthesis, electrons pass from a donor to an acceptor, the two forming a redox pair or couple. The ability of electrons to transfer is determined by the energy required, called the redox potential. The redox potential is determined by comparison with a standard hydrogen electrode which has, by definition, a voltage (E_0) of 0 volts under standard conditions and pH of 0. Biological reactions take place in solution close to pH7, so that the redox potential of biological redox substances is measured at pH7 and is called E'; E' is -0.42 V compared to E_0. The relation between redox potential and concentration of oxidized and reduced substances, [ox] and [red], is given by:

$$E' = E_0 + \frac{RT}{nF} \ln \frac{[ox]}{[red]} \qquad\qquad [1.11]$$

when n is the number of reducing equivalents (e^-) transferred, R is the gas constant, T the temperature and F the Faraday constant ($F = 96\,485$ C mol^{-1}). Redox potentials depend on concentration and are given as a midpoint potential when the two forms are equal in concentration.

Substances with more negative redox potential are energetically able to reduce (donate e^- to) others of more positive potential; it is impossible in the reverse direction unless energy and a mechanism are available. The maximum useful energy from a reaction is given by the difference in redox potential ΔE; the free energy of a reaction, ΔG, is related to redox potential by $\Delta E = -\Delta G/nF$.

Nature of light-gathering process and generation of reductant and ATP

Equation 1.2 includes a pigment; in eqns 1.3 and 1.4 this is identified as bacteriochlorophyll (bchl) and chlorophyll (chl), respectively but ignores the many complex oxidation–reduction processes. Absorption of light by bchl or chl leads to ejection of an electron from a special form of the pigment; the electron is then captured in a chemical form as a reduced acceptor (a quinone molecule) in an electron transport chain. Electrons pass from a higher to lower energy state along a chain of electron acceptors and donors, which are alternatively reduced and oxidized in the process, until a stable reduced compound is formed. In higher plants this is ferredoxin, which passes electrons to inorganic compounds, such as nitrate ions, or to secondary reductants, such as oxidized pyridine nucleotide (NADP$^+$). Energy transformations in living organisms take place at nearly isothermal conditions and involve multiple steps with rather small energy changes between them, which are rather easier to control than large 'jumps', and contribute to achieving optimum rates of processes and efficiency. It is not possible to obtain both maximum rates and efficiency in a system; organisms may have evolved to maximize the energy output at rather lower efficiency.

Energy is required to carry out the biological catalytic reactions leading to the reorganization of the primary reactants into complex products and is provided by the hydrolysis of ATP or other related phosphorylated adenylate compounds. ATP has three phosphate groups. When the bond joining the terminal group is hydrolysed it provides energy and releases phosphate, which may be donated to other compounds, activating them; phosphorylation is an essential step in many biochemical reactions.

$$ATP + H_2O \rightarrow ADP + P_i + energy\ (-31\ kJ\ mol^{-1}) \qquad [1.12]$$

ATP is resynthesized by photophosphorylation, a process in which the movement of electrons along the chain of electron carriers also pumps H^+ across the cell membrane (in bacteria) or the chloroplast thylakoid membranes (in other photosynthetic organisms), producing a concentration gradient of H^+ which is coupled to ATP synthesis (Ch. 6). The essential nature of the photosynthetic process is therefore a chemical oxidation–reduction reaction driven against the thermodynamic energy gradient by the energy of light captured by pigments. The light-transducing mechanism generates reductant in the form of organic molecules which pass it to biological reactions, themselves independent of the direct effect of light. In addition, light energy drives the synthesis of ATP, which is also essential for biochemical reactions. The study of photosynthesis considers the atomic and molecular processes underlying capture of light energy and conservation, and the relation between the production of energetically-favourable compounds and the assimilation of inorganic molecules. Ultimately, the fundamental processes of photosynthesis are related to performance of the organism at an ecological level.

Occurrence of photosynthesis

Photosynthesis is the only biological process that accumulates energy. It occurs in organisms as diverse as bacteria and trees (Table 1.1) and is similar in all, with light capture by a pigment and conversion of the energy to chemical form. Differences between organisms reflect their evolution; the greatest difference is between the photosynthetic bacteria, a very diverse group unable to oxidize water, and blue–green algae, algae and higher plants which oxidize water and evolve O_2.

The physical nature of light, but not its intensity or spectral distribution, has been a constant feature of the environment since the origin of the earth, and it is to be expected that the different pigment systems for light capture share many features; bchl and chl occur in several forms but are basically similar in all organisms and related biosynthetically with light absorption

Table 1.1 Main groups of photosynthetic organisms, their structure and photosynthetic characteristics

Prokaryotes	Eukaryotes
Single cell or little complexity	Mainly multicellular, complex intercellular interactions
Cell nucleus without membrane	Cell nucleus with membrane
Photosynthesis in vesicular membranes not in discrete compartment	Photosynthesis in vesicular membranes in discrete compartment — chloroplast
Anoxygenic forms	Anoxygenic forms: none
Examples: Photosynthetic bacteria (purple sulphur — Rhodospirillaceae, green sulphur — Chromatiaceae)	
Process Do not evolve O_2. Many obligate anaerobes. Source of reductant hydrogen sulphide, sulphur, thiosulphite, hydrogen, organic compounds, never water. Can reduce gaseous nitrogen	
Oxygenic forms	Oxygenic forms
Examples: Blue–green algae	*Examples:* Algae (green — Chlorophyceae, red — Rhodophyceae) Higher plants (Bryophyta, Angiospermae)
Processes: Evolve O_2. Source of reductant water. Reduce gaseous nitrogen	*Processes:* Evolve O_2. Source of reductant water. Do not reduce gaseous nitrogen

and transport of electrons taking place in membranes. In photosynthetic bacteria there is only a single light-driven process, whereas in all other organisms there are two, one of which oxidizes water. Bacteria use a variety of substrates to provide reductant suggesting that many substrates were available at early stages of evolution but later the abundance of water, and perhaps shortage of other sources of reductant, made it the preferred source.

Electron transport coupled to ATP synthesis was an early feature of photosynthetic systems, requiring a closed membrane vesicle separating regions of high and low proton concentration which are needed for ATP generation. The gradient is from high proton concentration, $[H^+]$, outside to low $[H^+]$ inside the photosynthetic bacterial cell (light energy is used to pump H^+ out of the cell) and from high $[H^+]$ in the thylakoids to low

[H$^+$] in the cytosol of blue—green algae or in the chloroplast stroma of higher plants. Thylakoid membranes, although internal, are equivalent to the external membrane of bacterial cells. The space within the thylakoid is equivalent to the medium surrounding a photosynthetic bacterium. Thus, the mechanisms driving ATP synthesis are comparable despite the great structural differences between groups. Photosynthetic bacteria are important as relatively simple organisms for analysis of photosynthesis and for indicating evolutionary process.

Types and structures of photosynthetic organisms

As Table 1.1 indicates there is a fundamental distinction between prokaryotes and eukaryotes with photosynthesis occurring in both but with anoxygenic and oxygenic forms occurring in the prokaryotes and oxygenic only in the eukaryotes. There are also many structural and functional differences within these main classifications.

Photosynthetic bacteria

The simplest known form of 'photosynthesis' is that of the halophilic bacterium *Halobacterium halobium* which has patches of purple pigment-protein molecules, bacteriorhodopsin, in its outer membrane. This pigment captures light and the energy drives transport of H$^+$ from inside the cell (Fig. 1.2a) to oustide; the pH gradient so formed provides (together with the electrical charge on the membrane) the energy for synthesis of ATP when the H$^+$ diffuses back into the cell through an ATP synthesizing enzyme called coupling factor. There is no electron transport associated with this type of photosynthesis.

Other photosynthetic bacteria (Fig. 1.2b) have a rather more complex form of photosynthesis in which electrons are transported and used to drive the formation of the pH gradient for ATP synthesis. The outer cell membrane is invaginated forming extensive multilayered membrane systems containing the photosynthetic pigments; these membranes are only indirectly in contact with the external medium. The bchl pigment and associated proteins of the photosystems capture light energy which leads to the formation of a reduced acceptor. At the same time the electrons reduce quinones in the membrane which carry H$^+$ from the cytosol to the external medium, in this case the space bounded by the membranes; this is equivalent to the cell exterior. A pH gradient results and the diffusion of H$^+$ back into the cell through a coupling factor drives ATP synthesis within the cell.

The prokaryotic cyanobacteria (blue—green algae) have oxygenic photosynthesis and a photosynthetic membrane structure resembling the more complex bacteria, without chloroplasts but with an extensive membrane system arranged around the cell periphery in parallel sheets. The red algae (Rhodophyta) are similar. The membranes have regularly arranged

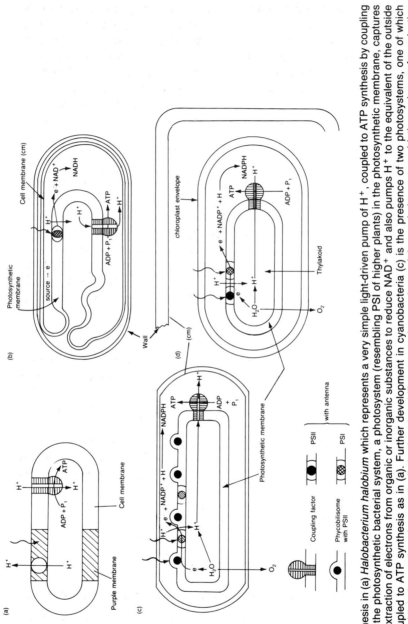

FIG. 1.2 Photosynthesis in (a) *Halobacterium halobium* which represents a very simple light-driven pump of H$^+$, coupled to ATP synthesis by coupling factor enzyme. In (b) the photosynthetic bacterial system, a photosystem (resembling PSI of higher plants) in the photosynthetic membrane, captures light and drives the extraction of electrons from organic or inorganic substances to reduce NAD$^+$ and also pumps H$^+$ to the equivalent of the outside of the cell; H$^+$ is coupled to ATP synthesis as in (a). Further development in cyanobacteria (c) is the presence of two photosystems, one of which has an arrangement of light-harvesting pigments in distinctive phycobilisomes on the membrane. It is coupled to water oxidation, using e$^-$ for reduction of NADP$^+$, H$^+$ for ATP synthesis and releasing O$_2$. The higher plant system (d) of the thylakoids has photosystems I and II embedded in the photosynthetic membrane — oxidation of water, H$^+$ pumping and ATP synthesis.

particles, called phycobilisomes (Fig. 1.2c) alternating on each side of the membranes which form a continuous system, the inside of which is equivalent to the external medium. The phycobilisomes contain a complex of pigments linked to the reaction centre. There are two types of photosystems in the membranes, one associated with electron transport (PSI), and the other with water oxidation and electron transport (PSII). As with the bacteria, transport of electrons leads to synthesis of reductant and also to formation of a pH gradient which drives the synthesis of ATP.

In the eukaryotic algae, bryophytes, ferns, gymnosperms and angiosperms, all with oxygenic photosynthesis, the photosynthetic membranes are enclosed in an envelope in a body called the chloroplast. The inside of the membranes (thylakoids) is equivalent to the outside of the cell (Fig. 1.2d) and the membranes are differentiated into stacked (granal) and unstacked (stromal) forms in many of these plants. There are two photosystems as in the cyanobacteria but of different form, particularly the light gathering pigments, without the phycobilisomes, and these are buried in the membrane not on its surface. However, as in the cyanobacteria, electron transport leads to the reduction of an acceptor, the formation of the pH gradient and synthesis of ATP via the coupling factor in the chloroplast.

Photosynthesis in relation to other plant functions

Respiration of the assimilates produced in the photosynthetic reactions provides a way of using the energy of light in periods of darkness or in parts of the organism not exposed to light. It is also exploited by heterotrophic organisms which cannot feed themselves but rely on photosynthetically competent ones to provide energy.

If the respiratory and photosynthetic processes are compared with the starting materials on the left-hand side of eqn 1.13 and the products on the right-hand side, the net result is energy conversion and closed cycles of carbon and water.

$$\text{Photosynthesis: } CO_2 + H_2O + \text{light energy} \xrightarrow[\text{catalysts}]{\text{biological}} (CH_2O) + O_2$$

$$\text{Respiration: } (CH_2O) + O_2 \xrightarrow[\text{catalysts}]{\text{biological}} CO_2 + H_2O + \text{heat energy} \qquad [1.13]$$

Net result of respiration and photosynthesis: light energy $\rightarrow$ heat energy

Thus, photosynthesis and respiration work in opposition, forming a cyclic, closed system for matter. The physical energy of light is converted into chemical energy and ultimately heat.

Photosynthesis and respiratory processes are very carefully regulated by complex biochemical mechanisms which link the processes in order that the products of photosynthesis are not consumed in a futile cycle of assimilation/respiration and so that cell, organ and organism growth are achieved in such a way that the organism is capable of continued growth and reproduction. Plants are composed of individual subsystems (cells, organs and processes) which are controlled by their internal conditions and outside factors but are highly integrated. Photosynthesis is the driving factor for all processes, yet is related to the other cellular functions in complex ways which are not yet well understood.

Evolution of photosynthesis

As a biochemical process based on proteins and organic molecules which are rapidly decomposed, photosynthesis has left little direct trace in the geological record. Yet enough is now known of the comparative biochemistry to suggest a plausible hypothesis of the development of this essential process. Here only the barest outline is attempted, and articles by Schopf (1978) and Bendall (1986) should be consulted. The earth was formed (Fig. 1.3) some 4.6×10^9 years ago and for the first 0.5×10^9 years cooled and solidified. Because of the earth's distance from the sun and its size, which

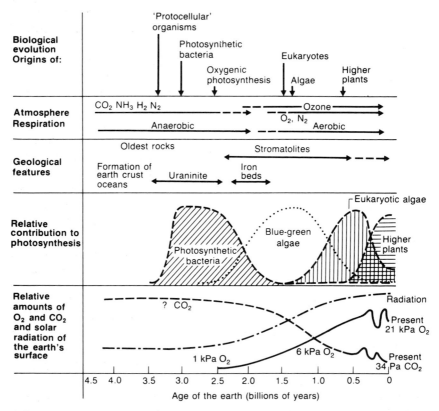

FIG. 1.3 Probable sequence of events in evolution of photosynthesis and the relation to some other geological and biological processes; highly schematic.

determined the heat received and the force holding gases on the surface, both liquid water and an atmosphere were retained. The primitive atmosphere was highly reducing, with methane (CH_4), H_2, H_2S, CO_2, CO, NH_3, etc., but there was no free O_2; it was anoxic. This is considered crucial to the evolution of life because oxygen destroys organic molecules. Also there was no ozone, which today forms a thin layer in the upper atmosphere, absorbing ultraviolet (UV) rays. This energetic form of radiation, together with high temperatures and abundant gases evolved by volcanic activity, provided conditions for synthesis of organic molecules such that the primitive oceans resembled 'hot dilute soup' to quote Haldane. How self-replicating biological systems were formed and evolved under these conditions is beyond the scope of this book and is still unresolved. However it occurred, organisms resembling present-day bacteria in size and cell structure are present in rocks 3.5×10^9 years old. A form of light-driven metabolism probably developed early, as derivatives of carotenoids have been detected in rocks of that age, although contamination by later material is possible. A primitive organism might have synthesized ATP by a light-driven proton pump such as the bacterium *Halobacterium halobium*. Organic carbon from deposits of this age show discrimination against the heavy isotope of carbon (^{13}C; see p. 251) suggesting some form of photosynthesis occurred very early in evolution. Such organisms would have been independent of the supply of preformed ATP and have had an evolutionary advantage by exploiting an almost limitless source of energy.

Many processes were linked to the light reactions and electron flow and ATP, including N_2, CO_2 and S assimilation. However, water was not split in early photosynthesis until 3.5×10^9 years ago so no photosynthetic O_2 was formed and the atmosphere was reducing. It was only after evolution of the water-splitting process, driven by light through two linked photosystems, that water could be oxidized and O_2 (the waste product) escaped to the atmosphere. This time scale is suggested by much evidence (Schopf 1978). Extensive beds of fossil limestone, called stromatolites, were formed 3×10^9 years ago and contain blue−green algae as do present-day stromatolites. However, oxygen production probably started before that time. Many geochemical processes would have consumed O_2; for example, iron (Fe^{2+}) reacts to form insoluble Fe_3O_4. Oceans would have been slowly depleted of Fe^{2+} with deposition of iron ores. This may be the origin of the 'red beds' formed $2.2 \times 10^9 - 1.7 \times 10^9$ years ago. To borrow Schopf's expression 'the world's oceans rusted'. A non-photosynthetic origin of O_2 by UV-splitting of water was too slow for such a massive chemical process. Uraninite (UO_2) is a uranium ore, insoluble at O_2 concentrations above 1 per cent; no deposits younger than 2×10^9 years have been found. Thus, between 3.5×10^9 and 3.0×10^9 years ago photosynthesis developed using H_2O as source of reductant and atmospheric O_2 increased. Deposition of reduced carbon (deduced from the change in $^{13}C/^{12}C$ ratio of ancient carbon deposits) may also have contributed to the removal of

C and the increase in O_2 in the atmosphere: CO_2 concentration may have been many hundred-fold greater than today. By 1.5×10^9 to 1.0×10^9 years ago aerobic conditions were established as the chemical buffers were exhausted and O_2 exceeded 1 kPa. Oxygen in the upper atmosphere formed an ozone layer, which absorbed UV radiation, allowing evolution of higher organisms and invasion of the land. Oxygen, by acting as a terminal receptor for respiratory processes, greatly (ten-fold) increased the amount of energy to be obtained by respiration of organic substances. Most of the present-day biota, including man, owe their existence to the waste product of photosynthesis!

Eukaryotes, with large internally compartmented nucleated cells, may have evolved early in earth's history, although evidence, e.g. the presence of steranes (molecules derived from sterols thought only to be made by nucleated, i.e. eukaryotic, cells) in rocks, only supports an age of 1.7×10^9 years and multicellular eukaryotic seaweeds occur in strata 1.4×10^9 years old. They evolved rapidly from about 1×10^9 years ago, forming multinucleate, macroscopic organisms (both plants and animals) perhaps linked to changes in the climate about 900−600 million years ago associated with tectonic and volcanic activity and large loss of C by burial in sediments and the onset of global climate changes (including glaciation). Evidence from the structure and function of the nucleic acids of chloroplasts and mitochondria of higher plants suggests that these organelles arose from the invasion of non-photosynthetic eukaryotic cells by bacteria and blue−green algae. Photosynthesis evolved more complex biochemistry, separation of respiration and photosynthesis and their regulation. Photosynthesis shaped the biosphere both directly and through its effects on the earth's climate and geology. Carbon from photosynthesis was sequestered in oil, coal and gas decreasing the atmospheric CO_2 and increasing the ratio of O_2 to CO_2. This may have been unfavourable for photosynthesis as the CO_2-fixing enzyme ribulose bisphosphate carboxylase is less efficient under these conditions. On land, prevention of water loss from the plant by a thick cuticle also reduced supply of CO_2. Evolution of different types of photosynthesis, C4 and CAM (Chapter 9), based on the earlier metabolic forms but of greater efficiency, is probably a response to an environment of decreasing CO_2/O_2 ratio and drier atmosphere, with intense radiation. Present human activity is increasing the CO_2 concentration of the atmosphere by burning fossil fuels. This may improve plant growth in the short term but it will also affect the world's climate (p. 234). Humankind's future is closely linked to the photosynthetic production of food, fuel and fibre and increasing population will lead to greater demands on the efficiency of the process.

References and Further Reading

Arnon, D.I. (1977) Photosynthesis 1950−75: Changing concepts and perspectives, pp. 7–56 in Trebst, A. and Avron, M. (eds), *Encyclopedia of Plant Physiology* (N.S.), Vol. 5, *Photosynthesis I*, Springer-Verlag, Berlin.

Bendall, D.S. (ed.) (1986) *Evolution from Molecules to Men*, Cambridge University Press, Cambridge.

Benedict, C.R. (1978) Nature of obligate photoautotrophy, *A. Rev. Plant Physiol.*, **29**, 67–93.

Briggs, W.R. (ed.) (1989) *Photosynthesis*, Allan R. Liss, Inc., New York.

Budyko, M.I. (1974) *Climate and Life* (English edn, ed. D.H. Miller), Academic Press, New York.

Clayton, R.K. (1980) *Photosynthesis: Physical Mechanisms and Chemical Patterns*, I.U.P.A.B. Biophysics Series, Cambridge University Press, London.

Cogdell, R. and **Malkin, R.** (1992) An introduction to plant and bacterial photosystems, pp. 1–15 in Barber, J. (ed.), *Topics in Photosynthesis Vol. II, The Photosystems: Structure, Function and Molecular Biology*, Elsevier, Amsterdam.

Danielli, J.F. and **Brown, R.** (eds) (1951) Carbon dioxide fixation and photosynthesis, *Symp. Soc. Exp. Biol.*, V, Cambridge University Press, Cambridge.

Danks, S.M., Evans, E.H. and **Whittaker, P.A.** (1983) *Photosynthetic Systems: Structure, Function and Assembly*, Wiley, New York.

Davis, G.R. (1990) Energy for planet earth, *Sci. Amer.*, **263**, 20–27.

Dennis, D.T. and **Turpin, D.H.** (1990) *Plant Physiology, Biochemistry and Molecular Biology*, Longman, Harlow.

Dose, K. (1983) Chemical evolution and the origin of living systems, pp. 912–24 in Hoppe, W., Lohmann, W., Markl, H. and Ziegler, H. (eds), *Biophysics*, Springer-Verlag, Berlin.

Giese, A. (1964) *Photophysiology, 1, General Principles: Action of Light on Plants*, Academic Press, New York.

Gifford, R.M. (1982) Global photosynthesis in relation to our food and energy needs, pp. 459–95 in Govindjee (ed.), *Photosynthesis, 2, Development, Carbon Metabolism and Plant Productivity*, Academic Press, New York.

Gregory, R.P.F. (1989) *Biochemistry of Photosynthesis* (3rd edn), Wiley, Chichester.

Hall, D.O. and **Rao, K.K.** (1981) *Photosynthesis* (3rd edn), Studies in Biology, No. 37, Edward Arnold, London.

Halliwell, B. (1981) *Chloroplast Metabolism: The Structure and Function of Chloroplasts in Green Leaf Cells*, Clarendon Press, Oxford.

Olsen, J.M. and **Pierson, B.K.** (1986) Photosynthesis 3.5 hundred million years ago, *Photosynthesis Res.*, **9**, 251–59.

Porter, G. (1989) Solar energy from photochemistry, pp. 3–7 in Barber, J. and Malkin, R. (eds), *Techniques and New Developments in Photosynthesis Research*, Plenum Press, New York/London.

Rabinowitch, E.I. (1945) *Photosynthesis*, Interscience Publishers Inc., New York.

Schopf, J.W. (1978) The evolution of the earliest cells, *Sci. Amer.*, **239**, 84–103.

Van Gorkom, H.T. (1987) Evolution of photosynthesis, pp. 343–50 in Amesz, J. (ed.), *Photosynthesis. New Comprehensive Biology*, Vol. 15, Elsevier, Amsterdam.

Woese, C.R. (1987) Bacterial evolution, *Microbiol. Rev.*, 221–171.

CHAPTER 2

Light — the driving force of photosynthesis

Characteristics of light

Photosynthesis is driven by the energy of light, therefore the physical nature of light and its interaction with matter are described in a much simplified, qualitative way, in order to understand the biological processes. The quantitative characteristics of light in relation to photochemistry are discussed in the books by Clayton (1980) and Hoppe *et al.* (1983). Light is electromagnetic radiation, emitted when an electrical dipole (a paired positive and negative charge, separated by a small distance) in an atom oscillates and causes a change in the field of force. The dipole produces an electrical and a magnetic vector, which are in phase but at right angles. Fluctuations in the field strength of these vectors are perpendicular to the direction of travel of the wave and, hence, light is a transverse wave. The electromagnetic wave is characterized by both wavelength, λ (in metres) which is the distance between successive positive or negative maxima on the sine wave, and by frequency, ν, the number of oscillations per unit time (s^{-1}).

Frequency is determined by the oscillations of the dipole. Wavelength and frequency are related by the velocity of propagation of the wave, v (m s^{-1}):

$$v = \lambda \nu \qquad\qquad [2.1]$$

The velocity of light (c) is 3×10^8 m s^{-1} *in vacuo*. Table 2.1 gives the approximate wavelengths and frequencies of the main groups of electro-magnetic waves. Photosynthesis is driven by radiation of $400-800$ nm and the term 'photosynthetically active radiation' (PAR) is applied to that spectral distribution. Frequencies between 7.5 and 3.8×10^{15} s^{-1} (wavelengths $400-700$ nm) are visible to the human eye and are called light. However, the simplicity of the word and similarity between the wavelengths of visible light and those used in photosynthesis, (but not the response of the processes to them), justify the use of the term 'light' to describe radiation used by photosynthesis.

Table 2.1 Electromagnetic radiation, wavelength, frequency and energy

Type of radiation	Wavelength	Frequency* (s^{-1})	Energy per photon† (eV)	(J)	Energy 1 mol‡ photon (J)
Radiowaves	10^3–10^{-3} m; for 1 m	3×10	1.24×10^{-6}	19.86×10^{-26}	11.96×10^{-27}
Infra-red	800 nm	3.8×10^{14}	1.55	25.16×10^{-20}	15.2×10^4
Visible red light	680 nm	4.4×10^{14}	1.82	29.13×10^{-20}	17.5×10^4
Visible green	500 nm	6.0×10^{14}	2.50	39.72×10^{-20}	23.9×10^4
Visible violet–blue	400 nm	7.5×10^{14}	3.12	49.65×10^{-20}	29.6×10^4
Near ultraviolet	200 nm	1.5×10^{15}	6.25	9.93×10^{-19}	59.5×10^4
Ultraviolet	10 nm	3.0×10^{16}	123.0	19.86×10^{-18}	119.6×10^5
X-rays	0.01 nm	3×10^{19}	1.24×10^5	19.86×10^{-15}	119.6×10^8

* Calculated from eqn 2.1.
† Calculated from eqn 2.2.
‡ Energy of 1 photon multiplied by Avogadro's number of photons.
Note: 1 nm = 10^{-9} m = 10 Ångströms = 1 millimicron (mμ).

Electromagnetic radiation passes through space without matter to transmit it, in contrast to wave propagation in solids, liquids or gases. The wave form of light is shown by interference phenomena and the slower transmission of light in dense media. Light has, however, the characteristics of both wave and particle. Emission of electrons from metal surfaces caused by light, called the photoelectric effect, and radiation of energy from atoms at distinct frequencies rather than as a continuous spectrum, show that light is particulate.

In 1900 Planck resolved the conflict between the wave and particle concepts by considering that light behaves as discrete particles of energy, called quanta, which can only be absorbed or emitted by matter in indivisible units. Thus, processes involving light are quantized, that is, 'all or nothing'. The quantum of energy is carried as the oscillating force field of the electromagnetic wave. The particle carrying a quantum of energy is called a photon, which has no rest mass. Quantum and photon are distinct concepts; the former is the energy carried by the photon. Changes in the state of atoms or molecules caused by light involve a transition in the energy state of electrons within the substance. This transition can only take place if all the energy of a photon is transferred to the electron; if the quantum is larger or smaller than the energy required for the transition, then the photon will not be 'captured'. The energy of a photon (ϵ) depends on the frequency of the electromagnetic wave, which is related to the wavelength, and is given by:

$$\epsilon = h\nu = hc/\lambda \qquad\qquad [2.2]$$

where h is Planck's constant (6.62×10^{-34} J s) which has the units of energy $\times$ time or 'action'. The greater the frequency and, from eqn 2.1, the smaller the wavelength, the larger the energy of the photon; characteristics of selected wavelengths are given in Table 2.1. Where the energy of light is to be related to a photochemical effect, as in spectroscopy, the wave number ($\bar{\nu} = 1/\lambda$; with units of cm^{-1} by convention) is employed as it is directly proportional to the energy from Planck's law (eqn 2.2) with $\epsilon = hc\bar{\nu}$. The Système International (SI) unit of energy is the joule (symbol J). Much of the older literature uses the calorie and energy levels of molecular orbitals, ionization potentials, etc., are often expressed as electron volts (eV) which is the energy acquired by an electron falling through a potential difference of 1 volt; conversion factors for different units of energy, which have been used in the literature of radiation biology are shown in Table 2.2.

A single photon is a small unit in biological terms; at noon on a bright day the earth's surface receives a maximum of about 1.3×10^{21} photons m^{-2} s^{-1}, so a larger unit of radiation, the mole of photons (more usually referred to as a mole of quanta) is used. It is the number of photons corresponding to Avogadro's number of particles (6.023×10^{23}), and may be thought of as the number of photons required to convert a mole of a

Table 2.2 Conversion factors for energy units used in the photosynthetic literature

1 electron volt (eV)	$= 1.602 \times 10^{-19}$ J
1 watt	$= 1$ J s^{-1}
1 kWh	$= 3.6 \times 10^6$ J
1 joule	$= 0.239$ calories
	$= 6.242 \times 10^{18}$ eV
1 calorie	$= 4.184$ J
1 kJ mol quantum^{-1}	$= 1.036 \times 10^{-2}$ eV
Planck's constant	$= 6.62 \times 10^{-34}$ J s
	$= 4.136 \times 10^{-15}$ eV s

1 mol contains as many elementary particles as there are carbon atoms in 0.012 kg of ^{12}C, or Avogadro's number, 6.023×10^{23}, of particles. 1 mol quantum of photons ($=$ 1 einstein)

Energy of 1 photon (J) $= 1.986 \times 10^{-16}/\lambda$

$$\text{With } \lambda \text{ in nm 1 mol quanta} = \frac{1.986 \times 10^{-16} \times 6.023 \times 10^{23}}{\lambda}$$

$$= \frac{119.616 \times 10^6 \text{ J}}{\lambda}$$

$$\text{1 mol quanta m}^{-2} \text{ s}^{-1} = \frac{1.2 \times 10^8}{\lambda} \text{ J m}^{-2} \text{ s}^{-1}$$

substance to another form with a 100 per cent efficiency, if captured in a single discrete step; a mole quanta is often called an 'einstein' although the unit is not permitted in the Système International. The relationship between wavelength, frequency and energy of individual photons and a mole quantum of photons is given in Table 2.1.

Confusion may arise over the many ways of measuring light and the units of expression because either the number of quanta, or their energy (or both) may be determined. There are also measures of the illuminance, that is, the visual impression to the human eye. To study quantitative and kinetic aspects of the response of chemical and biological processes to light only photon number or energy in defined spectral regions are useful. Photon number incident on a surface normal to the beam is given by photon flux (mol m^{-2} s^{-1}). Energy is given by the radiant flux (J m^{-2} s^{-1}; as 1 J s^{-1} = 1 watt (W) this is equivalent to W m^{-2}). Illuminance is given by the luminous flux, measured in lux ($=$ lx $=$ lumen m^{-2}). In the older literature the foot candle (1 fc = 1 lumen per square foot = 10 764 lx) was used extensively but is no longer acceptable. The biological literature uses the term photon flux density (PFD) as equivalent to photon flux; this term is not used in the Commission International de l'Eclairage which recommends use of photon flux and photon irradiance. Photon number (mol), photon flow (mol s^{-1}) and photon flux or photon irradiance (mol m^{-2} s^{-1}) may be qualified as normal or as a spherical flux (depending on the spatial distribution of the light) is preferable to the often misused term. Radiation from about 400 to 700 nm is used in higher plant photosynthesis and is called photosynthetically active radiation or PAR and is measured by quantum sensors. Quantum sensors include germanium diodes and lead sulphide resistors; their sensitivity changes with wavelength so that at different parts of the spectrum a true photon number is counted. Energy detectors or radiometers respond to energy independent of wavelength; the instruments include thermopiles and bolometers which can measure in mono- or polychromatic light. Detectors must have correct geometrical characteristics to detect light coming from different directions; most used in photosynthesis studies are cosine corrected.

Spectroradiometers are instruments which measure the energy of light in narrow wavebands over the whole spectrum and are used to determine spectra of light sources. Spectrophotometers are used to determine the response of photochemical and biological processes to wavelength. Illuminance is measured by light meters, which respond to light with a similar spectral response to the human eye. As the human eye is most sensitive to green light (*c.* 550 nm wavelength) measurements of illuminance should not be used in studies of photosynthesis.

Electronic states of matter, photon emission and photon capture

Light is emitted by matter undergoing changes in its energy state as a result of heating, which excites electrons from the ground state to higher excited states. Radiation is emitted when the electron drops back to the lower energy state (Fig. 2.1), at discrete wavelengths corresponding to the energy difference between the ground and excited states, and giving rise to line spectra because the energy levels are distinct. Excited molecules, particularly the more complex, may emit radiation at several wavelengths, often close together giving broader emission spectra. This is due to more, closely spaced, molecular orbitals from which electronic transitions occur.

The wavelength of the radiation emitted depends on the energetic characteristics of the states, thus, the temperature of the material determines the wavelength of light emitted. Lamps, for example, have a colour temperature at which a given spectrum of light is produced according to the particular material emitting it.

Absorption of radiation in the visible spectrum depends on the electronic states of the atoms and molecules in the absorbing substance. Substances absorbing visible light are called pigments and absorb particular wavelengths of light, the light not so absorbed being reflected or transmitted; thus chlorophyll absorbs light at about 450 nm and 650 nm, corresponding to the light detected by the average human eye as the colours blue and red, respectively. However, the light between those two wavelengths is relatively

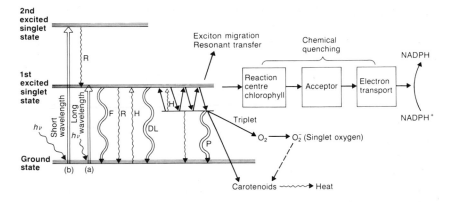

FIG. 2.1 Concept of absorption of photons ($h\nu$) by an atom, energizing an electron to an excited state (a) and its subsequent decay with release of energy. Capture of a more energetic photon (b) results in higher energy level orbitals being filled and then decay by radiationless transition (R). Heat (H) may also raise an electron to higher energy level and the energy is emitted when the electron drops back to the ground state. The main energy-dissipating processes are by radiationless transition (R), prompt fluorescence (F), delayed light emission (DL), phosphorescence (P), and by chemical reactions, for example, which are, in photosynthetic organisms assimilation of CO_2, and transfer, for example, of triplet energy to oxygen or carotenoids or of excitation energy to other chlorophyll and pigment molecules.

poorly absorbed, so the colour of vegetation appears green to the eye. As photon capture in photosynthesis is predominantly by large, complex molecules, brief consideration will be given to molecular orbitals and how electrons are arranged in them.

To capture a photon of visible light, the energy levels of the atomic or molecular orbitals must have a difference in energy corresponding to the absorbed quantum. The electronic orbitals of atoms are analogous to molecular orbitals. The main energy levels are referred to as the ground state (lowest energy), and the first and second (or higher) excited states. The energy levels are designated by a total quantum number. Within each level the number of possible orbitals depends on the magnetic motion and orientation of the electrons in relation to the nuclei, as described by quantum mechanics. Electrons spin within an orbital according to the prevailing magnetic field. Spin may be parallel to the field or antiparallel to it, and is designated by a spin quantum number, S, which has values of $+\frac{1}{2}$ or $-\frac{1}{2}$. Two electrons can only occupy the same orbital if their spins differ. When two electrons of opposite spin occupy the lowest energy orbital the configuration is a stable electronic ground state, S_0. Adding the spins of electrons together gives the total spin S and the spin multiplicity of the electrons is given by $2S + 1$. When all spins balance out $S = 0$ and the spin multiplicity is 1, which is called a singlet state. If there is spin reversal, then S equals 1 and $2S + 1$ becomes 3 giving a triplet. For triplet states spin reversal must occur; as this is a relatively infrequent event it has a low probability. Thus, the formation of a triplet is uncommon and the life-time may be long. Oxygen is an example of a ground state triplet. With an electron absent from an orbital, S equals $\frac{1}{2}$, and the spin multiplicity is 2 or a doublet. Singlet and triplet states are important in photochemistry, and the doublet in free radicals, which may cause photochemical damage in photosynthesis. The chemical and physical texts listed provide quantitative descriptions of molecular structure and quantum mechanics.

With a given electronic configuration, a molecule absorbs light of particular wavelength, smaller or larger quanta are not absorbed and cannot, by the Grotthus–Draper Law, cause a physical or chemical change. If the oscillating electronic vector of a photon causes an electron in a molecule to resonate (i.e. to vibrate at the same frequency) then the energy of the photon will be captured. The direction of the electrical and magnetic fields of the photon must be in correct orientation to the electronic oscillations to cause resonance. Capture of a quantum is rapid by all criteria of 'normal' biological interactions; an electronic transition from one energy level to another occurs in a time inversely proportional to the frequency of the wave, $(1/\nu)$. For red light (Table 2.1) this is about $2.3 \times {}^{-15}$ s which is much faster than nuclear vibrations (10^{-13} s) and so does not influence the nuclear configuration of the molecule.

In complex molecules, with many nuclei and electrons, the possible combination of orbitals is greatly increased, and the energy levels are split into vibrational and rotational levels as electrons within the molecule are influenced by magnetic and electrical forces. Molecules absorbing visible light exhibit differences in energy of about $1-5$ eV between ground and excited states corresponding to wavelengths between 1000 and 200 nm and vibrational energy differences of approximately 0.1 eV; rotational energy levels are not observed at these wavelengths in solid state, only in gases.

Larger molecules usually have more complex energy levels because the outer electrons which provide the bonding are 'delocalized' over the whole molecule. These π electrons travel in the extended π orbitals, even in the ground state and may undergo transition to an excited orbital, $\pi^{\star}$, in the same way that electrons in other, more 'rigid' orbitals can. The $\pi^{\star}$ orbitals are delocalized; the electrons are not orientated in the same way in π and $\pi^{\star}$ states; they are, respectively, bonding and antibonding. As π electrons are free to move in large volume the energy levels are smaller and, therefore, the binding energy is greater than for a system of double and single bonds and the molecule is more stable. Also, the energy required for the ground to excited state transition is low and the capacity to absorb light of longer wavelength is increased.

Delocalization is of great importance in organic molecules, including the photosynthetic pigments which have extensive orbitals over large molecules. Extensive π systems confer high efficiency of energy capture, together with stability under normal conditions *in vivo*. Also, orbitals of different energies provide for absorption of different wavelengths of light. Groups of atoms in a molecule which are responsible for absorbing light energy are called chromophores. In the 200−800 nm range chromophores always have loosely bound electrons, in π orbitals, the size of which determines wavelengths absorbed. For example, the peptide bond in proteins absorbs short energetic wavelengths (190 nm), the nucleic acid bases which have a larger π system absorb at 260 nm and β-carotene (absorbing at 400−500 nm) and chlorophyll (absorbing between 400 and 700 nm) have increasingly larger π systems. The absorption maxima become more dependent on the environment of the molecule as the size of the π system increases because the probability of interaction between the molecule and environment is increased and the small changes in energy can affect the properties of the electronic system substantially. Energy levels of a molecule depend on the environment, on intramolecular rearrangement, binding to other compounds, etc., which alter the absorption of particular wavelengths. Chlorophyll *in vivo* is associated with protein, which 'tunes' the absorption of light over a range of wavelengths.

Chemical reactions involve reorganization of the outer shell or valency electrons which form chemical bonds. For an excited electron to be used chemically it must be held in a configuration which is stable for long enough to transfer to a chemical acceptor directly or via an exchange of energy (not electrons). In photosynthesis the light-harvesting pigments, which capture the energy, transfer excitation via other pigment molecules to special reaction centres, composed of a form of chlorophyll, which convert the excitation energy to chemical energy.

The wavelength and number of photons captured determines the maximum energy available for biological processes and therefore governs the overall energetics. However, the efficiency of energy capture and of all the linked conversion processes limits the actual energy which can be obtained. It is possible to overcome the inefficiency of a photochemical process if the energy from the capture of several photons can be 'stored' or gathered in some way and then used, by multistep processes, for photochemical conversions. This occurs in photosynthetic organisms.

Dissipation of excitation energy

Electrons remain in the excited state for a period called the 'lifetime', dropping back to the stable ground state in an exponential decay. Processes dissipating the energy of the excited state are illustrated in Fig. 2.1. Thermal relaxation (also called radiationless transition) between the closely spaced vibrational energy levels leads to loss of excitation as heat within 10^{-12} s and is the normal pathway of energy loss from the second to first excited states. Excitation energy may be transferred to a chemical acceptor; this is the central event in photosynthesis. Electrons also decay from the excited singlet state to the ground state emitting 'prompt' fluorescence in 10^{-9} s. Electrons in a lower ground state at normal temperatures reach the centre of the excited state when excited, but decay rapidly by thermal relaxation to a lower level of that excited state, before finally dropping to an energy level in the ground state of different energy from which they started. The excitation energy that is required to excite a molecule is therefore greater than the energy emitted, i.e. the wavelength of the photon absorbed is shorter than that of the photon emitted and the fluorescence spectrum is shifted (Stoke's shift) towards longer, red wavelengths. If the molecule is excited to the second excited state then the absorption spectrum shows two bands. However, radiationless transition from the second to the first excited state allows fluorescence only from the first excited state, and only one red-shifted fluorescence band is detected (Fig. 2.2).

When electrons decay from the excited triplet before returning to the ground state and releasing a photon, the process is called phosphorescence. The photon is of longer wavelength than fluorescence due to a smaller energy difference between the triplet and ground states. Phosphorescence may be much delayed as triplet states have a very long lifetime, from milliseconds to tens of seconds, due to the change in spin which accompanies the triplet transition and occurs with low probability. Electrons in the triplet, when energized to the excited single by thermal energy, decay to ground state releasing a photon in delayed light emission (also called luminescence, delayed fluorescence and 'after glow'); the wavelength of the photon is shifted like fluorescence as both come from the same energy level. Delayed light emission is strongly dependent on temperature and on a small energy

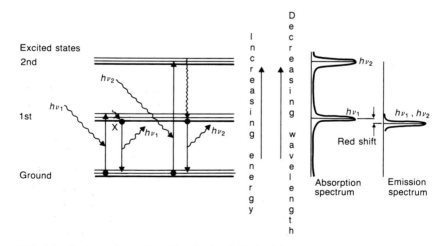

FIG. 2.2   Capture of photons of low ($h\nu_1$) or high ($h\nu_2$) energy, shown by the absorption spectrum, raises an electron to the first and second excited states. Electrons drop from the second to first excited state by radiationless decay, and photons of similar wavelength are emitted from the first excited state, giving a single peak in the emission spectrum. Radiationless decay from the higher to lower energy levels of the first excited state (at X) redshifts the emitted photons.

gap between the triplet and excited singlet. Fluorescence of chlorophyll in the thylakoid membrane is an important characteristic. Prompt fluorescence comes from chlorophyll which cannot pass excitation onto another molecule, either because it is not joined to it or because the system is already overexcited by light. This fluorescence may arise because the reduced acceptor of energy in the reaction centres (p. 99) passes energy back to the chlorophyll which then fluoresces. However, energy is lost from the chlorophyll matrix over a long period by delayed fluorescence and the rate depends on the state of the complete system. Fluorescence from chlorophyll *in vivo* is a very sensitive indicator of the energy status of the photosynthetic system, indicating how the multiple processes of energy absorption and utilization interact.

The rate of decay of the excited state depends on radiationless transitions, fluorescence and photochemistry. The processes have rate constants for de-excitation, respectively, K_d, K_f and K_p so that the overall rate constant is $K = (K_d + K_f + K_p)$. With n_0 excited states initially, the decrease to n excited states in time, t, is given by $n = n_0 \, e^{-Kt}$ where e is the base of natural logarithms. The number of excited states decreases exponentially to $(1 - 1/e)$ of the initial number of excited states in a time $= 1/K$ which is the lifetime, the time required for 63 per cent of the excited electrons to decay. The exponential decay process can also be characterized by the 'half-life', the time needed for half the original population of excited electrons to decay. Single excited states are relatively short lived, that of chlorophyll being about 5×10^{-9} s.

Transfer of excited states

Decay of excited electrons to the ground state, either by radiationless decay or fluorescence, wastes the energy for photosynthesis and *in vivo* the main mechanism which captures the energy of the excited pigment is transfer to a special form of chlorophyll at a 'reaction centre' (RC); this is a chemically reactive form of bacteriochlorophyll or chlorophyll which can pass an excited electron to an acceptor molecule and thus start the chemical reactions of photosynthesis. It is a gate between the world of physical energy and biological chemical energy and thus the heart of the whole process of life.

Photosynthetic organisms have evolved complex biophysical—chemical processes. The systems have multistep approaches to energy capture and transduction. Thus, energy capture by a chl molecule does not pass directly from an excited pigment molecule to the RC but the energy may be transferred via several intermediate molecules (which transfer energy but do not react chemically) before reaching the RC. However, the essential feature of the mechanism is shown in eqn 2.3. The pigment (P) is excited by light (P*) and donates the energy to an acceptor A, which is thus excited and after a number of such transfers excites the RC:

$$P + h\nu \rightarrow P^\star; \quad P^\star + A \rightarrow P + A_1^\star; \quad A_1^\star + A_2 \rightarrow A_2^\star + A_1 \rightarrow A_2^\star$$
$$+ A_n \text{ repeated}; \quad A_n^\star + RC \rightarrow A_n + RC^\star;$$

$$RC^\star + \text{chemical acceptor} \rightarrow RC + \text{reduced chemical acceptor} \quad [2.3]$$

The pigments D and A may be of the same type, e.g. chl a or different, e.g. chl a and chl b (see p. 32) and may be aggregated to form large energy collecting pigment groups or 'antennae' to maximize light capture.

The mechanisms of energy transfer depend on the types of molecules, their size, energy levels of the electronic orbitals, etc., and on concentration and orientation. In solids or concentrated solutions, orbitals 'overlap' and form extensive 'superorbitals'. When excited, an electron enters the delocalized conduction bands and leaves a 'hole' (a positive charge) in the ground state; electrons migrate through the pigment matrix. This semi-conductor type of mechanism leads to photoconductivity, which is observed in dried chloroplasts but is not thought to play an important role *in vivo*. However, it has been suggested to occur in closely bound chlorophyll groups. Another 'strong' interaction occurs between closely packed molecules at $1-2$ nm spacing; it is rapid (10^{-12} s) and depends on ($1/$distance) and there is no radiationless decay. None of these mechanisms is thought to be important in photosynthesis. More important is repeated energy transfer between donor and acceptor leading to energy migration between groups of the same and different molecules at distant spacing. This incoherent transfer with weak coupling is the only one relevant to photobiology. It means that after the system is excited by a very short pulse of light to excite all the pigments there is no uniform propagation of the energy through the system, rather there is rapid randomization of the excitation and there are only a few steps in the transfer. Dipole interaction causes inductive resonance in the acceptor, so that the excitation is passed to it. There is no mass or electron transfer, the excitation migrates as a spin-coupled electron−hole pair and is localized on a definite molecule. This 'weak interaction' involves rates of transfer of $10^{-12}-10^{-14}$ s and is called a Förster mechanism, after the discoverer. The interaction decreases as R_0^{-6}, where R_0 is the Förster distance, defined as the distance between molecules at which energy transfer rate equals the losses due to fluorescence and radiationless decay. R_0 is the mean distance between the dipole oscillating centres in each donor and acceptor molecule and, for photosynthetic systems *in vivo*, is between 4 and 10 nm for chlorophyll with an average of about 6 nm for efficient transfer; it minimizes the number of vibrations undergone by the donor and is temperature dependent.

Transfer is primarily via singlet excited states. Orientation between dipoles is important; transfer is zero with perpendicular orientation and maximal with parallel orientation. Energy levels are most critical; transfer only occurs if the fluorescence bands overlap because the electron drops by radiationless decay to the lowest vibrational level of the excited state from which it decays as a fluorescence photon if its energy is not transferred. The acceptor absorption band is not at the lowest vibrational level but in the centre of the energy band and will take excitation of the correct energy, that is, similar to fluorescence.

As energy is lost at each transfer, a donor—acceptor chain of decreasing ground state to first excited state energy difference favours funnelling of energy in a particular direction and increases the rate of exciton transfer. There may be transfer between similar molecules (homogeneous) by random walk with excitation passing in the pigment matrix at random but, as this leads to energy loss, transfer between different types of molecules (heterogeneous) is more likely. It is faster, requiring fewer steps (perhaps $10-100$), and very efficient. However *in vivo* there may be different transfer mechanisms in different parts of the pigment antenna within the many different types of antenna systems in photosynthetic bacteria, cyanobacteria and higher plants.

Light absorption and absorption spectra

When light passes through a solution of a compound, some wavelengths are tranmitted, that is, passed through without alteration, some scattered and the remainder absorbed in proportion to the concentration of absorbing substance. The same processes occur in gases and solids and in the complex state of living organisms such as the green leaves of plants. Absorbed light determines the rate of photosynthesis. The number of photons of a given wavelength is the difference between the incident and transmitted and scattered (reflected) photons. The number of photons absorbed at different wavelengths constitutes an absorption spectrum, which gives much information on molecular configuration, electronic transition and energy levels. Substances have characteristic spectra from which they may be identified and quantified. Absorption spectra are measured with a spectrophotometer; radiation of the required wavelength band or monochromatic light is passed through a known thickness of solvent (in which the substance to be studied will be dissolved) contained in a cuvette of material transparent to the wavelength to be measured (e.g. glass for visible light, quartz for ultraviolet light). This is a reference or standard solution to correct for reflectance and transmission characteristics of the cuvette. A solution of the substance is then substituted for the solvent and the decrease in transmitted light is measured. The book by Clayton (1980) and several contributions in the volume edited by Hoppe *et al.* (1983) consider techniques.

With I_0 photons per unit area and time passing into the cuvette of thickness l (cm) and a fraction I not absorbed and passing through it, the fraction absorbed is $I_a = I_0 - I$. Thus, dI photons are absorbed in a thin layer dl of area A by n molecules, uniformly distributed per unit volume. The average cross section, σ, of each molecule is the area of the molecule for photon capture. With a probability, p, that a photon will be captured by a molecule, the absorption cross section, k, is equal to $p\sigma$ (units, cm^{-2}).

Thus, the absorption of photons is given by:

$$-\mathrm{d}I = kn\mathrm{d}lI \qquad\qquad [2.4]$$

which integrates to give the Beer–Lambert law:

$$I = I_0 e^{-knl} \qquad\qquad [2.5]$$

where knl is the absorbance. This may be expressed as

$$I = I_0\, 10^{-\epsilon cl} \qquad\qquad [2.6]$$

where c (units, molar) is the molar concentration of absorber and ϵ is the molar extinction coefficient or molar absorptivity (unit, $M^{-1}\,cm^{-1}$), characteristic of the absorber in a particular solvent at a given wavelength. Transmittance, I/I_0, is related to optical density (OD), also called absorbance or extinction as

$$OD = \log I_0/I = \epsilon cl \qquad\qquad [2.7]$$

Absorbance increases linearly with concentration in dilute solution and is often quoted as $A_{1\,cm}^{1\%}$ or absorbance of a 1 per cent concentration in a 1 cm layer.

It is possible to measure the concentration of several components (which do not react) in a solution if their extinction coefficients are known at different wavelengths, as at any one wavelength

$$OD = \epsilon^A c^A l + \epsilon^B c^B l \qquad\qquad [2.8]$$

and by measuring at different wavelengths (at least as many as there are components) the equations can be combined to give the concentrations. This is the basis for Arnon's much used method of measuring chlorophyll a and b in the same extracts (see Neubacher and Lohmann 1983).

Global photosynthesis

The earth's surface receives about 5.2×10^{21} kJ year^{-1}, about 50 per cent of which is of wavelengths used in photosynthesis. Only about 0.05 per cent of this energy (3.8×10^{18} kJ year^{-1}) is captured in organic molecules, the rest is re-radiated into space as heat. Almost 50 per cent of the total photosynthesis is by marine organisms. The turnover of CO_2 is about 3×10^{12} tonne year^{-1} and of nitrogen about 5 per cent of this. The enormous scale of this most important photochemical reaction is comparable to geological processes, such as mineral weathering. Photosynthesis produces 5×10^{11} tonne of organic matter per year and the earth's standing organic matter is estimated at $10^{12}-10^{13}$ tonne dry matter. At present, the rate of destruction of forests and burning of fossil fuels is about 15 per cent of the rate of photosynthesis so that the CO_2 content of the atmosphere, which is small ($\sim$ 340 volumes per million volumes, or 34 Pa partial

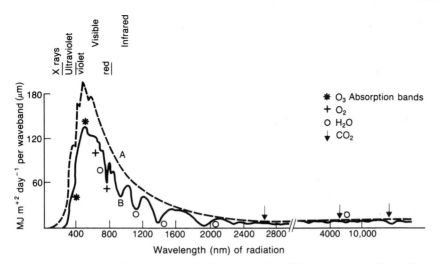

FIG. 2.3 Solar radiation above the earth's atmosphere (A) and at the earth's surface (B). The energy (in megajoules per day) for each wavelength is shown. Areas marked H_2O, O_3, O_2 show absorption of radiation by these components of the earth's atmosphere (recalculated from Gates (1962)).

pressure) turns over rapidly, with a half-life of 10 years. Oxygen turnover is slower, 6500 years, as the concentration is much greater (21 volumes per 100 volumes) and water turns over in 3 million years, owing to the huge global water reserves.

All the energy for global photosynthesis comes from the sun (although light of suitable wavelength from any source, electric lamps, for example, may be used by plants). Radiation is lost from the sun's corona as if from a 'black body' (a perfect radiator and absorber of energy) with a temperature of about 5800 K, in agreement with the Stefan−Boltzmann law. The distribution of the wavelengths of the solar spectrum is concentrated between 400 and 1200 nm with the peak around 600 nm (Fig. 2.3) in the yellow−orange. At the earth's surface the spectrum is modified by selective absorption of wavelengths by constituents of the atmosphere such as carbon dioxide and water vapour.

References and Further Reading

Broda, E. (1978) *The Evolution of the Bioenergetic Processes*, Pergamon Press, Oxford.
Clayton, R.K. (1980) *Photosynthesis: Physical Mechanisms and Chemical Patterns*, I.U.P.A.B. Biophysics Series, Cambridge University Press, London.
Gates, D.M. (1962) *Energy Exchange in the Biosphere*, Harper and Row, New York.
Giese, A. (1964) *Photophysiology, 1, General Principles: Action of Light on Plants*, Academic Press, New York.

Govindjee and **Whitmarsh, J.** (1982) Introduction to photosynthesis: Energy conversion by plants and bacteria, pp. 1–18 in Govindjee (ed.), *Photosynthesis*, Vol. 1, *Energy Conversion by Plants and Bacteria*, Academic Press, New York.

Häder, D.-P. and **Tevini, M.** (1987) *General Photobiology*, Pergamon Press, Oxford.

Hager, A. (1980) The reversible, light-induced conversions of xanthophylls in the chloroplast, in Czygan F.C. (ed.), *Pigments in Plants*, (2nd edn), Fischer, Stuttgart.

Hames, B.D. and **Rickwood, D.** (eds) (1987) *Spectrophotometry and Spectrofluorimetry, A Practical Approach*, IRL Press, Oxford.

Hart, J.W. (1988) *Light and Plant Growth*, Unwin Hyman, London.

Hipkins, M.F. and **Baker, N.R.** (1986) *Photosynthesis Energy Transduction a Practical Approach*, IRL Press, Oxford.

Hoppe, W., Lohmann, W., Markl, H. and **Ziegler, H.** (1983) *Biophysics*, Springer-Verlag, Berlin.

Neubacher, H. and **Lohmann, W.** (1983) Applications of spectrophotometry in the ultraviolet and visible spectral regions, pp. 100–9 in Hoppe, W., Lohmann, W., Markl, H. and Ziegler, H. (eds), *Biophysics*, Springer-Verlag, Berlin.

Nobel, P.S. (1991) *Physiochemical and Environmental Plant Physiology*, Academic Press Inc., San Diego.

Parson, W.W. and **Ke, B.** (1982) Primary photochemical reactions, pp. 331–85 in Govindjee (ed.), *Photosynthesis*, Vol. I, *Energy Conversion by Plants and Bacteria*, Academic Press, New York.

Salisbury, F.B. (1991) Système International: The use of SI units in plant physiology, *J. Plant Physiol.*, **139**, 1–7.

CHAPTER 3

Light harvesting and energy capture in photosynthesis

Three functions of the light-harvesting apparatus contribute to the utilization of light quanta to produce a chemical intermediate of higher energy state:

- light absorption: the energy of a photon is captured by an antenna pigment molecule and an electron is excited,
- energy transfer: excitation energy moves through the antenna to a reaction centre, a special form of pigment, in which it excites an electron,
- electron transfer: an energized electron from the reaction centre passes to a chemical acceptor and the oxidized reaction centre is reduced by an electron from organic or inorganic molecules (anoxygenic) or water (oxygenic).

The processes are controlled by the physical and chemical characteristics of the pigments, particularly chlorophyll (chl) in higher plants and bacteriochlorophyll (bchl) in photosynthetic bacteria. However, there are many other pigments involved in photosynthesis in different organisms.

Light-harvesting pigments

Several types of pigments harvest the energy of light (Table 3.1). Only special forms of chlorophyll a (chl a) and bacteriochlorophyll (bchl) form reaction centres (see p. 37, chapter 5) in higher plants and photosynthetic bacteria, respectively. All other pigments are therefore accessory pigments, forming groups arranged to extend the size of the light-capturing unit, thus acting as antennae (like a radiotelescope dish) for capturing photons which donate excitation to the reaction centres. A group of many antenna molecules of different type donates energy to a single reaction centre complex. Pigments absorb in different parts of the spectrum and, in combination, enable an organism to absorb light of different wavelengths. This is of great importance ecologically for organisms may then exploit a greater energy supply or grow in different radiation environments. For example, phycobilins of red algae absorb blue light, which predominates in deep water

Table 3.1 Pigments of photosynthetic organisms: only some of the more important pigments and groups of plants in which they occur are given. Primary pigments are those involved in the photochemical process, accessory pigments function only in light harvesting

Organism	Primary pigment	Accessory pigment(s)	Wavelength absorbed (nm) (approximate)
Prokaryotes			
Purple bacteria	bchl *a*	bchl *a*	Blue-violet−red 470−750
		bchl *b*	Blue-violet−red 400−1020
Green sulphur bacteria	bchl *a*	bchl *c* = *Chlorobium* chl	Blue−red 470−750
Blue−green algae	chl *a*	Phycocyanin	Orange 630
		Phycoerythrin	Green 570
		Allophycocyanin	Red 650
Eukaryotes			
Red algae	chl *a*	Phycocyanin	Orange 630
		Phycoerythrin	Green 570
		Allophycocyanin	Red 650
Brown algae	chl *a*	chl *c*	Violet-blue−red
Higher plants	chl *a*	chl *b*	Violet-blue− orange-red 454−670
Most higher plants, algae and bacteria		Carotenoids α, β in different groups Xanthophylls	Blue-green 450

where the plants grow. Figure 3.1 illustrates the pigments, wavelengths absorbed and composition of the light-capturing system for different organisms.

Photosynthetic bacteria have several forms of pigments donating energy to the reaction centre bacteriochlorophyll. In higher plants chl *b* is an auxiliary pigment passing excitation to chl *a*. In all O_2-evolving organisms excitation moves to reaction centres composed of forms of chl *a* which pass the energy on as an excited electron to chemical reactions.

Chlorophylls

Chlorophylls are probably the most abundant biological pigments and the world appears green to the human eye due to their absorption of blue and red light. Leaves may contain up to 1 g of chlorophyll m^{-2} of surface area but this varies with species, nutrition (particularly nitrogen fertilization), age, etc. Chlorophyll is extracted with fat solubilizing solvents (e.g. ether,

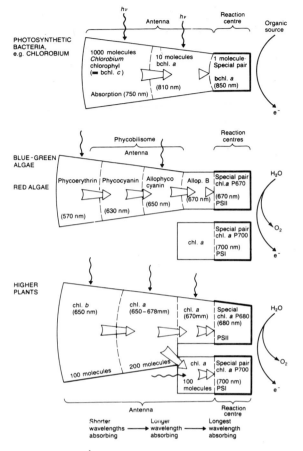

FIG. 3.1 Energy absorption (♪) and excitation transfer (⇒) between pigments in the light-harvesting antenna and to the photochemical reaction centres of different photosynthetic organisms.

acetone) for it is a lipophilic molecule only found in membranes containing lipid where it is bound to hydrophobic protein. Chl *a* and chl *b* and other pigments may be separated by chromatography.

Chemically chlorophylls are chlorin macrocycles with four-fold symmetry, derived from porphyrin. Chl *a* (Fig. 3.2) is a conjugated macrocyclic molecule (mass 894) with a planar 'head' of four pyrrole rings; it is about 1.5 × 1.5 nm but the overall size is much greater due to a phytol group, a terpene alcohol chain some 2 nm long, which positions the molecule in membranes. The chemical groups and H^+ on the outer edge of the pyrrole unit confine the electrons to a single plane, increasing absorption of red wavelengths, an advantage for plants of land or surface water. A non-ionic magnesium atom, bound by two covalent and two co-ordinate bonds in the centre of the molecule, co-ordinates the rings. Magnesium is crucial to capture of light energy; an iron atom (as in related pigments) cannot do

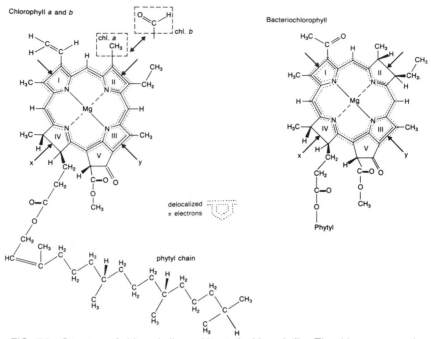

FIG. 3.2 Structure of chlorophyll *a* and bacteriochlorophyll *a*. The side groups on ring II are responsible for the difference between chlorophyll *a* and *b*. Axes x and y are the principal electronic transitions.

this. Magnesium is a 'close shell' divalent cation which changes the electron distribution and produces powerful excited states; chl *a* is a good donor of electrons (at a large negative potential), and excited chl *a* at the reaction centre is a very strong oxidizing substance able to accept electrons (indirectly) from water. The large size and extensive ring structures of the chlorophylls, with ten double bonds, allows electrons to delocalize in the π orbitals over the 'head' of the molecule, increasing the area for capture of a photon, giving many redox levels which are important for efficient energy capture and transfer, and producing a complex absorption spectrum. In chl *b* a formyl (—CHO) group replaces the methyl (—CH₃) group on ring II (Fig. 3.2), which increases the blue and decreases the red absorption maxima and alters solubility; chl *b* is less soluble in petroleum ether than chl *a* but more soluble in methyl alcohol, for example. Bchl *a* (Fig. 3.2) is similar to chl *a* but has an acetyl instead of a vinyl group on ring I, and ring II is saturated with hydrogen instead of unsaturated. This loss of the double bond alters the π system. Such differences, which are probably later evolutionary developments, increase absorption at longer wavelengths, but bchl cannot generate a sufficiently strong oxidant to remove electrons from water. Bchl *g*, which is very similar to the chl *a* of higher plants, has been described from *Heliobacterium chlorum* and may represent a stage in the evolution of chl *a*. Possibly the evolution of chl *a*, essential for the development of the oxygenic photosystem and water splitting, took place in bacteria together with the development of the special binding to the structural proteins which is essential for the modification of the electronic energy states to achieve particular absorption characteristics and sufficient oxidizing power to reduce (indirectly) water.

Absorption spectra of chlorophyll

Measured in organic solvents after extraction from the plant (Fig. 3.3), chl a absorbs most strongly at 430 nm (Soret band) and 660 nm and chl b at 450 and 640 nm. The absorption maxima 'shift' with the solvent; in forty different solvents the red absorption of chl a is between 660 and 675 nm. Polar solvents, such as acetone, cause strong dipole–dipole interactions, weaken London dispersion forces and change hydrogen bonding, altering the electronic configuration of the molecule and hence absorption. Aggregation of chlorophyll also causes a shift; crystalline chl a has its long wave absorption minimum at 740 nm.

Differences in absorption spectra and molar extinction coefficients enable chlorophylls to be distinguished and measured spectrometrically in unpurified solutions. Chl a has a molar extinction coefficient of 1.2×10^5 M^{-1} cm^{-1} at 430 nm; a 10^{-5} M solution is intensely coloured and absorbs some 80 per cent of the incident light. Chlorophyll is a very efficient pigment with a cross-section absorbing area per molecule of 3.8×10^{-16} cm^2. On a bright noon day at the earth's surface (a photon flux of 2×10^{-3} mol quanta m^{-2} s^{-1}) a chlorophyll molecule will capture about 45 photons s^{-1}.

The maxima in light absorption correspond to different energy levels in the molecule (Fig. 3.3). The highest energy level is the second (or higher) singlet, excited at 430 nm. The excited electrons make the transition in 10^{-12} s from the second to the first excited singlet state, to which electrons are excited by red light. Thus, the energy of blue light is dissipated and a blue photon is no more effective in photosynthesis than a red.

The lifetime of the lowest excited singlet state is about 5×10^{-9} s before it decays to the singlet ground state. In a solution of monomeric chlorophyll there is little radiationless dissipation to the ground state but in aggregated chlorophyll it dominates. Absorption spectra of chlorophylls show the electronic transitions along axes of the molecule. Polarized light and paramagnetic or electron spin resonance (ESR) are used to analyse the transitions, which are very important in understanding the mechanism of energy transfer (Clayton 1980 gives details of the methods). The x-axis of chlorophyll is through the nitrogen (N) atoms of rings II and IV and the y-axis through the N–N atoms of I and III (Fig. 3.2). The two main absorption bands in the blue and two in the red are called B and Q, respectively, due to $\pi \rightarrow \pi^*$ transitions. The polarizations of the transitions along the axes are called x and y, and may be from the lowest vibrational

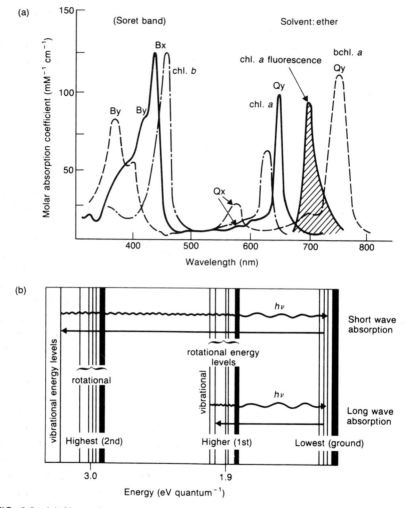

FIG. 3.3 (a) Absorption spectra of chlorophylls *a* and *b* and bacteriochlorophyll *a* in ether showing principal electronic transitions in blue (Soret) and red wavelengths. An energy level diagram (b) for chl *a* is related to the main absorption peaks.

energy state (called 0) or the next higher energy state (called 1). Thus, absorption at 430 nm is a Bx (0,0) transition and at 660 nm it is a Qx (0,0). The Qy transition is most altered by solvent (e.g. water bound to Mg) and association in membranes which increases the red absorption.

Chlorophyll fluorescence

Electrons in the higher levels of the first excited singlet, S, state (energy E_a) decay by radiationless transition (R) to the lower levels and, if not used in photochemistry or transferred to other molecules, decay to the singlet ground state, S_0, by emission of 'prompt' fluorescence of lower energy (E_f) than the exciting light, as $E_f = E_a - R$. Thus, a solution of chlorophyll irradiated with blue light emits red fluorescence. In chl a of the thylakoids prompt fluorescence is emitted at a peak of 685 nm. It shows the accumulation of excitation energy in the antenna and is inversely related to the use of electrons; it indicates the state of electron transport and biochemical processes relative to energy capture (Ch. 99).

The chlorophylls at the reaction centre

Normal absorption spectra cannot measure small changes in absorption occurring in a few molecules if the bulk of the pigments have overlapping absorption bands. Most bchl in bacteria and chl a and b molecules in higher plants form the antenna, and only a few bchl or chl a form reaction centres which pass electrons to acceptor molecules. Antenna chlorophyll 'swamps' the absorption even in preparations enriched in reaction centre chlorophyll. To overcome this, difference spectra compare absorption between light and dark with repetitive flashes or with chemical oxidation and reduction (see Witt 1975) enabling the chemically reactive form of chlorophyll to be studied. ESR and electron nuclear double resonance (ENDOR) techniques are also used to detect changes in paramagnetism caused by the transfer of electrons to an electron acceptor, independent of changes in the non-paramagnetic bulk chlorophylls and other pigments.

In isolated spinach chloroplasts from which most of the chlorophyll had been removed, difference spectra showed a decrease in absorption (called photobleaching) near 700 nm following illumination (Fig. 3.4). The signal was altered by the redox state of the chlorophyll; when oxidized there was a loss of absorption at 680 and 690 nm and an increase at 686. The form of chl a responsible is called P700 (P for pigment and the wavelength of the absorption change) and is the reaction centre for photosystem I (PSI) (p. 91). Only one in 300–400 chlorophyll molecules is in the special form, P700. Difference spectra also show fast change (60 ps) in absorption at 680 nm, separate from, but equivalent to the P700; it is associated with the reaction centre chl a of photosystem II (PSII) and is called P680.

Duysens observed, in bacterial chromatophores, a small (2 per cent) decrease in absorption at about 870 nm wavelength caused by illumination or chemical oxidation; the absorption then increased slowly in the dark. By removing the major part of the bchl with detergent the signal from P870, the bacterial reaction centre, was enhanced (see Clayton 1980).

Changes in absorption with illumination or chemical modification show charge separation during transfer of electrons from reaction centre chl a (or bchl) to an acceptor, A

$$\text{chl } a + A \xrightarrow{\text{light}} \text{chl } a^+ + A^- \qquad [3.1]$$

P700, P680 and P870 convert light energy to chemical form. This step is

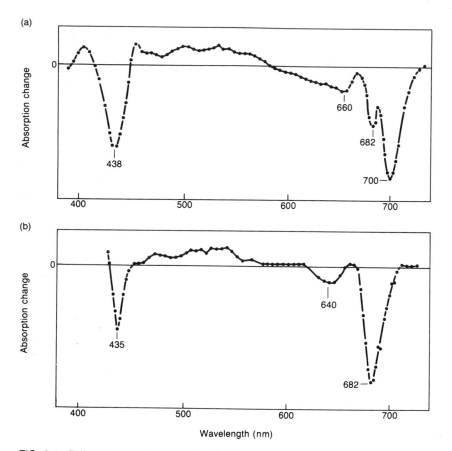

FIG. 3.4 Difference spectrum of spinach chloroplasts between light and dark; (a) the change at 700 nm is due to absorption by PSI reaction centre chl *a* (after Kok, 1961). (b) The reaction centre chl *a* absorption of PSII (P680) (after Döring *et al*, 1969).

the true 'photochemical act' in photosynthesis and thus has a unique role, not simply a special one.

The organization and function of higher plant reaction centres and comparison with bacterial reaction centres, which are similar to PSI, is discussed by Clayton (1980). The structure of the higher plant reaction centre is also considered by Okamura *et al.* (1982). Many of the concepts developed in the earlier work on the reaction centres have been substantiated by the detailed structural analysis of the bacterial reaction centre (see p. 74).

The molecular structure and arrangement of P700 and the other reaction centres are unknown in detail. Photo-oxidation is accompanied by an ESR signal identified as an unpaired electron delocalized over a large π-system. The ESR signal and photobleaching of P700 is 1 : 1, that is, an excitation produces one free radical and ejects one electron from chl *a*. The structure of P700 is, on evidence from ESR and ENDOR, probably a 'special pair' of chl *a* molecules joined covalently as a dimer in a specific way. In photosynthetic purple bacteria a special bacteriochlorophyll dimer complex (linked to protein and electron acceptors) acts as electron donor. The electron is distributed over the π system of both molecules of the dimer. Several models of the PSI reaction centre have been proposed. A recent model based on the X-ray structural analysis and on calculations of the molecular orbitals and electron spin densities of the dimer components is shown in Fig. 3.5. The methyl group (CH_3-) and the $-C-CH_3=O$ group on ring I of the bchl linked to the L subunit of the proteins forming the 'frame work' of the reaction centre (see p. 79), and interacts with the other bchl of the dimer bchl of the RC on the M protein. There is a strong influence of the electronic state of the atoms in each half of the dimer on the other half. The distant methyl groups on rings III in each bchl affect the spin density of the molecule and cause an imbalance so that the dimer is not symmetrical. Possibly this is necessary to achieve a directional effect, enabling the RC to eject an electron on excitation, as required for the chemical reaction with the acceptor molecule. This could give rise to signals which appear to be from a monomeric chlorophyll. Clearly the orientation of the molecules, their x—y transitions and close association (0.4 nm) is determined by the supporting protein. The spin densities of the dimer components will be influenced by the environment as well as in the molecule itself. However, the electronic state transition involves only the bchl dimer and not the proteins. The arrangement of the dimer makes a very extensive and stable but finely tuned state possible with extensive π orbitals. In membranes the Qy transition

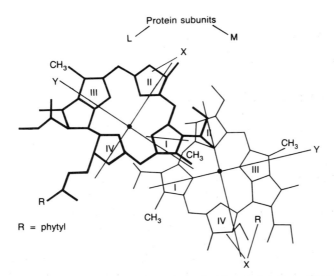

FIG. 3.5 Structure of the reaction centre bacteriochlorophyll *a* dimer P865 based on X-ray structure analysis, ENDOR and other techniques and calculated molecular orbitals (after Lendzian *et al* in Michel-Beyerle, 1990).

of special pair chl *a* is less closely orientated and chl *b* is possibly at angles greater than 35°.

The characteristics of bchl or chl *a* as a special pair dimer would be that the Qy transition is shifted to the required longer wavelength absorbing form, the energy levels are correct for sharing the unpaired electron and it is stable, allowing an excited electron to transfer to an acceptor (with which it forms a radical pair) whilst holding the electron for sufficient time for reaction to occur. Also the cation free radical formed on oxidation is not as reactive as the monomer, thus reducing the chances of back reactions with the acceptor. The type of model suggested for the bacterial RC is possibly applicable to the PSI and PSII RCs, although the details of structure and the absorption characteristics are probably different. There is still much to be done in establishing the features of this most important part of the photosynthetic process. Thus, the dimer model may not be fully acceptable for the PSI reaction centre, as more refined ESR and ENDOR spectroscopy favours a monomer of chl *a*, which has a more realistic redox potential. Thus, there is no generally accepted model of PSI reaction centre structure or function. The molecular arrangement of the PSII reaction centre, P680, is also not yet characterized, but new models of the structure of the reaction centre of the photosynthetic bacteria largely agree with the earlier generalizations and have greatly extended understanding of the molecular arrangements (Barber 1987). P680 may be a dimer of two chl *a* molecules close together and interacting, depending on the state of the RC, e.g. the 8g line width of P680$^+$ in time-resolved EPR spectra is probably from a single chl radical possibly one of the pair with the charge density on it rather than uniformly distributed over the dimer, however, it is probably a dimer in the ground state. Events at the reaction centre are of great importance and interest. The mechanism of the oxidation processes is one of exciton dissociation, with the dissociated negative charge transferred (via electron tunnelling) over distances of 2−3 nm, into an acceptor pool. This pool must be reduced at a rate of two orders of magnitude greater than the rate of photon absorption and the back reactions must be of the same order as the photon capture. These conditions give an efficiency of the PSI reaction centre of about 20 per cent of the solar radiation energy, although losses, outside the reaction centre, in the electron transport and related processes give an efficiency of 5 per cent; such efficiency has been measured for rapidly growing plants under optimal conditions. Detailed understanding of the mechanism may help to improve the efficiency of energy transduction in photochemical processes of many types. Detailed description of the bacterial photosystem complex is given in Chapter 5 together with discussion of the PSII of oxygen producing organisms.

Absorption spectra of chlorophylls in membranes

The absorption of light by chl *a* and chl *b* has been discussed (p. 35) for the pigments in solution. However, the situation in the thylakoid membranes, which are lipid hydrophobic environments, is very different with the pigment bound to proteins.

Aggregation of chlorophylls in membranes is analysed mathematically (deconvoluted) by fitting curves of a normal (Gaussian) distribution to absorption spectra of thylakoids. Chl *a* shows up to ten absorption maxima, the main ones absorbing at about 660, 670, 678, 685 and 689 nm. A 650 nm absorption is due to chl *b*. Light-harvesting chlorophyll−protein (LHCP) complex (p. 68) has six bands, including that at 678 nm. Chl *b* in LHCP may be in groups exchanging by the strong excitation mechanism and linked to chl *a* by the Förster mechanism. The chl *b* antenna shows strong interaction but not chl *a*.

The antenna, light harvesting and energy use

The pigment antenna is a mechanism for accumulating quanta in dim light to drive a process requiring several quanta. The effective cross-sectional area of the reaction centre is increased more than a hundred-fold, and by combining pigments, different wavelengths of light are used. In bright light excited chlorophyll can 'hold' energy for the next unoxidized reaction centre; however, there are many factors *in vivo* which are not well understood. Energy movement between closely aggregated chlorophylls is possibly by the 'rapid' (exciton) mechanism in homogeneous transfer (p. 26). These groups interact with less organized pigments by the slow Förster mechanism. Fluorescence occurs in 10^{-9} s and the Förster mechanism in 10^{-10} or 10^{-11} s and radiationless relaxation in 10^{-12} s. Therefore, excitation may pass by random walk between many molecules before it reaches a reaction centre, which can accept an electron. Theoretically many steps (perhaps up to 1000) may occur between antenna and reaction centres, but heterogeneous transfer directs excitation movement so there are many fewer; 200−300 steps. This is shown by, for example, loss of fluorescence polarization induced by polarized light. In dim light practically all photons are tunnelled to reaction centres and transfer is about 100 per cent efficient. To achieve this the chlorophyll molecules are arranged at 4−10 nm distances. Reaction centres of PSI trap the excitons from the antenna in 60−70 ps but PSII is slower, 200−500 ps; these rates are derived from fluorescence lifetime measurements. Models of energy transfer between groups of chlorophylls are discussed in relation to the light-harvesting chlorophyll−protein complexes in thylakoids (p. 68).

Formation of a reduced acceptor is the most important way of dissipating chlorophyll excited states and requires a particular structure of the reaction centre−acceptor complex, correct molecular orientation and overlap of molecular orbitals in order that net transfer is faster than back reactions caused by thermal excitation. In photosynthesis excitation drives an electron from the reaction centre over an energy barrier to an acceptor, the pigment acting as a catalyst or sensitizer.

Accessory light-harvesting pigments

Pigments (Table 3.1) other than bchl or chl *a* form a major part of the light gathering antenna. In organisms such as algae, accessory pigments are important for they capture light of the wavelengths not absorbed by chlorophyll and pass the energy to the reaction centre.

Bacteriorhodopsin and bacteriochlorophyll

In the simplest form of transduction of light energy, that of *Halobacterium halobium* (see p. 9 and Ch. 5) only one pigment, bacteriorhodopsin is involved. However, the 'true' photosynthetic bacteria contain a wide range of accessory pigments, some of which are mentioned in Table 3.1. Bchl *a* and bacteriopheophytin (bpheo) absorb light at significantly longer wavelength than chl *a* and chl *b*; the RC absorbs in the infra-red region between 870 and 960 nm with the Soret bands below 400 nm in the UV-A region (Fig. 3.3a). Such a shift in light absorption is very important to the ecological performance of these bacteria for they inhabit oxygen-poor environments often in the shade of green algae and plants which cannot use the wavelength so enabling the bacteria to fill a special niche.

The green bacteria (Chlorobiaceae and Chloroflexaceae) contain bchl *b*, *d* or *e* in addition to bchl *a* and carotenoids. The pigments are often arranged into chlorosomes, antennae structures bound to the cytoplasmic membranes which contain four types of polypeptides. One of these is probably a dimer of molecular mass 3.7 kDa and 10−16 bchl molecules are bound to it. In the Chloroflexaceae bchl *a* is bound to a 5.8 kDa hydrophobic protein which absorbs at 792 nm and donates the energy to the RC.

Phycobiliproteins and phycobilisomes

Phycobiliproteins of the red algae and blue−green algae are composed of a chromophore, a bilin pigment (phycocyanin, phycoerythrin or allophyco-cyanin) attached to a protein, characteristic of the organism. The chromophores are straight chain tetrapyrroles related to porphyrins and therefore chlorophyll, but are water soluble. Phycocyanin and phycoerythrin absorb mainly at 630 and 550 nm, respectively and allophycocyanin at 650 nm; they have high molecular extinction coefficients. Phycobiliproteins transfer energy to reaction centres with almost 100 per cent efficiency. The chromophore groups are bound to the proteins with thioether bridges and form a variety of different wavelength absorbing complexes, phycoerythrobilin, phycobilin, phycourobilin and phycobiliviolin. Energy absorbed by the shorter wavelength phycoerythrin is transferred to the phycocyanin and allophycocyanin which absorb at progressively longer wavelengths. This enables the energy between 480 and 630 nm to be exploited. The energy is then transferred to chl *a*, probably by a Förster mechanism with efficiency approaching 100 per cent.

In the cyanobacteria and red algae, the phycobilin pigment−protein

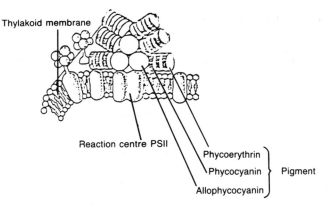

FIG. 3.6 The phycobilisome antenna complexes of a red alga on the thylakoid membrane (a) and details of its subunit structure and attachment to the membrane in the cyanobacterium *Mastigiocladus laminosus*, with three central cylinders to which rods of pigment−protein complexes are attached.

complexes are organized into phycobilisomes. These are arranged very regularly on the thylakoid membranes and are anchored to PSII particles (Fig. 3.6). The centre of the phycobilisome is a nucleus of allophycocyanin organized in three cylindrical complexes of stacked subunits. Onto this nucleus, radially arranged, are three or four rods, also made up of segments of pigment−protein complexes. Next to the allophycocyanin is phycocyanin and towards the extremity of the rods phycoerythrin or phycoerythrocyanin. The individual segments are joined by linker protein (L) of molecular mass 25−30 kDa. The subunits are $\alpha-\beta$ heteromers, organized into cyclic trimers, thus producing a hexamer, $(\alpha-\beta)_6L$. The α subunit of 17 kDa binds 1 or 2 chromophores and the β subunit 1−4 chromophores. The α and β subunits are always in the 1 : 1 ratio. In addition, large subunits occur. The composition of the phycobilisomes of some algae is strongly influenced by the wavelengths of light they experience during development, a phenomenon called chromatic adaptation. This allows an additional and more effective exploitation of light with ecological benefits for survival and reproduction of the organism.

Carotenoids

Carotenoids, that is the carotenes and xanthophylls which are synthesized by the same synthetic pathway except for the final steps, have the dual functions of accessory pigments in energy capture and of dissipating energy and excited states of O_2 in plant membranes. The main carotene of leaves is β-carotene and lutein is the principal xanthophyll, although violaxanthin and neoxanthin are also very important. Carotenes are fat soluble orange pigments (maximum absorption at 530 nm); they are long (3 nm) chain hydrocarbons of isoprene units (five-carbon) with alternating double bonds, nine or more in different positions in the different photosynthetic

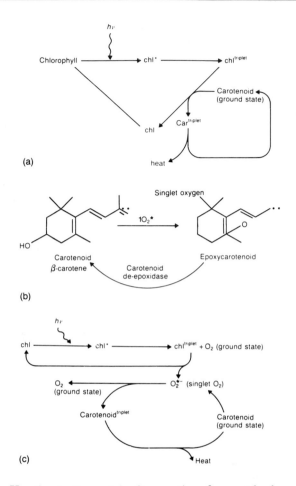

carotenoids. Xanthophylls contain O_2, are therefore not hydrocarbons and they differ in the position of the oxygen in the terminal ring structure.

The most important function of carotenes in higher plants is not as accessory pigments, for they are only 30 to 40 per cent efficient in transferring energy to reaction centres, but in dissipation of the excess energy of chlorophyll and of 'detoxifying' reactive forms of O_2. Carotenoids occur in all photosynthetic organisms that evolve O_2, and are closely associated with the reaction centres. Plants lacking carotenoids, because of chemical inhibition of synthesis or mutation, are damaged during photosynthesis. The photosynthetic bacterium *Rhodopseudomonas sphaeroides* is also destroyed in the light in the presence of oxygen when a mutation prevents carotenoid production. In the light, ground state oxygen reacts with excited state chlorophyll giving singlet oxygen which is very reactive, oxidizing (bleaching) chlorophyll, purines in nucleic acids and polyunsaturated fatty acids, which form lipid peroxides. Superoxide $(O_2^{\cdot -})$ with a free radical (an unpaired electron) is produced in photosynthesis; it is both a reducing

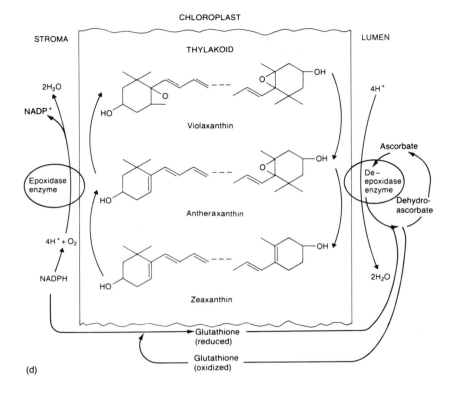

(d)

FIG. 3.7 Mechanisms of energy regulation in photosynthesis by carotenoids. (a) The direct reaction with excited chlorophyll; (b) the reaction with excited states of oxygen by epoxidation; (c) reaction of carotenoids with excited states of oxygen; (d) reaction with NADPH and O_2 in the violaxanthin cycle.

and oxidizing agent. Hydrogen peroxide, which is formed when O_2 reacts with electrons from photosynthesis, reacts with $O_2^{\cdot -}$ to form the hydroxyl (OH) radical, which is very reactive indeed and damages most biological material (see p. 97). It is therefore essential to prevent formation of these compounds or destroy them quickly, before the photosynthetic system is damaged. In light, oxygen becomes toxic and photosynthesis self destructive!

Peroxide and $O_2^{\cdot -}$ are destroyed by catalase and superoxide dismutase (SOD), respectively, but carotenoids dissipate $O_2^{\cdot -}$ and also quench excited chlorophyll, preventing reaction with O_2. Carotenoids form excited triplets, by energy migration from the chlorophyll triplet state, which are dissipated by radiationless decay (Fig. 3.7a), transferring energy to the medium. For the triplet energy migration the distance between carotenoid and chlorophyll must be very small, one order of magnitude smaller than the Förster distance so that close association between pigments is essential for π-electron systems to overlap. Excited O_2 is destroyed in an epoxidation reaction involving the ring structure of the carotenoids and an enzyme. Singlet O_2 is also removed, via the carotenoid triplet state to give ground state oxygen. The reaction with β-carotene is by an electron exchange transfer mechanism. There are differences between carotenoids in their efficiency: all *trans*-lycopene quenches (destroys) $^1O_2^*$ faster than β-carotene. Carotenoids also absorb 380–520 nm radiation, which is rather energetic and may damage biological systems; they offer protection against a wide range of damaging cellular products not only in photosynthesis but in general metabolism in plants and animals. Carotenoids themselves are destroyed only with excessive energy load. The structure with nine or more double bonds is essential for efficient energy dissipation, seven double bonds or less being ineffective.

The xanthophyll cycle

Xanthophylls are involved in a number of energy and reductant regulatory processes in plants and are closely associated with photosynthesis. It has long been known that when *Chlorella* is illuminated violaxanthin, a di-epoxide xanthophyll, is converted (in a de-epoxidation reaction) via antheroxanthin (a mono-epoxide) to zeaxanthin (no epoxide group). In dim light or darkness the zeaxanthin is epoxidized back to the violaxanthin; ATP or an energy rich state is required. This cycle is replaced with one based on diadinoxanthin and diatoxanthin in *Euglena*. The cycle does not occur in photosynthetic bacteria, so it suggests that it is related to protection from photochemical damage. The cycle (Fig. 3.7d) summarizes the system (see Krinsky 1978). In the light, as the level of reductant increases relative to ATP, the reduced NADPH must be turned over to provide a 'sink' for electrons and to allow photochemistry to proceed. This also protects the membranes from damage. The xanthophyll (violaxanthin) cycle involves two enzymes on the thylakoid membrane. In the light de-epoxidation occurs; the enzyme (with a molecular mass of 54 kDa) is on the lumen side of the membrane and has an optimum pH of 5.2 and is inactive at pH above 7. Thus, it is active in the light when the lumen pH falls (see Ch. 5). Ascorbate is also required as well as reduced glutathione (see p. 99) and both will increase concentration in the lumen. The epoxidation occurs at the stromal side of the membrane; the reaction is catalysed by a mixed-function oxygenase which works optimally at pH 7.5, the condition of the illuminated stroma. The reaction consumes NADPH producing water and making $NADP^+$ available to accept more electrons. As the two enzymes are located in the two compartments and have different pH optima, maximum activity of the xanthophyll cycle depends on the light-induced proton gradient across the thylakoid. Under these conditions ATP is synthesized. It is likely that a major function of the xanthophyll cycle is to regulate the ratio of NADPH to ATP, reducing the possibility of photo-damage and optimizing conditions for membrane function and chloroplast biochemistry. Importantly, chloroplast envelopes are rich in violaxanthin in darkness and zeaxanthin in the light and may regulate chloroplast stromal conditions relative to the cell cytosol. Probably zeaxanthin also regulates the dissipation of excess excitation in the antenna chl of the photosystems, as indicated by the blockage of de-epoxidation by thiol compounds such as dithiothreitol which inhibit a large part of the 'high energy state quenching' of chlorophyll fluorescence by radiationless dissipation in the antenna. Probably zeaxanthin interacts with other mechanisms of regulation.

Carotenoids are thus an important 'safety valve' dissipating the excess energy of the excited pigments and the products of the reaction with O_2. This limits the danger of photodestruction of tissues in intense light, when O_2 concentration is high and the normal acceptor of electrons is deficient.

Action spectra and two light reactions

The absorption spectrum is the amount of light captured by photosynthetic pigments as a function of wavelength and the action spectrum is the rate of photosynthesis (e.g. O_2 evolution or CO_2 absorption) or other response resulting from the capture. An action spectrum of O_2 release by algal cells is given in Fig. 3.8. Action spectra show the efficiency of energy use. From the Stark–Einstein Law of equivalence the amount of a reaction is

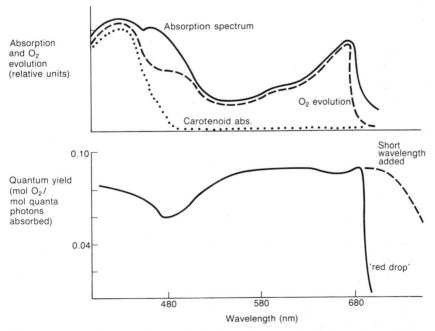

FIG. 3.8 Absorption and action spectra for O_2 evolution by the alga *Chlorella*. Red light is ineffective in O_2 evolution ('red drop'). Supplementation of red light (>690 nm) with shorter wavelength light (<690 nm) increases the O_2 evolution (---). In the blue region, where carotenoids absorb, efficiency of O_2 production is poor.

proportional to the number of photons absorbed, so that efficiency may be determined from an action spectrum. However, if the action spectrum is the product of several reactions, as in photosynthetic CO_2 assimilation or O_2 release, the action spectrum of the basic reactions may be obscured. Ideally a process directly linked to energy consumption is measured, independent of other processes. For example, measurement of photosynthetic action spectra by O_2 or CO_2 exchange of whole leaves, algae, etc., is complicated by respiration and photorespiration which must be assessed in deriving the true action spectrum. Also the response at different wavelengths can only be compared when light intensity is limiting (i.e. the process is linearly dependent upon the amount of light captured) not at saturating light intensity where other factors limit.

If all pigments captured energy and delivered it to reaction centre chlorophyll with equal efficiency, and each wavelength was equally effective in promoting photosynthesis, then the chemical process would be the same at all wavelengths, and the action and absorption spectra would have the same shape. However, several pigments may capture energy with different efficiencies at different wavelengths so that absorption and action spectra do not match. Thus, in Fig. 3.8 there is close agreement between the action and absorption spectra for chloroplasts between 570 and 680 nm, but in the blue region where carotenoids absorb, less O_2 is produced per unit of light absorbed, as energy transfer from carotenoids to chlorophyll is inefficient.

Algae evolve O_2 at different wavelengths of light (Fig. 3.8) with an almost constant quantum yield (mol O_2 per mol photons absorbed) up to about 650−690 nm, at which point the rate decreases precipitately despite absorption of photons by chl *a*. This phenomenon is known as the 'red drop'. Adding short wavelengths together with red light greatly increases (enhances) O_2 evolution compared with the two wavelengths given separately (Fig. 3.8) — this is the enhancement or Emerson effect. It shows that there is co-operation between two pigment systems, called photosystems (abbreviated PS), one absorbing short, the other long wavelengths. The two photosystems are arranged in series, the short-wave absorbing photosystem (PSII) preceding the long (PSI). The order of wavelengths is important. Where the 'red drop' occurs only PSI operates and the PSII does not. The long-wavelength absorbing PSI 'primes' the system and the shorter-wavelength absorbing PSII delivers energy to it, with the photosystems acting in series. Identification of the two photosystems was a major achievement and forms the basis of much of our understanding of the biophysics of photosynthesis. Before discussing the characteristics of PSI and II in Chapter 5, evidence for groups of chlorophylls functioning as semi-discrete, energy gathering units, called photosynthetic units, will be considered.

The concept of photosynthetic units developed from studies of oxygen production. Algae (*Chlorella*) were illuminated with brief (10^{-5} s) flashes of intense light to saturate photosynthesis. This permitted only a single photon capture per chlorophyll molecule ('single turnover flash') and prevented the energy being used chemically during the flash. A dark period of about 40 ms at room temperature was required for maximum O_2 production per flash. Thus, photosynthesis was divisible into light and dark reactions which together gave a maximum efficiency of light utilization. In the cold a longer (0.4 s) dark period was needed for maximum O_2 evolution due to slower chemical processes. However, the maximum O_2 yield per flash was the same despite the different rates. When O_2 evolution was measured following flashes of different energy, separated by a dark period greater than 40 ms (at 25 °C) O_2 per flash increased to a maximum of one O_2 molecule for about 2400 molecules of chlorophyll activated. The ratio of O_2 evolved to photons absorbed (the quantum yield) in dim light is about one O_2 molecule per eight or ten photons absorbed, but at saturating light 2400 chlorophylls are excited to produce one O_2. Therefore, to evolve one O_2, 2400 chlorophyll molecules co-operated to collect eight photons, *c*. 300 chlorophylls per photon, and supply the energy to the chemical processes.

The group of chlorophylls, reaction centre and other pigments have been called a 'photosynthetic unit', although since the complexity of the system has been realized and the fact that the different photosystems occur in variable ratios, the term 'unit' is inappropriate. However, the linkage between photosystems implied by the term must still be considered. The eight quanta are captured by the two photosystems, so the energy of four photons per photosystem must be transferred to each reaction centre for 1 O_2 molecule, 4 e^- and 4 H^+ to be evolved and liberated from 2 H_2O molecules. A series coupling of two photosystems is suggested; each e^- is energized twice, first by a short wave photon (PSII) and again by a longer wave photon (PSI). Some antenna chlorophylls gather photons and transfer the excitation to the reaction centre, increasing the effective area per chlorophyll and allowing quanta to be absorbed in very dim light. This explains the gush of O_2 in experiments where the chance of capturing a photon was small; the eight quanta must first be 'accumulated' before electrons can be removed from water. However, energy gathered in one photosynthetic unit can be passed to other units or photosystems ('spill-over') under normal conditions, so that units are not functionally distinct. The photosynthetic unit corresponds loosely to the chlorophyll−protein complexes in thylakoid membranes (p. 69) associated with an individual photosystem. A thylakoid disc 500 nm in diameter may contain 10^5 chlorophylls associated with some 200 electron transport chains, each with one PSI and one PSII unit, 250 chlorophylls per photosystem. Chlorophyll *b* passes excitation energy on to chl *a* and then to P680 (PSII) or P700 (PSI) reaction centres. However, present concepts of photosynthetic organization are more dynamic; the size of the antenna and number of reaction centres and the proportion of PSI and II per chloroplast thylakoid are not fixed, but vary with species and conditions. Also, the photosynthetic unit is not the functional unit for all thylakoid processes; photophosphorylation, for example, is related to the whole thylakoid.

Quantum yield (also called quantum efficiency) measures the ability of photons to produce chemical change and is the number of O_2 molecules evolved (or CO_2 fixed) per quantum of light absorbed. Its reciprocal is the quantum requirement, that is, the number of quanta needed per O_2 produced or CO_2 consumed. The quantum yield of photosynthesis has been controversial; Warburg and associates claimed one O_2 evolved per four quanta absorbed but from eight to ten quanta is now accepted. The value of quantum yield is very important as it provides a basis for understanding the energetics of photosynthesis and to interpret the mechanism. To reduce one CO_2 requires 460 kJ mol^{-1} of quantum energy. Three quanta of red light (680 nm, 174 kJ mol^{-1}) would suffice if efficiency of capture and conversion approached 90 per cent, which is thermodynamically impossible (Clayton, 1980). Even four quanta of red light would require 66 per cent efficiency at minimum. Blue light, although more energetic, is no more effective than red light. Four quanta per CO_2 fixed or O_2 evolved conflicts with experiment and the concept of two light reactions. It is now accepted that eight photons are needed per O_2 as a theoretical minimum and more may be required, depending on conditions. The energy available from eight red photons is $8 \times 2.9 \times 10^{-10}$ J or 1.4 MJ mol^{-1}. Energy fixed in carbohydrates is 470 kJ mol^{-1} so the efficiency is $(0.47 \times 10^6)/(1.4 \times 10^6) \times 100$ or 34 per cent; ATP formulation increases efficiency to 38 per cent, a good efficiency for many energy conversions. However, under intense natural illumination with only about 50 per cent of radiation as PAR (400–700 nm), overall efficiency is less than 5 per cent; algae in dim light achieve 10 per cent efficiency.

References and Further Reading

Alscher, R.G. (1989) Biosynthesis and antioxidant function of glutathione in plants, *Physiol. plant*, **77**, 457–64.

Barber, J. (1987) Photosynthetic reaction centres: a common link, *TIBS*, **12**, 321–26.

Barber, J., Malkin, S. and **Telfer, A.** (1989) The origin of chlorophyll fluorescence *in vivo* and its quenching by the photosystem II reaction centre, pp. 227–39 in Walker, D.A. and Osmond, C.B. (eds), *New Vistas in Measurement of Photosynthesis. Phil. Trans. R. Soc. Lond.*, **B323**, 225–448.

Butler, W.L. (1978) Energy distribution in the photochemical apparatus of photosynthesis, *A. Rev. Plant Physiol.*, **29**, 345–78.

Clayton, R.K. (1980) *Photosynthesis: Physical Mechanisms and Chemical Patterns*, I.U.P.A.B. Biophysics Series, Cambridge University Press.

Cogdell, R.J. (1983) Photosynthetic reaction centers, *A. Rev. Plant Physiol.*, **34**, 21–45.

Döring, G., Renger, G., Vater, J. and **Witt, H.J.** (1969) Properties of the photoactive chlorophyll in photosynthesis, *Zeitschrift für Naturforschung*, **B24b**, 1139–43.

Dörr, F. (1983) Photophysics and photochemistry. General principles, pp. 265–88 in Hoppe, W., Lohmann, W., Markl, H. and Ziegler, H. (eds), *Biophysics*, Springer-Verlag, Berlin.

Drews, G., Kaufmann, N. and **Klug, G.** (1985) The bacterial photosynthetic apparatus: molecular organization, genetic structure, and biosynthesis of the pigment–protein complexes, pp. 211–22 in Steinback *et al.* (eds), *Molecular Biology of the Photosynthetic Apparatus*, Cold Spring Harbor Laboratory.

Gantt, E., Lipschultz, C.A. and **Redlinger, T.** (1985) Phycobilisomes: A terminal acceptor pigment in cyanobacteria and red algae, pp. 223–29 in Steinback *et al.*, (eds), *Molecular Biology of the Photosynthetic Apparatus*, Cold Spring Harbor Laboratory.

Glazer, A.N. (1983) Comparative biochemistry of photosynthetic light-harvesting systems, *A. Rev. Biochem.*, **52**, 125–57.

Glazer, A.N. (1985) Phycobilisomes: Structure and dynamics of energy flow, pp. 231–40 in Steinback *et al.*, (eds), *Molecular Biology of the Photosynthetic Apparatus*, Cold Spring Harbor Laboratory.

Golbeck, J.H., Lien, S. and **San Pietro, A.** (1977) Electron transport in chloroplasts, pp. 94–116 in Trebst, A. and Avron, M. (eds), *Encyclopedia of Plant Physiology* (N.S.), Vol. 5, *Photosynthesis I*, Springer-Verlag, Berlin.

Holzwarth, A.R. (1991) Excited state kinetics in chlorophyll systems and its relationship to the functional organisation of the photosystem, in Scheer, H. (ed.), *The Chlorophylls*, CRC Handbook, Boca Raton, CRC Press.

Kok, B. (1961) Partial purification and determination of oxidation–reduction potential of the photosynthetic chlorophyll complex absorbing at 700 nm, *Biochim Biophys. Acta*, **48**, 527.

Krinsky, N.I. (1978) Non-photosynthetic functions of carotenoids, *Phil. Trans. R. Soc. Lond. B.*, **284**, 581–90.

Mathis, P. (1981) Primary photochemical reactions in photosystem II, pp. 827–37 in Akoyunoglou, G. (ed.), *Photosynthesis III. Structure and Molecular Organisation of the Photosynthetic Apparatus*, Balaban International Science Services, Philadelphia.

Mathis, P. and **Paillotin, G.** (1981) Primary processes of photosynthesis, pp. 98–161 in Hatch, M.D. and Boardman, N.K. (eds), *The Biochemistry of Plants*, Vol. 8, *Photosynthesis*, Academic Press, New York.

Mauzerall, D. (1977) Porphyrins, chlorophyll and photosynthesis, pp. 117–24 in Trebst, A. and Avron, M. (eds), *Encyclopedia of Plant Physiology* (N.S.), Vol. 5, *Photosynthesis I*, Springer-Verlag, Berlin.

Michel-Beyerle, M.-E. (ed.) (1990) *Reaction centres of photosynthetic bacteria*, *Springer Series in Biophysics*, Vol. 6, Springer-Verlag, Berlin.

Okamura, M.Y., Feher, G. and **Nelson, N.** (1982) Reaction centers, pp. 195–272 in Govindjee (ed.), *Photosynthesis*, Vol. 1, *Energy Conversion by Plants and Bacteria*, Academic Press, New York.

Parsons, W.W. (1987) The bacterial reaction centre, pp. 43–61 in Amesz, J. (ed.), *Photosynthesis. New Comprehensive Biochemistry*, Vol. 15, Elsevier, Amsterdam.

Renger, G. (1983) Photosynthesis, pp. 515–42 in Hoppe, W., Lohmann, W., Markl, H. and Ziegler, H. (eds), *Biophysics*, Springer-Verlag, Berlin.

Truscott, T.G. (1990) The photophysics and photochemistry of the carotenoids, *J. Photochem. Photobiol. B: Biol.*, **6**, 359–71.

Witt, H.T. (1975) Energy conservation in the functional membrane, pp. 495–554 in Govindjee (ed.), *Bioenergetics of Photosynthesis*, Academic Press, New York.

CHAPTER 4

Architecture of the photosynthetic apparatus

Structural components of photosynthesis form a hierarchy of organization 'levels' of different size and complexity which co-operate. In eukaryotes, and particularly vascular plants, they are:

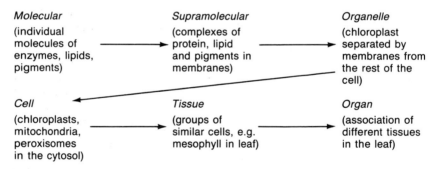

Molecular
(individual molecules of enzymes, lipids, pigments)

Supramolecular
(complexes of protein, lipid and pigments in membranes)

Organelle
(chloroplast separated by membranes from the rest of the cell)

Cell
(chloroplasts, mitochondria, peroxisomes in the cytosol)

Tissue
(groups of similar cells, e.g. mesophyll in leaf)

Organ
(association of different tissues in the leaf)

The structures developed in photosynthetic organisms, from bacteria to higher plants, for capturing and converting light energy to chemical form, are similar, particularly at the molecular and supramolecular levels (p. 10, Ch. 1). In all photosynthetic systems, a membrane containing the light capturing and energy conversion components separates the inside of the cell from what is, in terms of the developmental morphology of the photosynthetic membrane, effectively the outside of the cell. The basic structure may be regarded as that of *Halobacterium* (Fig. 1.2). In many photosynthetic bacteria extensive invagination of the external cell membrane into the cell's interior has produced, during evolution, large surface area for light capture, H^+ diffusion and ATP synthesis. Progression of organization from simple membranes to organelles and cells is marked by great variation in structure. Particularly compartmentation of the chloroplast from the rest of the cell is a major distinction between prokaryotes and eukaryotes. Further differences at tissue and organ level are very apparent in more complex algae and vascular plants.

Photosynthetic prokaryotes are a diverse group, differing in structure as well as pigment composition (Ch. 3) and in the source of reductant for CO_2 assimilation. The photosynthetic bacteria, as a group, have photosynthetic membranes of different extent and form, ranging from simple, flat membranes as in *Rhodopseudomonas viridis* to extensive flattened sacks in *Chromatium vinosum* which probably form a continuous system, to large vesicles. In *Chlorobium limicola* these vesicles contain up to 10 000 bchl molecules each. The membranes are 5 nm thick, contain up to 50 per cent protein in the form of granular patches and 5 per cent pigments.

The cyanobacteria lack membrane bound organelles but have extensive double thylakoids arranged in parallel on which the pigment–protein antenna complexes, the phycobilisomes, are regularly distributed. They are described in Chapters 3 and 5, respectively. Clearly the great similarities between different photosynthetic organisms reflects evolution from a single type of basic mechanism or parallel evolution in different groups.

Photosynthetic eukaryotes include algae and higher plants. The algae are a large, diverse group differing in structure and pigment composition (Ch. 3) but all have thylakoid membranes enclosed in a chloroplast envelope and so are effectively compartmented from other parts of the cell. Structure of their thylakoids and chloroplasts resembles that of vascular plants and is not further considered. These algae often form very complex thalli with extensive laminae for light capture.

Leaves are the site of most higher plant photosynthesis and provide the necessary conditions to maintain it, for example, epidermis and cuticle minimize water loss. Leaf structure, shape and cell distribution are genetically determined but change, within limits, with growth conditions, allowing adjustment to environment. A semi-quantitative estimate of the number and size of component parts of an 'average' higher plant leaf (Table 4.1) is based on data for spinach, tobacco and wheat.

The mesophyll tissue is differentiated in many species (Fig. 4.1) into palisade and spongy mesophyll cells, both with extensive air passages contiguous with the substomatal cavities. Rate of exchange of gases between cells and the atmosphere is related to their surface area and to the geometry of the intercellular spaces of the mesophyll (see p. 245); capture of light also depends on mesophyll structure, for example, the number and density of cell layers. Photosynthetic processes appear to be the same in palisade and spongy mesophyll cells of C3 plants but in C4 plants (Ch. 9) the mesophyll is clearly differentiated into morphologically distinct tissues with different photosynthetic functions.

Individual photosynthetic cells in algae and most higher plants perform all the primary, and many secondary, reactions of photosynthesis, requiring only water and nutrients. Mesophyll cells may be isolated from leaves by enzymes which break the bonds holding together the walls of the adjacent cells. Protoplasts may also be freed from the cell wall by enzymes which lyse bonds in the wall. Cells and protoplasts will photosynthesize actively when given CO_2, light and nutrients, if the osmotic potential of the

Table 4.1 A semi-quantitative analysis of the photosynthetic system in an 'average' C3 plant leaf

Leaf characteristics

Area of leaf	1 m^2
Thickness of leaf	$3 \times 10^{-4} \text{ m}$
Volume of (1 m^2) leaf	$3 \times 10^{-4} \text{ m}^3$
Fresh mass of (1 m^2) leaf	0.17 g
Dry mass of (1 m^2) leaf	0.04 g
Volume of cells (assuming density of dry matter = $1.2 \times 10^3 \text{ kg m}^{-3}$)	$2.2 \times 10^{-4} \text{ m}^3$
Air space in leaf = total–cell volume	$0.8 \times 10^{-4} \text{ m}^3$
As % of total volume	27%

Leaf cell number per m^2 leaf

Leaf has: 1 layer of palisade mesophyll cells in transverse section and in parallel section contains	4×10^9 cells
5 layers of spongy mesophyll cells in transverse section and in parallel section each layer contains	0.6×10^9 cells
Total spongy mesophyll cell number	3×10^9 cells
Total mesophyll cell number	7×10^9 cells

Cell dimensions

Palisade cell	
'average' diameter	$2 \times 10^{-5} \text{ m}$
'average' length	$8 \times 10^{-5} \text{ m}$
total volume (for cylinder with hemispherical ends)	$2.9 \times 10^{-14} \text{ m}^3$
total volume of all palisade cells	$1.2 \times 10^{-4} \text{ m}^3$
Spongy mesophyll cell	
'average' diameter of cell	$3 \times 10^{-5} \text{ m}^3$
volume of cell	$1.4 \times 10^{-14} \text{ m}^3$
total volume of all spongy mesophyll cells	$4.2 \times 10^{-5} \text{ m}^3$
Total volume of mesophyll cells	$1.6 \times 10^{-4} \text{ m}^3$

Chloroplasts in cells

Assume that each palisade cell contains 100 chloroplasts: total chloroplasts m^{-2}	4.0×10^{11}
Assume each spongy mesophyll cell contains 50 chloroplasts: total chloroplasts m^{-2}	1.5×10^{11}
Total number of chloroplasts m^{-2} leaf	5.5×10^{11}

Chloroplast

A chloroplast approximates a hemisphere of diameter	$5 \times 10^{-6} \text{ m}$
Volume of 1 chloroplast	$3.3 \times 10^{-17} \text{ m}^3$
Total chloroplast volume in 1 m^2 of leaf	$1.8 \times 10^{-5} \text{ m}^3$
Chloroplasts occupy $\approx$ 8% of cell volume	

Table 4.1 (continued)

Vacuoles occupy 80% of the cell volume, therefore the cytoplasm volume is	4.8×10^{-5} m^3 m^{-2} leaf
and the chloroplasts occupy 38% of cytosol volume	
Chloroplast envelope area chloroplast^{-1}	5.9×10^{-11} m^2
Chloroplast envelope area m^{-2} leaf	32 m^2
Chloroplast envelope area mg chlorophyll^{-1}	0.064 m^2
Total volume of chloroplast stroma	1.2×10^{-5} m^3

Thylakoid system

A typical chloroplast contains 60 grana stacks and each has 15 thylakoids/granum, i.e. 30 membranes. Diameter of 1 'end' of granum is	4.5×10^{-7} m
Area of vesicle (not allowing for ends)	1.6×10^{-13} m^2
Total area of membranes in one stack	5.0×10^{-12} m^2
Total area of grana membranes in one chloroplast	3×10^{-10} m^2
If four stromal thylakoids the width of the granum pass across the diameter of chloroplast the stromal thylakoid area is	2.4×10^{-10} m^2
Total thylakoid area, grana + stroma per chloroplast	5.4×10^{-10} m^2
Total thylakoid area m^{-2} leaf	300 m^2 (other estimates 835 m^2)
Thickness of thylakoid membranes (approx.)	7.5×10^{-9} m
Volume of thylakoid membranes	4.1×10^{-18} m^3
Volume of thylakoid lumen	
area of end surface of a granum	1.6×10^{-13} m^2
space between membranes is approx.	8.0×10^{-9} m
volume of individual vesicle	1.3×10^{-21} m^3
volume of lumen in one stack	1.9×10^{-20} m^3
volume of grana lumen in a chloroplast	1.2×10^{-18} m^3
volume of stromal thylakoids chloroplast^{-1}	9.0×10^{-19} m^3
total volume of thylakoid lumen chloroplast^{-1}	2.1×10^{-18} m^3
total volume of thylakoid lumen m^{-2} leaf	1.2×10^{-6} m^3
total volume of membrane + lumen in a chloroplast or ~ 20% of chloroplast volume	6.2×10^{-18} m^3

Chlorophyll content of leaves

Average content of chlorophyll m^{-2}	0.5 g
molecular mass of chlorophyll *a*	894
chlorophyll content	5.6×10^{-4} mol m^{-2}
or	1.9 mol m^{-3} leaf
Chlorophyll/chloroplast	9×10^{-13} g
or	1.0×10^{-15} mol
Vol. thylakoid lumen (g chlorophyll)$^{-1}$	2.3×10^{-6} m^3 g^{-1}
Vol. chloroplast (g chlorophyll)$^{-1}$ or vol. mol^{-1}	3.6×10^{-5} m^3 g^{-1} 0.033 m^3 mol^{-1}
Vol. thylakoid membrane (g chlorophyll)$^{-1}$	4.6×10^{-6} m^3 g^{-1}

Table 4.1 (continued)

Area thylakoid membrane (g chlorophyll)$^{-1}$	600 m^2 g^{-1} (other estimates 1670 m^2 g^{-1})
Conc. chlorophyll in thylakoid	2.2 $\times$ 10^5 g m^{-3}
or	2.4 $\times$ 10^2 mol m^{-3} (0.24 M)
No. chlorophyll molecules chloroplast^{-1}	6.7 $\times$ 10^8
Area membrane (chlorophyll molecule)$^{-1}$	1.2 $\times$ 10^{-18} m^2
Area per 'head' of chlorophyll molecule	2.2 $\times$ 10^{-18} m^2
Miscellaneous values	
Chlorophyll molecules per photosystem I & II	300
Conc. of NADP reductase in stroma	8 $\times$ 10^{-5} M
Coupling factor (CF$_1$)	0.45 g (g chlorophyll)$^{-1}$
Molecular mass of CF$_1$	325 kDa
CF$_1$ granum^{-1}	200
ATP content illuminated chloroplasts (nmol ATP mg chlorophyll^{-1})	40
ADP content (nmol ATP mg chlorophyll^{-1})	12
ATP conc.	1.5 mM
ADP conc.	0.5 mM
ATP/ADP ratio	3
NADPH conc.	0.1 mM
RuBP carboxylase/oxygenase conc. of enzyme sites	4 mM
Electron transport chains per thylakoid 'disc'	200
Dry mass per chloroplast	20 $\times$ 10^{-12} g
Vol. water to total vol.	75%
Protein content (% dry mass of chloroplast)	60%
Lipid content (% dry mass of chloroplast)	20%
Chlorophyll (% dry mass of chloroplast)	4%
Carotenoids (% dry mass of chloroplast)	0.9%
Nucleic acids (% dry mass of chloroplast)	2.5%
Soluble products of photosynthesis	7.5%
Mg conc. in chloroplast stroma (light)	30 mM
Mg conc. in chloroplast stroma (dark)	15 mM
Inorganic phosphate in chloroplasts (mol P$_i$ mg chlorophyll^{-1})	3
Inorganic phosphate conc.	100 mM
RuBP conc.	0.1–2 mM

medium is correct. Cells and protoplasts from C3 plants function independently. However, in C4 plants co-operation has evolved between photosynthetic cells and their autonomy has been lost.

The ultrastructure of photosynthetic cells is shown in Figs 4.2–4.4. In the thin layer of cytoplasm around the large central vacuole are chloroplasts, mitochondria and peroxisomes, which are all involved in aspects of

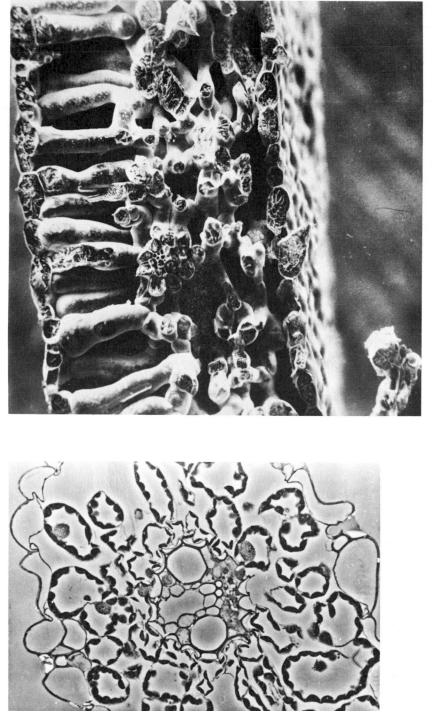

FIG. 4.1   (a) Transverse section of wheat (*Triticum aestivum* L.) leaf (× 400) showing the loosely packed mesophyll cells arranged around a central vascular bundle. The cells contain chloroplasts lining the cell wall. (b) Scanning electron micrograph of a transverse section of a bean (*Vicia faba*) leaf (× 216). Electron micrographs courtesy of R. Turner, Rothamsted Experimental Station, Harpenden, Herts, UK.

(a)

(b)

photosynthetic metabolism, together with inclusions such as pigment granules and starch grains. The number of chloroplasts per cell is very variable, approximately 20–100 in higher plants. Chloroplast distribution in cells differs with species and changes with conditions, such as illumination. C3 plants have typically flattened spherical or lens-shaped chloroplasts some 3–10 μm in greatest dimensions (Fig. 4.2). Mitochondria are prominent, with characteristic membranes, the inner often folded. Peroxisomes have a single limiting membrane and granular contents, without distinctive features in electron micrographs, and are often closely associated with chloroplasts. The number of mitochondria and peroxisomes varies and may depend on the function of the cell and on the environment during growth. Peroxisomes metabolize products from the chloroplast in co-operation with mitochondria, for example, in photorespiration (Ch. 8).

Chloroplasts perform all the primary (e.g. light capture and electron transport leading to NADPH and ATP synthesis) and most of the secondary processes (e.g. synthesis of 3-carbon phosphorylated compounds from CO_2) in photosynthesis and their structure will be considered in detail. They are bounded by a continuous envelope and contain all the membrane bound light-harvesting chlorophyll and other pigments, proteins and redox compounds involved in transport of electrons, and the associated proteins, plus the soluble enzymes and substrates required for CO_2, NO_3^- and SO_4^{2-} assimilation by photosynthesis together with its products.

Chloroplasts are isolated by breaking the cell wall mechanically, in a buffer solution (pH about 7) with non-permeating osmotic substances (e.g. sorbitol or sucrose, 0.3 M) to maintain the osmotic potential and ions, for example, phosphate (30 mM) and magnesium (5 mM). Chloroplasts are separated from other organelles by centrifugation (90 s at 3000 g) and then suspended in buffered solution (see Leegood and Walker (1983) for methods).

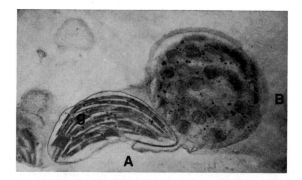

FIG. 4.2   Structure of higher plant chloroplasts, illustrated by an electron micrograph (× 10 000) of bean (*Vicia faba*). (A) shows in transverse section the double chloroplast envelope enclosing the stroma and the thylakoid membranes forming granal stacks (g) and unstacked stromal thylakoids; (B) has been sectioned at right angles to (A). The granal discs are clearly visible. Courtesy of Dr A.D. Greenwood, Imperial College, London.

However, mechanical damage and unphysiological conditions may impair the envelope membranes. Intact chloroplasts assimilate CO_2 faster than damaged ones because they retain enzymes and co-factors. The envelope is detected by the bright appearance of isolated chloroplasts in phase contrast microscopy. However, the envelope may appear intact yet have broken and resealed. Ferricyanide, an electron acceptor in electron transport, cannot penetrate intact membranes, so measuring its reduction ($Fe(CN)_6)^{3-} + e^-$ $\rightarrow (Fe(CN)_6)^{4-}$ at 420 nm indicates the state of the envelope. Hall (1972) has classified chloroplasts according to intactness of the outer envelope and retention of metabolites.

Manipulation of isolated chloroplasts has been very important in analysing their functions. Separated thylakoids and envelope may be disrupted by detergents or ultrasonically; the components are separated by column chromatography, density gradient fractionation, gel electrophoresis, etc., and chemically analysed to determine the components of light-harvesting systems and electron transport chains. The position of proteins or other components within the membranes is found by using specific antibodies. If an antibody only affects, for example, electron transport in disrupted thylakoids, then the antigen is inside the membranes. Similarly, chemical probes of known function are used to penetrate the membranes, changing processes and indicating the position of components. X-ray studies may be made on particles or membrane pieces to study the orientation of components. These techniques in combination provide a picture of chloroplast membrane and thylakoid vesicle structure and function.

Chloroplast internal structure

Electron microscopy shows the chloroplast to consist of an envelope enclosing a complex of membranes, the thylakoid system often joined or stacked into grana (Figs 4.2 and 4.3); the lipid membranes contrast with the background when stained with lipophilic electron dense osmium. The space between the envelope and thylakoid membranes is the chloroplast stroma. The envelope is composed of two membranes each about 5.6 nm thick separated by the intraenvelope space (*c*. 10 nm) with areas of high electron density which are possibly contact points between the membranes; they may be involved in transport, e.g. of proteins between cytosol and stroma. The membranes are lipid bilayers, of galactosyl glycerides and phosphatidyl choline, containing carotenoids but no chlorophyll. The membranes are not identical in structure or function. The outer (cytoplasmic layer) has a density of 1.08 g cm^{-3} and 1500–3300 particles μm^{-2}; the lipid/protein ratio is 2.5–3. The inner membrane has 7500–10 000 particles μm^{-2}, a correspondingly greater density (1.13 g cm^{-3}) and 1–1.2 mg lipid mg protein^{-1}. MGDG makes up 50 per cent of the inner envelope and thylakoid membranes, but has very different function in the two membranes. There are many polypeptides (more than 75 and possibly

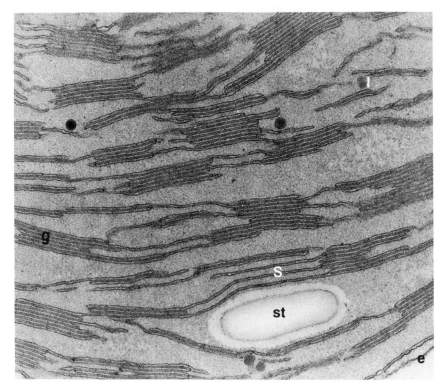

FIG. 4.3 Transverse section of spinach (*Spinacea oleracea*) chloroplast by electron microscopy (× 103 000); thylakoid membranes are associated into grana (g) joined by the stromal thylakoids (S). Osmophilic lipid globules (l) are apparent and a starch granule (st) also. The envelope (e) appears as a single membrane at the lower right hand corner. Courtesy of Dr A.D. Greenwood, Imperial College, London.

several hundred) of 10−120 kDa, with major components of 54, 37, 14 and 12 kDa, possibly involved in metabolite transport. The 30 kDa polypeptide is probably the triosphosphate : P_i translocator and the 37 kDa component is protochlorophyllide oxidoreductase. There are glycolipids giving rise to galactolipids and sulpholipids; up to 95 per cent of these are polyunsaturated fatty acids, e.g. linolenic acid (18:3), and others have 16:3 fatty acids but the significance of these differences is not understood. Phosphocholine is a major envelope phospholipid making up 30−35 per cent of the outer membrane glycerolipid. The outer cytoplasmic membrane allows may substrates to pass freely, whereas the inner (stromal) membrane is highly selective, allowing passage of only some solutes by special enzyme systems, called translocators. The cytoplasmic membrane has 9 nm diameter particles in both halves of the lipid bilayer at a density of 1.5×10^2 particles μm^{-2}. The stroma membrane has 7 nm and 9 nm diameter particles at a density of 1×10^3 and 1.8×10^3 μm^{-2} in the halves of the lipid bilayer next to the stroma and intermembrane space, respectively. The particles in both membranes are probably protein complexes, some associated with the translocators but constitute only 1 per cent of the total chloroplast protein.

The stroma contains indistinct granules and particles, mainly of proteins; the enzyme ribulose bisphosphate carboxylase (Rubisco) is the major soluble protein and may crystallize in unfavourable conditions such as water stress or air pollution. Other inclusions are products of the photosynthetic processes; for example, starch granules up to 2 μm long accumulate in the stroma and disturb the thylakoid membranes, and globules of lipids and plastoquinone accumulate; RNAs and DNA occur in chloroplasts which synthesize many of their constituent proteins.

The most noticeable feature of chloroplasts in electron micrographs (Fig. 4.2) is the thylakoid (from the Greek $\theta\upsilon\lambda\alpha\kappa\omicron\epsilon\iota\delta\acute{\varsigma}$ for 'sack like') membrane vesicle system. In transverse section the thylakoids appear as parallel pairs of continuous membranes separated by a space, the thylakoid lumen (Fig. 4.3) which is 5–10 nm wide and contains few identifiable features. Thylakoid membranes frequently associate into granal stacks, interconnected by pairs of membranes, called stromal thylakoids (or alternatively intergranal connections or frets), which are in contact with the stroma on both sides. The interface between the appressed membranes is the partition region. In C3 plants over 60 per cent of the thylakoid surface is typically in the grana. The end membranes of stacked thylakoids and the ends of the grana, but not the partition regions, have direct contact with the stroma (Fig. 4.4).

Several models of the three-dimensional structure of the thylakoid system have been suggested from analysis of serial sections of chloroplasts. Thylakoid membrane vesicles in grana are stacked and flattened, but not closed, sacs (Fig. 4.4) interconnected with the other membranes. The vesicles join the stromal lamellae at different points around the periphery of a granum. The structure derives from folding and joining of separate sheets of lamellae which are interconnected and probably originate from a single point (Fig. 4.4), the prolamella body, in the developing chloroplast (Staehelin and Arntzen 1979). The thylakoid system appears to be a single interconnecting giant closed vesicle with continuous lumen, a feature of great importance in electron transport and ATP generation. Composition of the lumen is not known; but proteins, of the water-splitting complex and the light-harvesting complex for example, may occupy part of the volume and it is unlikely to be a homogeneous aqueous solution of small molecules. Grana differ in extent and size between species, and with conditions during growth, for example, with bright illumination there is less granal stacking. A semi-quantitative summary of the size of an average thylakoid system is given in Table 4.1. Grana in isolated thylakoids stack and unstack, according to the ionic concentration and light quality.

The internal composition and structure of thylakoid membranes

A membrane is 5–7 nm thick and consists of a lipid bilayer (50 per cent of the mass) together with proteins, pigments and other major components

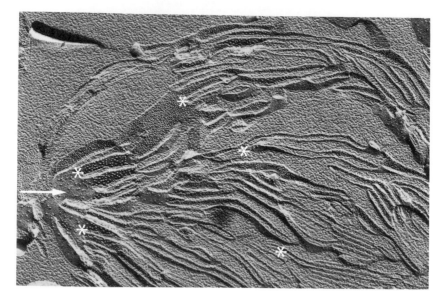

FIG. 4.4   Thylakoid membranes of an isolated spinach chloroplast after freeze-fracture, showing progressive branching of membranes (*) from a common point (→). Electron micrograph (× 50 000) by courtesy of Professor L.A. Staehelin, University of Colorado, Boulder, Colorado, USA.

which are vital for photosynthesis. Thylakoid lipids are a complex mixture; some 80 per cent is glycolipid containing galactose, such as monogalactosyl-diglyceride (50 per cent of total lipid on a molar basis) and digalactosyl-diglyceride (25 per cent) which have neutral hydrophobic heads. The remainder is mainly phospholipid (10 per cent) and sulpholipid (5 per cent) charged at pH 7. Synthesis and structure of lipids is considered in Chapter 8. The fatty acids of lipids are highly unsaturated. Linolenic acid (C18:3) is the predominant fatty acid and *trans*-3-hexadecanoic acid (C16:1) acylated to phosphatidyl glycerol is specific to thylakoid membranes; its function may be structural. The fatty acid tails form a non-aqueous, hydrophobic, central core to the membrane, whilst the hydrophilic heads are at the surface. The outermost half of the membrane, next to the stroma is some 3−5 nm thick; the inner, next to the lumen is 2−3 nm thick. As the two most abundant lipids are highly unsaturated, the membrane is very fluid at physiological temperatures with little cholesterol or other sterol to cause rigidity. Fluidity allows movement of pigment−protein complexes laterally through the membrane. The lateral diffusion coefficient of lipids is 10^{-10} m^2 s^{-1} and of proteins 5×10^{-11} m^2 s^{-1}. As the distances over which pigment−protein complexes move are small, displacements of the order of 10−100 nm will occur rapidly, particularly if the proteins are charged, enabling the thylakoid to change its structure and function as necessary for optimizing ion and water fluxes across the membrane and energy distribution between light transducing complexes within it.

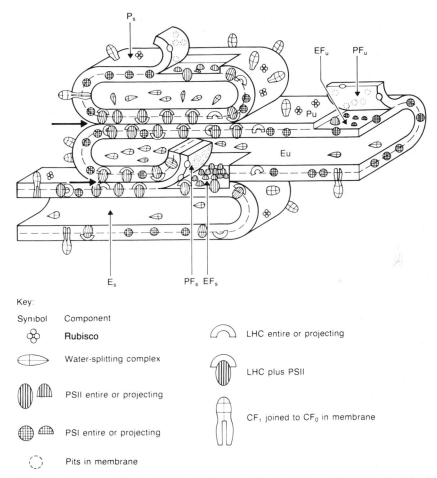

Key:

Symbol	Component

Rubisco

Water-splitting complex

PSII entire or projecting

PSI entire or projecting

Pits in membrane

LHC entire or projecting

LHC plus PSII

CF$_1$ joined to CF$_0$ in membrane

FIG. 4.5   Diagram of the thylakoids of a higher (C3) plant chloroplast. Part of a granal stack and stromal thylakoid is shown. Coupling factor (CF$_1$) and ribulose bisphosphate carboxylase (Rubisco) are attached but not in the partition region (→). In the lumen of the granal thylakoids are particles of the photosystem II, water-splitting complex. The partition region contains particles on the endoplasmic fracture face (EF$_s$) identified with PSII and the light-harvesting protein complex. Smaller particles on other membranes may be PSI and the base of coupling factor. See text for explanation of membrane surfaces and particles.

Particles in thylakoid membrane (Fig. 4.5)

Surface structure of membranes is observed by electron microscopy of isolated membranes. Inner membrane structure may be seen after freeze-fracturing thylakoids. Membranes are frozen and cut, during which the membrane bilayer separates along the line of weakness caused by the hydrophobic tails of the membrane lipids, exposing particles within the membrane.

The surfaces of membranes in electron micrographs of freeze-fractured thylakoids are denoted by their contact with the stroma, i.e. protoplasmic surface (PS) or the lumen, i.e. endoplasmic surface (ES) and their fractured surfaces are PF and EF, respectively. Membranes from stacked (granal) or unstacked (agranal) regions are shown by subscript s and u, respectively (Branton *et al.* 1975). On the outer surface of stromal (PS_u) and of granal thylakoids (PS_s) in contact with the stroma, are particles of Rubisco loosely attached and easily removed; most prominent are club-shaped 15 nm diameter particles of coupling factor, called CF_1 which synthesizes ATP. Neither Rubisco nor CF_1 which are extrinsic, (i.e. external) proteins occur in the partition region. Smaller (9 nm diameter) particles on PS are exposed parts of intrinsic proteins (i.e. occurring within the membrane) some possibly the base part of CF_1, called CF_0. Also, particles from within the membrane may project rather indistinctly out of the surface, and form a lattice with rows 8−9 nm apart and interparticle distances of 10 nm. On the very smooth inner lumen surface, ES_s, of granal thylakoids are rectangular (10 × 15 nm) particles of four (sometimes two or six) subunits each *c.* 5 nm in diameter (Fig. 4.6). In artificially unstacked thylakoids these large particles may be arranged in a lattice when they always have four units of 18 × 20 nm spacing on the ES_s face. The stromal thylakoid inner surface (ES_u) is much more textured than ES_s and has only a few smaller (4−6 nm diameter particles) just projecting above the surface. More than

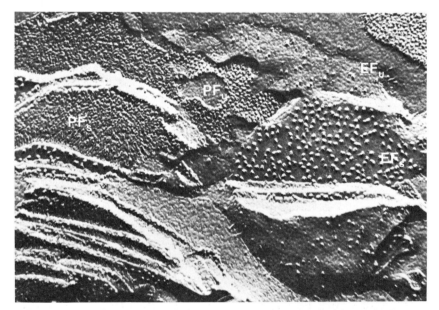

FIG. 4.6   Freeze-fractured granal stacks of isolated thylakoids of a spinach chloroplast, showing the EF and PF faces of stacked and unstacked areas. Note the characteristic large (15 nm) particles on EF_s. The stacks are linked by a sheet of unstacked membrane (↓). Electron micrograph (× 185 000) by courtesy of Professor L.A. Staehelin, University of Colorado, Boulder, Colorado, USA.

80 per cent of the large particles of EF faces occur in granal regions and 20 per cent or less in stromal areas.

Particle distribution on fractured membranes has been analysed mainly on spinach (Staehelin *et al.* 1977). Figure 4.6 shows an electron micrograph of the freeze-fracture faces. On EF_s are many (*c.* 1500 particles μm^{-2}) large particles in two populations of 15 and 11 nm diameter, 60–70 per cent and 30–40 per cent, respectively of the total population. Depending on the conditions (salt concentrations, etc.) these particles, which appear lobed, may be arranged in very regular arrays or lattices forming a uniform population about 16 nm diameter. There are also about 700 smaller particles μm^{-2}, 10 nm diameter, of two subunits. On the stromal face, EF_u, there are fewer (450–570 μm^{-2}) smaller (10 nm diameter) particles which are not arranged in a lattice, and are more scattered than on the EF_s face.

The PF faces in both stromal and granal thylakoids contain particles: PF_s has 3500–4500 particles μm^{-2}, of 8 nm average diameter (range 5–12 nm). PF_u has 3600 particles μm^{-2}, in two groups of 8 and 11 nm average diameter and less deeply embedded than those in the PF_s faces. Deep pits in the PF_s are left by the large particles on the EF_s face tearing out of the PF_s lipid bilayer during fracturing (Figs 4.5 and 4.6). Four (sometimes two or three) smaller PF_s particles are regularly arranged around pits and would be closely associated with large EF particles. The groups, separated by about 2.5 nm, form the lattices in granal areas when they penetrate the ES and PS faces.

Particles in membranes have been identified with PSI and II, and the light-harvesting chlorophyll–protein and cytochrome *b*–*f* complexes (see p. 89). Probably the large EF particle has a core of a PSII complex (8 nm diameter) associated with two, four or six units of light-harvesting complex in granal thylakoids, but only one in stromal membranes. Smaller particles in the PF_u and PF_s surfaces may be chlorophyll–protein complexes of PSI (see p. 91), light-harvesting complexes and CF_0. Variations in size may be due to association of light-harvesting complexes with different numbers of other complexes or with components of the electron transport chain.

Large particles span the lipid bilayer from inside to outside surfaces in stacked membranes. Unstacked membrane particles are not arranged in regular lattice arrangements and do not span the membrane. Perhaps forces on or in the membrane cause the large particles to stand upright, make the lipid layer thinner, and produce arrays. However, the smaller PF particles do not project into the thylakoid lumen.

Thylakoid membranes are 'sided' in construction, with the water-splitting complex in the lumen, a PSII chlorophyll–protein complex, a cytochrome *b*–*f* complex and light-harvesting complex spanning the membrane interspersed with the PSI chlorophyll–protein complex on the outer side, and finally enzymes of carbon metabolism and ATP synthesis on the outer surface. This sidedness allows thylakoids to transport electrons to the stroma from water in the lumen and accumulate protons in the lumen.

Membrane structure of C4 plants

Considerable differences occur between C4 species in chloroplast structure; here maize is taken as an example. Its photosynthetic tissue is divided into

(a)

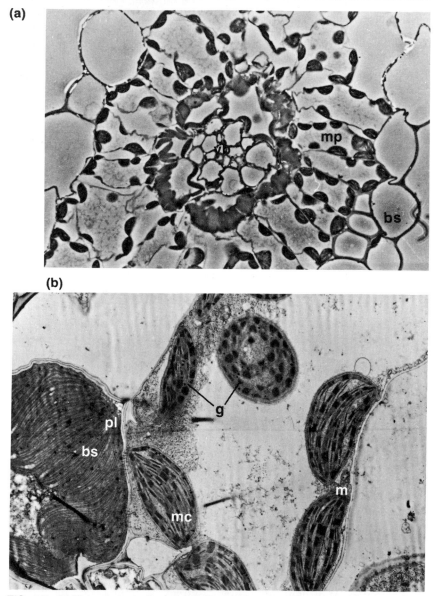

(b)

FIG. 4.7 (a) Transverse section of a maize leaf (× 440) with the vascular bundle (vb) surrounded by a bundle sheath (bs) of large cells with prominent chloroplasts. Cells of the mesophyll parenchyma (mp) have smaller, less densely packed chloroplasts. (b) Section of maize bundle sheath cell with chloroplast (bs) and mesophyll cell (m) with chloroplast (mc) showing chloroplast dimorphism. Plasmodesmata (pl) traverse the wall between the two cell types (g = grana). Electron micrograph (× 3560) by courtesy of Mr R. Turner, Rothamsted Experimental Station, Harpenden, Herts, UK.

bundle sheath and mesophyll (Fig. 4.7a and b) with different forms of chloroplast (Fig. 4.7b). Mesophyll cell chloroplasts are comparable to those of C3 plants already described, with many discoid chloroplasts about 8 μm diameter, double membrane envelope, thylakoids arranged in parallel and many obvious grana. Bundle sheath cells contain few, large (15−20 μm diameter) chloroplasts with many parallel, densely packed thylakoids with no obvious grana, although some contact occurs. A dense layer of tubules, the chloroplast reticulum, lines the envelope which is closely appressed to a dense layer of cytoplasm next to the cell wall; groups of plasmodesmata connect the bundle sheath and mesophyll cells.

Thylakoid membranes of the two types of chloroplast differ in macromolecular construction (Fig. 4.8). Granal and stromal areas of the C4 mesophyll chloroplast are similar to the equivalent areas of C3 plants.

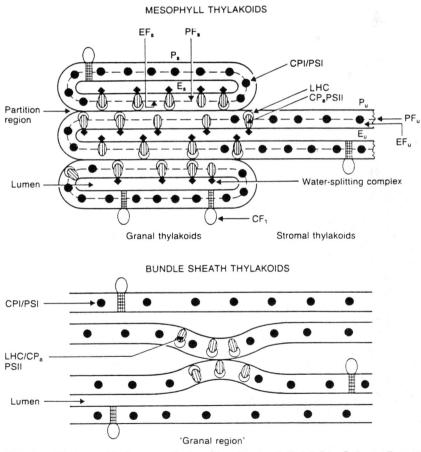

FIG. 4.8 Diagram of maize mesophyll and bundle sheath thylakoids. Only small areas of contact (grana) occur between bundle sheath lamellae compared with extensive grana in the mesophyll. Particles of PSII and associated light-harvesting complex are in grana of both types.

The EF_s face is extensive with large particles (15 nm diameter) on a smooth background. On the PF_s face the particles are smaller and more numerous. EF_u faces lack 15 nm particles and have 10 nm diameter particles, similar to the PF_u particles, but at much lower density.

Bundle sheath thylakoids lack obvious grana. However, the restricted contact areas have 15 nm EF_s particles and PF particles of 8.5 nm, only slightly larger than the mesophyll particles, and appear to be rudimentary grana, of comparable subunit structure to mesophyll and C3 chloroplasts. Particle density is similar in stacked mesophyll and in contact areas in the bundle sheath, but as there is much less stacked area in bundle sheath compared to mesophyll chloroplasts, the number of larger particles per thylakoid is only one-tenth. However, the PF particles are similar in number. Stromal lamellae of bundle sheath and mesophyll chloroplasts have 10 nm particles on EF_u and PF_u surfaces. The small PSII activity (p. 187) and few large EF particles in bundle sheath thylakoids suggest that the EF particle is a PSII unit together with the light-harvesting chlorophyll−protein complex which is responsible for chloroplast stacking. Thus, the basic structure of thylakoids is similar in different plants and tissues but with quantitative differences in composition related to function.

Chlorophyll and protein complexes in thylakoids

Chl *a* and chl *b* occur only in thylakoid membranes and may form 5 per cent of their total mass (Table 4.1). Chlorophyll is complexed with, but not covalently bonded to proteins; the hydrophobic phytyl group of chlorophyll may be between the membrane proteins and lipids, and the hydrophilic part of the porphyrin ring in the protein. This would orientate pigment molecules for efficient energy capture and transfer. The macromolecular structure of thylakoids has been determined from studies (Thornber 1975) in which chlorophyll−protein complexes are removed from membranes, with the anionic detergent sodium dodecyl sulphate (SDS), for example, and separated by electrophoresis on polyacrylamide gels containing SDS. As conditions of extraction and electrophoresis affect separation, a variable number of bands results, and size and composition of membrane components is difficult to interpret — for example, polypeptide monomers may associate into larger units (oligomers) which give distinct bands in gel separation. Although there is uncertainty about the chlorophyll−protein complexes in the membrane and their correspondence to membrane particles, three main complexes contain 90 per cent of the chlorophyll (Table 4.2). One corresponds to PSI and its antenna chl *a*; it is called P700 chlorophyll *a* complex or chlorophyll−protein complex I, CPI for short. A second is light-harvesting chlorophyll *a/b*−protein complex, now called light-harvesting complex, LHC, which has only antenna function and no photochemical activity. The third is less well resolved but contains PSII and its antenna chlorophyll *a*. It is called CP_a

Table 4.2 Chlorophyll–protein (CP) complexes and light-harvesting complexes (LHC) and some features of their composition in higher plant thylakoids

Name	Associated photosystem	Components	Chlorophylls (chl a/b ratio)	Chlorophyll content (% total)	Mol/mol protein	Main carotenoids (α/β carotenoid) ratio	No. subunits	Mass (kDa*)	
								Complex	Subunit
CPI	PSI	CPI	chl a, P700 ($a/b \simeq 20$)	30	12	β-carotene ($\alpha/\beta \simeq 0.7$)	2	100	50,60,70
		CPII							
CP$_a$	PSII	CP$_a$-1 (CP47)†	chl a, P680 no chl b	10	6	β-carotene ($\alpha/\beta \simeq 0.5$)	1	45–51	43
		CP$_a$-2 (CP43)†						40–45	
LHC	(PSII)	LHCII$_a$	chl a, chl b ($a/b \simeq 1.2$)	4	6 chl a	lutein, neoxanthin ($\alpha/\beta \simeq 50 \rightarrow 100$)	1	35	29
		LHCII$_b$		40	5 chl b		3	75	25
		LHCII$_c$		3	2.3 xanthophyll		1	30	30
		LHCII$_d$		3			1	24	21

* 1 dalton = 1.6605×10^{-27} kg, mass of hydrogen atom. A mass of 20 000 Da (20 kDa) is equivalent to a protein of mol mass 20 kg.
† Alternative designation of complexes.

and is the chlorophyll−protein complex serving as the internal chl *a* antenna of PSII.

CPI has a mass of 110 kDa, composed of two polypeptides of 50 or 60 and 70 kDa, and contains about 30 per cent of the total chlorophyll (no chl *b*), 10 mol chl *a* per mole of protein and 1 P700 molecule to every 100 chl *a* molecules. P700 is probably surrounded by a tightly bound inner and loosely bound outer chl *a* antenna, with about 5 β-carotenes and 2 xanthophylls. A quinone (naphthoquinone, phylloquinone or tocopherylquinone but no plastoquinone) and perhaps iron−sulphur centres are bound in the complex, and transport electrons. Some CPI may be composed of P700 and 5−10 monomers of protein−chlorophyll, a functional unit. CPI occurs throughout the plant kingdom in all organisms; photosynthetic bacteria have an equivalent and very similar photosystem complex. CPI predominates in membranes in contact with the stroma.

The CP_a complex occurs in all oxygen-evolving plants (and seems essential for O_2 evolution) and contains 10 per cent of the total chlorophyll, mainly or only chlorophyll *a*. It is composed of two different polypeptides, one molecule of each per unit, called CP_{a-1} and CP_{a-2} (alternative names are CPIII and CPIV, respectively, particularly in the older literature). P680 forms the core with 6 mol of chlorophyll *a*, β-carotene and electron carriers bound per mole of protein. Probably an antenna of chl *a* links CP_a to the LHC, thus preventing easy or consistent isolation. It has been suggested that there may be two types of CP_a units associated with PSII forming $PSII_\alpha$ and $PSII_\beta$. $PSII_\alpha$ contains more chl *a* in the antenna than $PSII_\beta$ and several $PSII_\alpha$ complexes may join together but the $PSII_\beta$ complexes are not combined together or with PSII. The more $PSII_\alpha$ the less grana stacking occurs. Possibly $PSII_\beta$ complexes are in the stromal membrane and $PSII_\alpha$ in the stacked regions; maybe these different forms are complexes in stages of synthesis or repair with the D1 protein (p. 80) being removed and inserted into the complex in stromal thylakoids before the repaired, fully functional unit moves into the granal region.

Light-harvesting complex, mainly found in granal thylakoids of plants with chl *b*, contains 50 per cent of chl *a* and all chl *b* (ratio *a*/*b*, 1:3) and approximately 1 carotenoid per 4−6 chlorophylls. Four forms of LHC are recognized, $LHCII_{a, b, c, d}$; the most abundant is $LHCII_b$ which is an oligomer of 75 kDa mass composed of three apoprotein subunits (mass 25 kDa each) making up 40 per cent of the chlorophyll and protein of the thylakoid total LHC. Within the LHC complex, groups of chl *b* molecules, probably arranged within exciton transfer distance, deliver energy with high efficiency to groups of chl *a* molecules more loosely arranged and transferring energy by the Förster mechanism. LHC passes energy mainly to CPII and hence to PSII but as light changes so does the association of LHC with CPI and CPII, altering the energy transfer between LHC and CPI and therefore the distribution ('spill-over') of energy between PSI and PSII.

The protein is composed of two abundant polypeptides, 25 and 23 kDa, found with chl *b* and chl *a* (it is not known if there is specific binding between polypeptides and pigments). LHC is not found in mutants or plants grown in intermittent light which lack chl *b*. If each polypeptide contains 6 chlorophyll molecules there will be 30 or more monomers aggregated to form LHC *in vivo*. Most of the LHC is hydrophobic and buried in the membrane. However, a segment (2 kDa) of 20 amino acids on the major polypeptides is exposed to the stroma, and a segment of each is exposed to the lumen. The exposed portion in the stroma may be removed by adding protease enzyme (trypsin). The amino acid sequence at the C-terminal of the peptide segment is (lysine$^+$ or arginine$^+$) lysine$^+$, arginine$^+$, serine, alanine, threonine$^+$, threonine$^+$, lysine$^+$, lysine$^+$, with positive charges as indicated. Despite these positive charges the thylakoid surface has a net negative charge (1 charge per 6 nm^2) so that the membranes are mutually repelling and dissociate unless cations (e.g. Mg^{2+}) are present to shield the negative charges. This allows the positive charged portion of LHC to join onto some negative charges and cause thylakoid stacking. The threonine residues on the 2 kDa peptides can be phosphorylated by a specific kinase, abolishing the positive charge and so altering the balance of charges on the membrane surface, changing stacking and distribution of particles in the granal and stromal thylakoids and also energy distribution.

Arrangement of complexes in thylakoids

Current models of particle organization are reviewed by Hiller and Goodchild (1981). A large LHC antenna of chl *a* and *b* is linked closely to the antenna chl *a* of CPII which adjoins the PSII core. LHC delivers energy preferentially to PSII and the CPI antenna transfers energy to P700. PSI may be almost restricted to membranes exposed to the stroma and absent from the interior of stacked membranes. Probably 80 per cent of PSII is in the granal regions, sheltered from the stroma (Staehelin *et al.* 1977; Barber 1982). Illumination causes PSI to associate with PSII and LHC, from which it receives more energy, at the edge of the granum. Groups of LHC, CPI and CP$_a$, perhaps 200 per granum, may function independently, only distributing energy within a single group (a 'puddle' model) or co-operatively (a 'lake' model) with excitation energy passing between groups, but efficient energy distribution under different conditions may require that complexes interact in various ways.

These groupings of LHC, CPI and CP$_a$ in the membrane allow greater flexibility in photosynthetic response to environment and to physiological conditions in the photosynthetic apparatus. They form the structural basis of the functional 'photosynthetic units', a group of 200−300 chlorophyll molecules co-operating in all the light reactions (see Ch. 3).

References and Further Reading

Anderson, J.M. (1986) Photoregulation of the composition, function and structure of thylakoid membranes, *A. Rev. Plant Physiol.*, **37**, 93−136.

Arntzen, C.J. and **Briantais, J.-M.** (1975) Chloroplast structure and function, pp. 52−133 in Govindjee (ed.), *Bioenergetics of Photosynthesis*, Academic Press, New York.

Barber, J. (1982) Influence of surface changes on thylakoid structure and function, *A. Rev. Plant Physiol.*, **33**, 261−95.

Bennett, J. (1991) Protein phosphorylation in green plant chloroplasts, *A. Rev. Plant Physiol Plant Mol Biol.*, **42**, 281−311.

Branton, D. *et al.* (1975) Freeze-etching nomenclature, *Science*, **190**, 54−6.

Douce, R. and **Joyard, J.** (1981) The chloroplast envelope: structure, composition and biological properties, pp. 187−98 in Akoyunoglou, G. (ed.), *Photosynthesis III, Structure and Molecular Organisation of the Photosynthetic Apparatus*, Balaban International Science Services, Philadelphia.

Hall, D.O. (1972) Nomenclature for isolated chloroplasts, *Nature New Biol. (Lond.)*, **235**, 125−6.

Hiller, R.G. and **Goodchild, D.J.** (1981) Thylakoid membrane and pigment organization, pp. 1−49 in Hatch, M.D. and Boardman, N.K. (eds), *The Biochemistry of Plants*, Vol. 8, *Photosynthesis*, Academic Press, New York.

Hinshaw, J.E. and **Miller, K.R.** (1989) A novel method for the visualization of outer surfaces from stacked regions of thylakoid membranes, pp. 111−14 in Barber, J. and Malkin, R. (eds), *Techniques and New Developments in Photosynthesis Research*, Plenum Press, New York.

Horton, P. (1983) Control of chloroplast electron transport by phosphorylation of thylakoid proteins, *FEBS Lett.*, **152**, 47−52.

Kaplan, S. and **Arntzen, C.J.** (1982) Photosynthetic membrane structure and function, pp. 65−151 in Govindjee (ed.), *Photosynthesis*, Vol. 1, *Energy Conversion by Plants and Bacteria*, Academic Press, New York.

Leegood, R.C. and **Walker, D.A.** (1983) Chloroplasts (including protoplasts of high carbon dioxide fixation ability), pp. 185−210 in Hall, J.L. and Moore, A.L. (eds), *Isolation of Membranes and Organelles from Plant Cells*, Academic Press, London.

Newcomb, E.H. and **Frederick, S.E.** (1971) Distribution and structure of plant microbodies (peroxisomes), pp. 442−57 in Hatch, M.D., Osmond, C.B. and Slatyer, R.O. (eds), *Photosynthesis and Photorespiration*, Wiley Interscience, New York.

Pierson, B.K. and **Olson, J.M.** (1987) Photosynthetic bacteria, pp. 21−42 in Amesz, J. (ed.), *Photosynthesis, New Comprehensive Biochemistry*, Vol. 15, Elsevier, Amsterdam.

Staehelin, L.A. and **Arntzen, C.J.** (1979) Effects of ions and gravity forces on the supramolecular organisation and excitation energy distribution in chloroplast membranes, pp. 147−75 in *Chlorophyll Organisation and Energy Transfer in Photosynthesis*, Ciba Foundation Symposium 61 (N.S.), Excerpta Medica, Amsterdam.

Staehelin, L.A., Armond, P.A. and **Miller, K.R.** (1977) Chloroplast membrane organization at the supramolecular level and its functional implications, *Brookhaven Symp. Biol.*, **28**, 278−315.

Thornber, J.P. (1975) Chlorophyll−proteins: Light harvesting and reaction centre components of plants, *A. Rev. Plant Physiol.*, **26**, 127−58.

CHAPTER 5

Electron and proton transport

In anoxygenic and oxygenic photosynthetic organisms, the excitation of the special pair chlorophylls and transfer of electrons to the primary acceptors produces a strongly reduced intermediate which can be used to reduce CO_2 and NO_3^-. Although the light reactions appear very similar, the source of electrons to replace those ejected from the special pair differs between organisms which do or do not produce O_2 and also within the anoxygenic types. Further, there are differences in the mechanisms of electron movement reflecting the diversity of organisms, particularly those with the anoxygenic mechanism. Oxygenic systems are more uniform.

Bacterial electron and proton transport

One of the simplest light capture and energy transduction systems in bacteria is a light-driven proton pump involving protein conformation changes but without e^- movement. It is linked to ATP synthesis. This pump (Fig. 5.1) occurs in *Halobacterium halobium* (Kushner 1985), a bacterium which uses light to drive protons out of the cytosol into the external medium. Diffusion of the H^+ back into the cell through an ATP synthase enzyme provides a part of *H. halobium*'s energy needs, but it is not fully autotrophic. The organism grows in brackish water in light; with deficient oxygen (but not if O_2 is completely absent) purple patches 1 μm long form over more than half of the cytoplasmic membrane. The patches are of bacteriorhodopsin, pigmented protein molecules similar to visual rhodopsin in the retina of the vertebrate eye; they are arranged in a crystalline lattice. Retinal, which absorbs light, is an aldehyde of vitamin A bound by a lysine group to a protein (an opsin-like polypeptide of 26 kDa) by a protonated Schiff's base. The protein crosses the membrane with 7 α-helical segments in a crescent with the N-terminal of 3–6 amino acids on the external side of the membrane and 17–24 forming the C-terminal at the cytoplasmic surface. Retinal is attached to the last segment about one-third of the way to the outer membrane. When the protonated chromophore in the 13 *cis*-all *trans*retinol

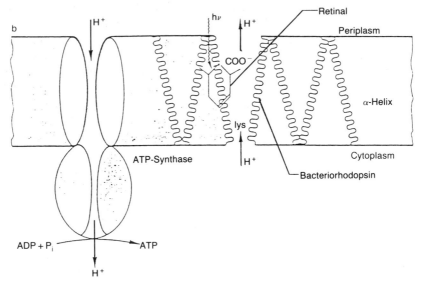

FIG. 5.1 The light-driven proton pump of the bacterium *Halobacterium halobium*, based on light capture by retinal combined with a protein, which spans the membrane. Protons are expelled from the cytoplasm into the external medium and diffuse back into the cell via an ATP synthase enzyme.

form absorbs light at 570 nm it loses a proton forming 13 *cis* all *trans*retinal and absorption changes to 412 nm. The H^+ is transferred by the protein (which alters conformation and has been called a 'proton wire' because it carries protons), from the inside of the cell to the external medium. The deprotonated pigment then absorbs a proton from the cytoplasm. Thus, the simple mechanism acts as a pump of H^+ analogous to other photosynthetic H^+ transport (see p. 109), but without electron transport. The H^+ from the medium diffuses back into the cell through an enzyme synthesizing ATP. The ATP supports metabolism, particularly when the respiratory activity in the bacterium is limited. This mechanism illustrates how photosynthetic systems may have evolved. Its presence in a 'primitive' bacterium suggests an early form of exploitation of radiation although it may be a later evolutionary addition.

Photosynthetic bacterial reaction centre

The photosynthetic reaction centre (RC) of the photosynthetic purple bacterium *Rhodopseudomonas viridis* is the first membrane protein complex to be crystallized (Deisenhofer *et al.* 1985). This was achieved by solubilizing membranes in solutions of particular detergents with long alkyl chains and small polar heads together with careful addition of small ampiphilic molecules such as benzamidine to prevent denaturation of these very hydrophobic proteins. The crystals were analysed by X-ray crystallography to 0.3 nm resolution. For describing the structure of the bacterial reaction centre Huber, Michel and Deisenhofer were awarded the 1988 Nobel Prize for chemistry. The detailed structure (Deisenhofer *et al.* 1985; Michel and Deisenhofer 1986) shows (Fig. 5.2) that the RC contains 4 bchl, 2 bphaeo *b* molecules and 2 quinones, and these are arranged around a two-fold rotation axis between the two bchl molecules and through a non-haem iron atom in a symmetrical manner. The bchl molecules form the 'special pair' on the periplasmic side of the membrane. The Fe atom is between the two quinones on the cytoplasmic side of the membrane. There are two electron

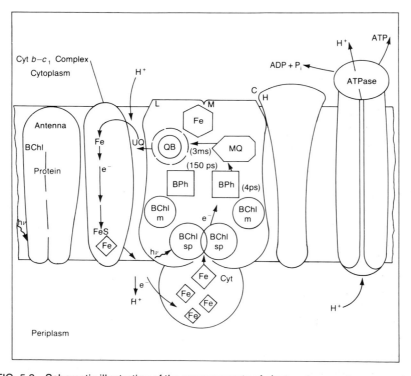

FIG. 5.2 Schematic illustration of the arrangements of electron transport components in the bacterial reaction centre (*Rhodopseudomonas viridis*). Two proteins, L and M, carry the bacteriochlorophyll special pair (BChl sp), which absorbs energy and ejects electrons. The electrons pass to a bacteriopheophytin (BPh), with another acting as a 'chaperone', and on to menaquinone (MQ) and to a bound ubiquinone (QB) before transport to the cyclic electron transport path in the Cyt *b*–*c*$_1$ complex. Other components of the system are: Antenna (low molecular mass polypeptides, < 10 kDa). Cyt *b*–*c*$_1$ complex. Reaction centre polypeptide subunits L (light, 28 amino acid residues); M (medium, 30 amino acids); H (heavy, 46 amino acids); mass of subunits 20–30 kDa. ATP synthase (CF$_0$–CF$_1$). Cyt. cytochrome. UB, ubiquinone; ⟨Fe⟩ haem iron. ⟨Fe⟩ Fe, ferrous iron atom magnetically coupled to MQ which is menaquinone linked to a histidine residue. BChlm, bacteriochlorophyll monomer. BPh, bacteriopheophytin, one of which is linked to the chain of MQ. Transfer times for electrons are given in brackets. Much simplified (after Deisenhofer *et al.* 1985; Michel and Deisenhofer, 1986, 1988).

paths from the special pair, each with a bchl, a bphaeo and a quinone. However, the quinones function in series, not parallel and only one path is used in photosynthesis; the other may have carried electrons early in evolution but its present role is perhaps to 'fine tune' the reaction centre. The accessory (i.e. not the RC) bchl is close to the carotenoid and may provide a path for quenching excited states of the bchl. The oxidized RC, after e^- ejection, is reduced again by electrons from two low and two high redox potential haem groups which are in a row. All these components are linked to the L and M subunits of the RC proteins. The subunits possess five helices of 19 or more amino acids, without charge and very hydrophobic, containing many S amino acids and histidines which bind the pigments. These helices cross the membrane and are arranged with two-fold symmetry. Because of the different distribution of amino acids in the segments of the polypeptides the periplasmic and cytoplasmic membrane surfaces have four negative and four positive charges, respectively. This may be important for organization and orientation of the complex in the membrane, and also for e^- transport and development of electrical potential across the membrane. The peptide segments on the membrane surface form flat planes orienting the complex and perhaps holding it steady. There is an H subunit with one helix crossing the membrane and its carboxy-end is bound to the L−M complex on the cytoplasmic side of the membrane. Pigments are bound into hydrophobic regions within the L and M subunits. The histidine residues join in five-fold co-ordination to the Mg of the special pair bchl. A few water molecules are found inside the membrane and may be important for hydrogen bonding in the protein. This very symmetrical and organized structure is important for e^- transport. The L and M subunits are homologous to D1 and D2 proteins of PSII (p. 80).

Electron transport in the anoxygenic photosynthetic bacteria

In Chapter 3 the structure of the purple bacterial light capturing system was described and the mechanisms of e^- transport are considered further here. Ejection of an electron from P870 to one of the bchl molecules occurs within a few picoseconds of light capture (Fig. 5.3). Then two e^- are transferred via bacteriochlorophyll to ubiquinone thus decreasing the back reactions. Ubiquinone picks up two protons from the inside of the cell and transfers them to the outer side of the membrane. These protons diffuse back into the cytoplasm through an enzyme complex, the ATP synthetase or coupling factor, where ATP is synthesized from ADP and inorganic phosphate (see Ch. 6). The electrons pass to cytochrome (type c_2) catalysed by an oxidoreductase of three subunits, a cyt b apoprotein, a cyt c_1 and an FeS protein. The apoprotein has 8 hydrophobic sections which span the membranes and contain 4 histidine residues which bind 2 haem molecules. These may act as transmembrane e^- carriers. Other hydrophilic subunits are found on the membrane surface (Dutton 1986).

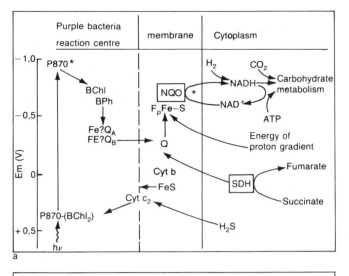

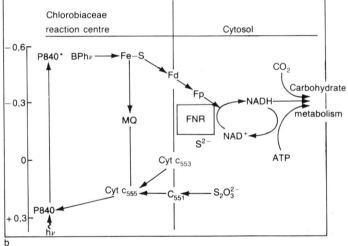

FIG. 5.3 Electron transport in the photosynthetic bacteria. (a) Purple bacteria, showing the flow of e^- in the reaction centre, reduction of quinones in the membrane and passage of e^- to CO_2 catalysed by NAD quinone oxidoreductase (shown at $*$). As the available energy is insufficient, energy from the proton gradient is coupled to the enzyme. Electrons from substrates are fed into cytochrome c_2 (cyt c_2) and via a $b-c_1$ complex (quinol cytoreductase). (b) Electron flow in Chlorobiaceae. Reduced Fe–S centres pass e^- to NAD^+ for reduction of CO_2 via the enzyme ferredoxin nucleotide reductase (FNR). Electrons from substrates reduce the oxidized reaction centre via cytochrome 551 (c_{551}). (After Dutton 1986.)

Electron flow in the photosynthetic bacteria differs between bacterial types. The purple bacterial and chlorobial forms are shown in Fig. 5.3a and b, respectively. In the purple bacteria, electrons from substrates reduce the cytochrome and are passed back to the oxidized reaction centres.

Alternatively electrons from metabolism of organic acids, e.g. conversion of succinate to fumarate by succinate dehydrogenase, may contribute to the reduction of quinone. The electrons from quinone reduce NAD^+, a reaction catalysed by NAD-quinone oxidoreductase. Because the redox potential for the reaction is unfavourable for the reduction of NAD^+ by ubiquinone, energy from the proton gradient formed in the light is coupled to the enzyme in order to achieve the reduction. NADH is the reductant for CO_2 (Clayton 1980).

The green bacteria (Chlorobiaceae and Chloroflexaceae) contain bchl *c*, *d* or *e* as main photosynthetic pigments organized in chlorosomes (see Ch. 4), forming a very large antenna, 1000 molecules to one reaction centre. One of the four polypeptides in chlorosomes is a 3.7 kDa protein which carries bchl *c*; it is probably a dimer and has 10−16 chlorophyll molecules attached. A water soluble protein, which binds bchl *a*, forms a basal plate between the chlorosomes and the reaction centre. The reaction centres of *Chloroflexus* are almost identical to those of purple bacteria but other green bacteria have different structures. Electrons from the reaction centre (Fig. 5.3b) pass to ferredoxin and these reduce NAD^+ via a flavin-NAD reductase enzyme on the cytoplasmic side of the membrane. Electrons from H_2S pass to a cytochrome on the opposite side of the membrane and reduce the oxidized reaction centre. Alternatively, electrons from ferredoxin pass to the cytochrome via the quinone−cytochrome complex, thus forming a cyclic electron pathway. The NADH is used in the cycle, together with ATP generated by the ATP synthetase, as described for purple bacteria but reduction of NAD^+ is achieved without the need for energy from the proton gradient. In *Chlorobium*, glycogen and hydroxybutyric acid are major products.

Electron transport in oxygen-producing plants

Photosynthesis of the prokaryotic cyanobacteria as well as that of eukaryotic algae and higher plants, produces oxygen and the basic process is similar in them all.

A form of the Hill and Bendall 'Z' scheme of the sequence of processes and electron transport leading from water-splitting through $NADP^+$ reduction is given in Fig. 5.4 and the structural components related to it in Fig. 5.5.

Photon capture by the photosystem antennae and excitation transfer to PSII and I provide the energy for oxidation of water and electron movement to acceptors, which donate e^- to biochemical processes, and for passage of protons into the thylakoid lumen, for synthesis of ATP. The electron transport system may be considered in five parts: (1) a water-splitting complex, (2) a photosystem II complex, (3) an electron carrier chain, (4) a PSI complex, and (5) a group of e^- carriers which reduce electron acceptors ($NADP^+$, O_2).

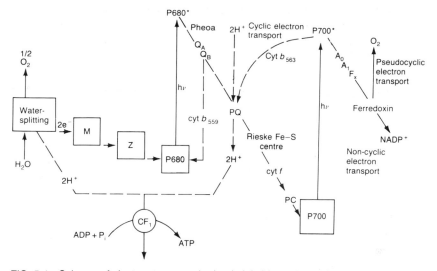

FIG. 5.4   Scheme of electron transport in the thylakoid membrane from water to NADP⁺ with two light reactions, PSII and I in series; S and Z are intermediates in water-splitting, pheophytin (pheo), quinone acceptors Q_A, Q_B and PQ (plastoquinone), cytochrome f (cyt f) and plastocyanin (PC) pass e⁻. A_0 A_1 and F_x are intermediates in e⁻ movement to ferredoxin. Proton transport leads to synthesis of ATP. Electrons also cycle from PSI back to the intermediate chain or pass to oxygen.

Electron transport starts with the capture of photons by chlorophylls and accessory pigments. Transfer of the energy to reaction centres of PSI and PSII excites the dimer chlorophylls and causes ejection of electrons to acceptors, starting e⁻ transport along the chain of redox components. Excitation of P680 of PSII results in an oxidized reaction centre P680⁺. PSII is defined as that part of oxygenic photosynthesis catalysing photo-induced transfer of e⁻ from water to plastoquinone (PQ):

$$4\,H^+ + 2\,PQ + 2\,H_2O \xrightarrow[\text{PSII}]{4hv} 2\,PQH_2 + O_2 + 4\,H^+ \qquad [5.1]$$

with transfer of H⁺ from the stromal to lumen side of the thylakoid membrane. This oxidized PSII is reduced by e⁻ from a water-splitting complex via intermediate states M and Z (Fig. 5.4) which are components of the water-splitting complex and electron carrier system between it and the reaction centre. The energized e⁻ passes, from more to less negative potential, to the primary acceptor pheophytin and then in sequence to the quinone acceptors Q_a, Q_b and PQ. Quinones are important carriers of e⁻ and H⁺ in many biological processes. From PQ the electron passes to cytochrome f and plastocyanin before reducing an oxidized PSI reaction centre. Here it is energized again by excitation energy derived from photon energy trapped in the chlorophyll matrix, and passed via intermediate states A_0, A and B to oxidized ferredoxin (Fd) and NADPH⁺, which are reduced and are able to enter into biochemical reactions in the chloroplast stroma.

Electron transport chains bridge the thylakoid membranes, allowing electrons removed from water held in the water-splitting complex of proteins, manganese ions and other components inside the thylakoid lumen to pass across the membrane to ferredoxin on the stromal side. Plastoquinone in the membrane is reduced by the electrons; the H^+ from the stroma attaches to reduced plastoquinone and is carried to the lumen, where it is released and the plastoquinone oxidized. Thus, electron transport is coupled to a plastoquinone cycle which carries ('pumps') H^+ from stroma to the thylakoid lumen in the reverse direction to electron transport, increasing H^+ concentration in the thylakoid lumen and forming the proton concentration gradient, the energy of which drives ATP synthesis (p. 109). Velthuys (1980) has reviewed aspects of electron transport and text by Clayton (1980) provides detailed discussion.

Photosystem II complex

This photosystem has been intensively studied and there is rapid development in understanding its structure and function at the molecular and atomic levels of organization, particularly since description of the purple photosynthetic bacterial reaction centre which is a valuable analogue of PSII. Thus, the protein subunits of the bacterial reaction centre are analogous to the proteins of higher plant and cyanobacterial PSII, but the bacterial structure lacks the oxygen-evolving complex, of course.

Considering the structure of PSII from higher plants, such as spinach, it may be regarded as consisting of a chlorophyll pigment antenna reaction centre complex with a molecule of P680 joined to structural proteins and components linking P680 to the water-splitting enzyme complex and to the electron acceptors of P680. Purified spinach reaction centre complexes have been analysed by methods such as separation of proteins by SDS polyacrylamide gel electrophoresis with urea to denature and dissociate the subunit structure, coupled with Coomassie blue staining to identify the bands. However, early preparations were complex, with antenna chlorophyll interfering with the optical, EPR and X-ray spectroscopic signals used to analyse the structure. Recently it has been possible to isolate a reaction centre core complex (the D1/D2/cyt b_{559} complex; see later) which contains very little chlorophyll and pheophytin but can still perform photochemistry; this has greatly facilitated analysis of reaction centre processes.

Table 5.1 lists the main components of PSII so far identified. Some 22 polypeptides are associated with PSII. Firstly the reaction core: it consists of two polypeptides called D1 and D2 (D for 'diffuse' appearance on gels) of similar mass. D1 is homologous to the L and D2 to the M subunit polypeptides of the bacterial reaction centre complex. The RC of PSII contains one each of D1 and D2, and is thus a heterodimer in construction.

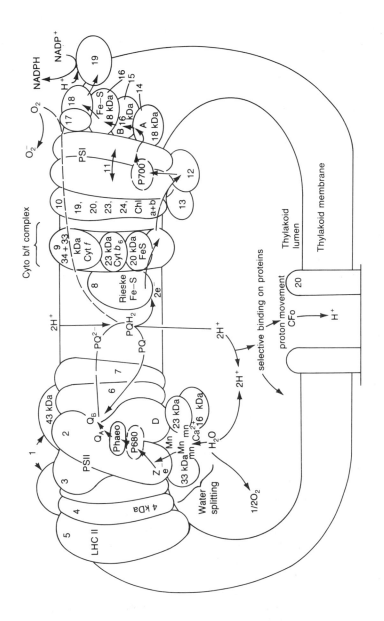

FIG. 5.5   Schematic representation of the light-harvesting and photosystem complexes and electron transport chain in the thylakoid membrane. The number of protein complexes and their relation is shown. The mass of components is indicated. The components are listed in Table 5.1.

Table 5.1 Components of the light-harvesting and electron transport complexes of the oxygenic plant thylakoid membranes as depicted in Fig. 5.5

1.	Antenna protein, pigment complex
2.	32 kDa, D1 herbicide binding protein of the reaction centre
3.	32 kDa, D2 reaction centre protein
4.	Cytochrome b_{559}, 9 kDa b_{559} type 1 and 4 kDa b_{559} type 2 proteins
5.	Light-harvesting antenna
6.	10 kDa docking protein
7.	22 kDa stablizing protein (intrinsic membrane protein)
8.	20 kDa Rieske Fe$-$S centre
9.	Cytochrome b_6-f complex with polypeptides
10.	Light-harvesting protein$-$pigment (chlorophyll a and b) complex of PSI and polypeptides
11.	PSI reaction centre, with two 70 (?) kDa polypeptides
12.	Plastocyanin, 10.5 kDa
13.	Plastocyanin binding protein (10 kDa)
14.	Fe$-$S proteins
15.	Fe$-$S proteins
16.	Fe$-$S proteins
17.	Ferredoxin binding protein
18.	Ferredoxin
19.	Ferredoxin, NADP oxidoreductase
20.	Coupling factor, CF_0, membrane subunits

Each polypeptide has five transmembrane segments but their functions are not equal. D1 only carries the oxidizing and reducing components of the RC.

Electron accepting side of PSII

Under physiological conditions P680$^+$ oxidizes the donor, Z, as the kinetics of Z$^+$ formation are in phase with the reduction of P680$^+$. In turn, Z$^+$ is reduced by electrons from the water-splitting complex M. It is now accepted that Z is a tyrosine residue Y_z (Y; the biochemical symbol for tyrosine) at amino acid position 160 on the D2 protein. This tyrosine gives rise to a special EPR signal having transients with different decay rates called signal II$_{very fast}$ (20 μs), and II$_{slow}$ (1 s); II$_{very fast}$ results from transfer of e$^-$ to P680$^+$. The signal is not from a quinone but is a radical located on the e$^-$ donor side of PSII. The II$_{very fast}$ signal is stopped by Tris, which removes Mn^{2+} (and possibly polypeptides of the water-splitting complex) and also by DCMU. Thus, the evidence points to tyrosine tightly bound to the reaction centre, close enough to donate electrons very rapidly. P680 and Y_z are separated by 1$-$1.5 nm and e$^-$ transfer occurs within the membrane. The signal II$_{slow}$, which is stable in the dark, is probably from another tyrosine at position 161 on the D2 protein.

The P680 donates e^- to a pheophytin molecule (generally called H_A as it is the primary acceptor) located on the D1/D2 heterodimer. There is a second pheophytin (H_B) not directly involved in e^- transport and on a part of the protein which is functionally distinct. H_B may have a role in modifying the electronic structure of the RC. There is a similar H_B in the bacterial RC with the function of modifying electron flow. Probably D1 and D2 can change their conformation according to whether the quinone binding site (Q_B) is filled and also both can be phosphorylated, which is a way of modifying the efficiency of the electron transport system.

Purple bacterial reaction centres (Fig. 5.2) contain a third polypeptide called the H subunit, which does not function in e^- transfer but probably conforms the L/M heterodimer. PSII may have a similar structure, called the quinone shielding protein which has homology in gene sequence and similar molecular mass. It has one transmembrane sequence and a very large N-terminus probably on the stromal side of the membrane. Another 22 kDa protein may also serve to regulate PSII. Reaction centres of PSII contain at least one tightly bound haem, cyt b_{559}, which has two subunits of 9 and 4 kDa, each with a histidine group which probably co-ordinates haem binding into the protein subunits. This cyt b_{559} can exist in two forms of different potential — high and low — but these do not correlate with PSII activity and probably have protective function in decreasing the chance of photoinhibition which damages the D1 protein particularly. It may also be involved in proton pumping during cyclic electron flow.

Within the core complex P680 is, of course, crucial to electron transfer capacity. Yet the structure of P680 is poorly understood, e.g. whether it is a monomer or dimer of chl a (Ch. 3). Probably P680 is a pair of chl a molecules which interact strongly, especially in the singlet state, and weakly in the triplet state. The chlorophyll is oriented with its macrocycle parallel to the membrane plane. Separation between components of the RC is probably crucial to the efficiency of energy transfer. The distribution of energy between P680, H_A and Q_A is approximately in equilibrium. Q_A^- is, on evidence of the electrochemical shift in the blue and green spectral region of pheophytin, closer to pheophytin than Q_B^-; the distance between the Fe and the Q_A quinone ring is 0.7 nm. Y_D, although not involved in electron transport, interacts with manganese in the M complex within distances of 3−4 nm and sits almost midway between the thylakoid surfaces.

A major characteristic part of PSII is the 'regulatory cap' which is involved in water-splitting. The cap is composed of three major extrinsic proteins (hence called EP) of 33, 23 and 16 kDa and a minor protein of 5 kDa mass, which are hydrophilic and bind in equimolar amounts, probably two copies of each, to the luminal surface of the thylakoid. The cap may regulate passage of ions (Ca^{2+} and Cl^-) and water into the water-splitting Mn complex. EP33 links to the RC and to the antenna and EP23 and EP16 may shield it from the lumen and link to the distal antenna, judged by antibody studies. The binding sites may be of different charge and they may be removed with loss of tetrameric complexes on the ES_s surfaces of thylakoids leaving two-fold structures, indicating that PSII is a dimer.

The antenna of PSII is made of two distinct, linked parts, the proximal antenna (nearest to the RC) and a more distant distal antenna. The first is made of two pigment–protein complexes, CP_{a-1} and CP_{a-2} (see p. 68, Ch. 4) subunits of which both have 7 transmembrane segments with about 12 histidine residues, probably the sites of chl binding and a large hydrophilic segment in the lumen. About 25 chl a and 5 β-carotene (oriented parallel to the membrane) are present. There is tight linkage between the proximal antenna and the RC and CP_{a-1} is essential to e^- transport, for if genes for the polypeptide are mutated assembly of the RC is inhibited. The distal antenna contains chl a with chl b and xanthophyll as accessory pigments. It is made up of LHC polypeptides as described in Chapter 4.

Electron transport in PSII reaction centre

A reaction centre will accept or 'quench' excitation if it is reduced (P680), i.e. contains an electron which can be ejected by excitation, but will not use excitation when oxidized (P680$^+$). In the latter case excitation may migrate to reduced reaction centres. Ejection of an electron occurs within nanoseconds; the fewer reduced reaction centres the longer the time that excitation dwells in the antenna and the greater chlorophyll fluorescence. If Q is reduced, fluorescence yield (p. 99) is about 3 per cent but increases to 12 per cent if Q is oxidized. Events at the reaction centre are summarized by the following sequence with Z the donor, P680 the reaction centre, pheophytin (pheo) the primary acceptor and Q and Q_B secondary acceptors:

$$\text{Z.P680.pheo.Q} \xrightarrow{h\nu} \text{Z.P680}^+.\text{pheo.}^-\text{Q} \xrightarrow{} \text{Z}^+.\text{P680.pheo.Q}^- \quad [5.2]$$

P680$^+$ is a very powerful oxidant ($+1.2$ V or greater) able to remove electrons from water ($+0.8$ V) and produce a relatively weak reductant (-0.6 V). The state of Z.P680.pheo.Q$^-$ is non-quenching, as electrons cannot be transferred even if P680 contains an electron. Possibly recombination of pheo$^-$ with P680$^+$ gives rise to the high variable fluoresence. Charge separation forms the radical pair P680$^+$.pheo$^-$, which can be detected by ESR (see Malkin 1982). For efficient separation, back reactions of the electron with P680$^+$ must be limited; this is achieved by loss of energy as heat, delocalization of the electron and formation of a triplet-like state, and most importantly, spatial separation by rapid transfer to the secondary acceptor, Q.

The secondary acceptor, Q, is a quinone bound to protein; it passes electrons to another acceptor Q_B, which is bound to a 32 kDa lysine-free polypeptide when oxidized but when reduced it diffuses into the PQ pool (Velthuys 1980). Q_B is a 2 e^- transfer molecule. Q_B^- is tightly bound to the protein but Q_B^{2-} is not bound. Inhibitors such as the herbicide DCMU may bind next to or at the site of Q_B attachment and thereby alter the energy charge at the site and prevent e^- transport. Different herbicides, for example atrazine and DCMU, bind to the same site or closely linked sites. The peptide probably regulates H^+ movement to Q_B^-, Q_B reduction by added reductants and also controls herbicide activity. Trypsin destroys the protein and influences granal stacking. Herbicides such as atrazine which bind to the protein inhibit electron transport, but in many species of plants mutants have arisen which are insensitive to these herbicides; the binding site amino acids and structure have changed so that the herbicides cannot block e^- transport.

The M complex and water-splitting

The nature of the water-splitting enzyme complex, that is, the type, structure and number of the proteins and associated electron donors and acceptors, particularly of manganese the most important metal in the complex, and their binding to structural components is poorly known and an active, challenging area of research. The complex forms a four-electron gate and governs the S states of the water-oxidizing process, passing electrons to Y_z to reduce P680 and holding the charge until four positive charges are 'accumulated' when water is split and O_2 is released. The water-splitting complex is on the lumen side of the thylakoid membrane exposed to the aqueous medium, and is designated M. Its role in the production of $4 H^+$, $4 e^-$ and O_2 from water is well characterized but the sequence of electron, proton and O_2 release is not yet clear and the detailed molecular processes and mechanisms remain to be described.

Manganese is an essential component of the complex with four atoms per PSII but their location on the PSII proteins or regulatory cap and how they function is speculative. In thylakoids two Mn fractions related to O_2 production have been identified: one fraction of two Mn atoms is tightly bound to PSII and probably involved in charge accumulation and the others more loosely bound. For very active O_2 production four Mn per PSII complex is essential. The labile fraction is removed by treatment with Tris buffer, chelating agents (EDTA), or by heating, which stops O_2 liberation but not electron transport. Mn^{2+} added back to thylakoids stimulates O_2 evolution only after chemical reduction or illumination. This 'photo-reactivation' requires phosphorylation and uses only PSII reactions; there is no interaction between PSII centres. Photoreactivation has two dark reactions: one changes Mn^{2+} to Mn^{3+}, the next Mn^{3+} to Mn^{4+} and it is possible that these are required so that the metal can bind to a site in the configuration needed for water-splitting.

Manganese is probably in clusters at two binding sites and may be in groups of one and three or in binucleate structures with one pair bound closely together (0.27 nm) the other less so, as judged from X-ray absorption and near infra-red optical spectroscopy and EPR. Manganese ions are bound to the O and N of the protein matrix and probably the changes in S states are not related to physical rearrangement of their position. The regulatory cap has been discussed; the EP33 shields the Mn ('Mn stabilizing protein') but its regulation is complex. Removal of EP33 prevents the functions of Mn but if high concentrations of ions are present changes of S states may still occur although it may slow or stop O_2 evolution. Two cysteines in EP33 form a disulphide bridge, which, if reduced, changes the protein conformation and this prevents active O_2 evolution. Oxidation of the sulphide bridge can restart the process. Calcium and chloride ions are essential co-factors for PSII and their presence in the M complex is probably controlled by the regulatory cap. For example, if EP16 and EP23 are removed then O_2 evolution may decrease by up to 50 per cent, but adding Ca and Cl stabilizes the processes probably by affecting the binding sites. Calcium is essential for O_2 evolution and there is a calcium binding calmodulin type of protein which influences behaviour. How chloride ions function is unknown; there may be several binding sites and not only is Cl effective but also bromide and nitrate ions, suggesting electrostatic interactions determine the function. Possibly the regulatory cap controls water entry and access of other components of the lumen to the active site of water-splitting.

Water molecules appear to bind directly to Mn judging from the use of isotopically labelled water and its analogues (NH_2OH, NH_3). Ammonia, which is isoelectronic with water, inhibits O_2 evolution, the S states and EPR signals. It is not known if water is continuously bound at all stages of the oxidative cycle to Mn ions or to the protein matrix close to them or if intermediate redox states involve Mn ions and proteins; perhaps the histidines on the protein regulate the redox changes and electron transfer.

Valency changes in the Mn clusters (and possibly associated proteins) are detected as EPR signals (at g approximately 2.0 and 4.1) which are probably both related to Mn in the S_2 state. The $g = 2.0$ multiline signal is due to a multinuclear Mn complex, either two different conformations of a 4 Mn cluster or a variable number of different valency Mn ions in the S_2 states. Changes in valency result in loss of the EPR signal. X-ray absorption-edge energies measurement suggests that the Mn oxidation state increases in the $S_0 \rightarrow S_1$ and $S_1 \rightarrow S_2$ but not in the $S_2 \rightarrow S_3$ states or it is hidden. Currently the sequence of valency changes and associated electron and proton transfers are not understood (see p. 93).

Manganese is a period IV transition metal with oxidation states of $+2$ (most common) up to $+7$. Two electrons lost from each of two linked Mn atoms would give four oxidizing equivalents ($4+$). Mn^{3+} is stable in complexes and could act as intermediary oxidant; Mn^{4+} is also stable and water-splitting could take place by single electron transfer steps. The reactions and stable state should not be too long-lived. The Mn^{2+} complex with O_2 is unstable, so that O_2 is rapidly released as required for fast water-splitting. Oxygen must bind without releasing singlet or other forms of O_2

which could damage the protein. Manganese fulfils these requirements; ESR studies at low temperature on previously heated thylakoids, which release Mn^{2+}, strongly suggest that manganese is involved in the S-states changes, and may be the physical entity corresponding to intermediate S. The molecular mechanism of H_2O binding to the complex and the role of ions such as chloride and nitrate which act around PSII and are required for O_2 evolution, are unknown.

It is important to know how water-splitting is carried out *in vivo* because it may show how to produce H_2 for fuel from an unlimited source — H_2O. Methods of doing this chemically, by reactions using ruthenium atoms in membrane systems, for example, are being explored. Altogether, the chemistry of water-splitting and the role of Mn are fascinating but unclear (Rutherford 1989).

Water-splitting and cycle of S states

When dark-adapted algae are illuminated with short flashes of bright light, separated by darkness, a characteristic pattern of O_2 evolution results (Fig. 5.6). The first two flashes evolve little or no O_2, the third a large 'gush' and the fourth a smaller amount of O_2 than the third but more than the first, that is, a periodicity of four. Oscillations are damped and after some 20 flashes yield per flash is constant. This pattern is characteristic of algae and chloroplasts. Oxygen production is slower just after illumination than with longer illumination and this 'lag' period is inversely proportional to light intensity $\times$ time, showing that activation of an intermediate of water-splitting is needed for O_2 production. The states induced by light flashes are not stable. Light captured by one photosystem cannot contribute to another photosystem. Four photons captured by PSII co-operate to dissociate water; they cause the reaction centre of PSII to eject four electrons. Four oxidation equivalents accumulate on an intermediate, S, before they accept four electrons from two molecules of water, releasing O_2. If S^{4+} is the oxidized component which reacts with water:

$$2\,H_2O + S^{4+} \rightarrow S + 4\,H^+ + O_2 \qquad\qquad [5.3]$$

and S is a 'charge'-accumulating chemical device. As photon capture is infrequent in dim light, the intermediate oxidized states must remain stable for sufficient time to enable four positive charges to accumulate, and allow water oxidation.

A model by Kok *et al.* (1970) explains the periodicity of O_2 as a cycle of S states. It proposes that S accumulates four oxidizing equivalents solely from P680 and that only O_2 is liberated in a single process. If the water-splitting and PSII complex, written as S.Z.P680.Q, is equivalent to S_0, then S_1 is S^+.Z.P680.Q^- with the reaction centre refilled with an electron and one oxidizing equivalent accumulated on S; further flashes cause the sequence shown in Fig. 5.7.

Events involving PSII reaction centre and water-splitting are therefore:

(1) Activation of the reaction centre chlorophyll, P680 and charge separation.
(2) Reduction of acceptor and rapid donation of an electron from the water-splitting complex, S, via Z to $P680^+$.
(3) Repetition until S_4 is formed releasing $4 H^+$.
(4) Removal of four electrons from two water molecules by S_4, and liberation of O_2.

To explain why most O_2 is released on the third flash (rather than the fourth as expected), it is assumed that in darkness S is 75 per cent in the state S_1 and 25 per cent in state S_0. The damped oscillations observed experimentally are thought to result from 5 per cent of the light flashes causing double hits and two oxidations and 10 per cent of the photons not hitting a target, so averaging the states (see Clayton 1980). The rate of conversion of each step $S_0 \rightarrow S_1$ and $S_1 \rightarrow S_2$ is probably about 50 μs, $S_2 \rightarrow S_3$ about 500 μs, but $S_3 \rightarrow S_4 \rightarrow S_0$ requires 1 ms so the O_2 evolving step is rate limiting. It is not clear if water oxidation limits the process or the rate of O_2 release. In (4) it is implied that all oxidation occurs at the $S_3 \rightarrow S_4 \rightarrow S_0$ step and this may be so judged on the spectroscopic and other evidence.

The S states decay in darkness; S_2 and S_3 are unstable and are deactivated to S_1 by addition of electrons, probably from reduced compounds via the electron transport chain and cyclic electron transport. Electrons flow from PQ via Q to P680 then to S_2, but not to S_3. States S_0 and S_i equilibrate in darkness and S_0 goes to S_1 if artificially reduced by ferricyanide or DCPIP and ascorbate. Hydroxylamine (NH_2OH) and ammonia bind tightly to the O_2-evolving complex and donate electrons,

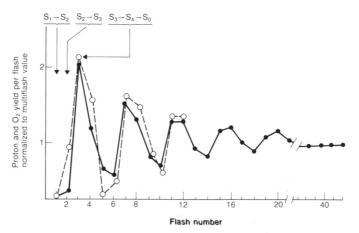

FIG 5.6   Oxygen evolution and proton release by chloroplasts given short (2 μs) intense flashes of light separated by darkness. The number per flash is expressed relative to the production after many flashes. ●——●, O_2 evolution; o --- o, H^+ evolution.

removing two oxidizing equivalents and increasing the lag in O_2 evolution which cannot be reversed by washing.

The S_2 and S_3 states are rapidly destabilized by a diverse group of very lipid soluble, weak acid anions with —NH and —OH groups, called ADRY agents from Acceleration of the Deactivation Reactions of Y (alternative sign for the water-splitting system), an example being Ant 2p (2(3-chloro-4-trifluoreomethyl) anilino-3,5-dinitrothiophene). ADRY reagents transport electrons from Q over the energy barrier to S.

The mechanism of O_2 evolution involves more than accumulating four oxidizing equivalents before O_2 release. Measurements of pH changes in thylakoids (treated with uncoupling agents (p. 95) to prevent accumulation of H^+ caused by the electron flow through the thylakoid membrane) show a complex release of protons (H^+) in relation to the light flashes (Fig. 5.7). Single protons are released at the S_0 to S_1 and S_2 to S_3 steps, whilst 2 H^+ are released from S_3 to S_0 via S_4. Release of H^+ neutralizes the charge accumulation on S, so it is not justified to call it a 'charge accumulating' device as previously stated (see Velthuys 1980).

Electron transport beyond the primary acceptor of PSII

The energized electron from P680 reduces Q_B (which is a bound plastoquinone) to the semiquinone and anionic plastoquinone which accepts

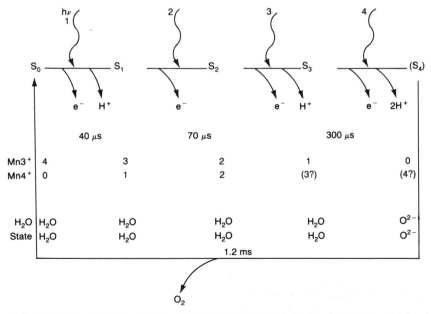

FIG. 5.7 Sequence of events in water splitting and O_2 evolution in PSII. The S states are related to the electron and proton release, manganese changes and states of the water molecules which are split (speculative, see text). The cycle is driven by light (v) capture.

protons to give hydroquinone. Reduced Q_B, in the Q_B^{2-} form, diffuses from the binding site and enters the plastoquinone pool. Quinones are important in many biological electron transport processes, carrying e^- and H^+. PQ is found almost exclusively in thylakoid membranes probably between the lipid layers (p. 61) because of its solubility. Several forms exist but PQ B and C are usually less than 20 per cent of the amount of PQ A and their structures are poorly known. PQ is a substituted benzoquinone with two methyl ($-CH_3$) groups attached to the ring and a long chain of nine isoprene groups; other types of PQ have three and four isoprene units. PQ A is colourless, absorbs light in the ultraviolet (260 nm) and is reduced at $+0.1$ V via plastosemiquinone to plastohydroquinone which absorbs at 290 nm. PQ is implicated as an electron carrier because e^- transport is inhibited when it is extracted with solvents, and restored when it is added back to the membrane, and from spectroscopic measurements of its oxidation and reduction in light and dark. In bright light most, but not all, of the PQ in thylakoids is reduced. The pool of PQ engaged in electron transport in the light is several times larger than that of other carriers. PQ has a molar ratio to chlorophyll of 1:10, with about 40 molecules of PQ per PSII reaction centre; this large pool accepts e^- from several PSII units, acting as a 'buffer' with PSI. PQ is oxidized by cytochrome f (cyt f) which itself is mainly oxidized under bright light because electrons are more rapidly removed by PSI than they are supplied from PQ, that is, the rate limiting step lies between PQ and cyt f. The large pool of PQ and rapid oxidation by cyt f ensure that PQ is not completely reduced in bright light; this enables PSII to function and donate e^-, even though the transport rate from Q to cyt f is slow.

PQ also has the important ability to transport protons across the thylakoid membrane (Fig. 5.4). Reduced (anionic) PQ on the stromal side picks up protons and carries them to the lumen, thus increasing the H^+ concentration inside as a consequence of electron transport. The method by which PQ 'turns over' in the membrane is not yet clear. Probably PQH_2, which is hydrophobic, diffuses laterally between PSII and cyt f (distances of the order of 100 nm) within the membrane, losing H^+ due to the electrical charge on the membrane. If PSII and PSI are unequally distributed between the granal and stromal thylakoids (p. 63) transport of electrons between them over some distance may be by diffusion of PQ.

The transfer of e^- and H^+ has been measured from absorbance at 320 nm (where PQ^- but not PQ or PQH_2 absorb) and of H^+ accumulation during a sequence of discrete millisecond light flashes using dark-adapted chloroplasts. PQ^- is formed by the first flash but H^+ is taken up only every second flash. Regeneration of PQ occurs after the second flash. Some H^+ is taken up with the first, third, fifth flashes, etc., as PQ is partially reduced, even in darkness. With time and increasing number of flashes the oscillations are damped. The cycle of PQ reduction and oxidation, which transfers H^+ across the thylakoid, is central to the generation of the pH gradient component of the proton motive force for ATP synthesis (p. 109) and for coupling to e^- transport. Mitchell (1975) suggests that PQH_2 loses e^- to Rieske $Fe-S$ centres and H^+ to the lumen of the thylakoid. The Rieske $Fe-S$ centres are part of the cytochrome $b-c_1$ or b_6-f complex in photosynthetic membranes and they are known to react with quinones from the ESR signals obtained when DBMIB is applied. The PQH then 'dismutates', giving e^- to cyt b_{563} and to cyt b_{559} low potential form, and losing H^+ to the lumen. The cytochrome cycles e^- back to reduce PQH on the stromal side of the thylakoid and picks up more H^+. This model (referred to as the proton motive 'Q cycle') provides a role for the cytochromes and could also give 2 H^+ per e^- transported. Presently, however, the experimental evidence is equivocal; 1 and 2 H^+ per e^- have been measured.

Electrons pass to cyt b_{563} (called b_6) and cyt f in a protein complex which has been isolated from thylakoids. The complex may be associated with the Rieske centres. The protein from cyt f is rather hydrophilic and is probably exposed at the lumen surface of the thylakoid. The redox characteristics of cyt f are not altered by incorporation in the complex. Electrons pass from PQ to cyt f via a Rieske $Fe-S$ centre, with characteristic ESR signal and mid-point potential of $+0.2$ V, which functions in a one electron transfer,

$$PQH_2 + Fe-S_{oxidized} \rightarrow PQH + H^+ + Fe-S_{reduced} \qquad [5.4]$$

The 2 e^- from PQH_2 go to separate $Fe-S$ molecules as the Rieske centres are single e^- carriers; it has been suggested that they carry H^+ also. Mutants of *Lemna* without the $Fe-S$ centre lack non-cyclic electron transport.

Cytochromes, porphyrin-proteins containing iron bound into the porphyrin as a haem, are important oxidation−reduction electron carriers. Cytochrome f (from *folium* = leaf) and two cytochrome bs are involved in thylakoid electron transport. Cytochrome f from higher plants has a molecular mass of about 65 kDa (although determinations differ) and an absorption band at 554 nm when reduced; it has a potential of $+0.37$ V. Cytochrome f donates single electrons to plastocyanin; it may move towards the outside of the thylakoid on illumination as the membrane changes shape and becomes thinner, increasing its ability to accept electrons from the stromal side during cyclic electron flow. However, as mentioned, cyt b_{563} and cyt b_{559} may link cyt f to the stromal side of the membrane. Cyt b_{559} has two forms: a low potential form of about $+0.02$ V which may have a role in e^- transport around PSII with PQ and cyt b_{563} and the high potential form of $+0.4$ V which may be involved in water-splitting. However, much evidence on cytochromes has accumulated without their role being clarified.

Plastocyanin (PC) a 40 kDa protein of four polypeptides, each with one Cu atom co-ordinated by two N atoms from histidine and two S atoms from cysteine, and a redox potential of 0.37 V, is found in thylakoid lamellae of oxygenic tissue, in the ratio 1 molecule to 400 chlorophylls. PC is blue when oxidized, with a major absorption band at 597 nm and is inhibited by treatment with mercury ($HgCl_2$) which blocks the $-SH$ groups, stopping all electron flow to PSI. PC accepts one electron from cyt f and may form a pool of electrons which can be passed to PSI.

Photosystem I

This photosystem is organized in chlorophyll–protein complexes, CPI of 130 kDa mass and CP_a of 200 kDa. It consists of ~ 200 antenna chl a with perhaps an inner core antenna of 100 chl a per P700 (the reaction centre chlorophyll); P700 is probably a chlorophyll dimer (see Ch. 3). Some 10–20 β-carotenes orientated with respect to the chlorophyll iron–sulphur proteins (non-haem iron) and proteins which bind and structure the units are found in the complex. The core of the photosystem, the P700–chl a-protein complex has two 62 kDa polypeptides and 100 chl a per P700 and there are several e^- acceptors (A_0, A, Fe–Sx) and one each of the smaller polypeptides. The main polypeptides, to which P700 and chl a attach, are of 58 and 62 kDa mass. These are specified by the chloroplast DNA psaA and psaB genes; from the amino acid sequences of the proteins the Fe–S centres are thought to link to cystine. The psaG gene codes for the 9 kDa subunit which binds the Fe–S group to the main complex. The 22 kDa subunits bind ferredoxin and the 19 kDa unit links the plastocyanin to the PSII complex inside the thylakoid lumen.

Excitation from the antenna leads to oxidation of P700, the reaction centre chlorophyll. Electron movement around PSI is so fast that there is little fluorescence from the antenna at room temperature, although at the temperature of liquid nitrogen (-196 °C) fluorescence at 730 nm is observed from PSI antenna chlorophyll. The primary acceptor, X, is possibly a pheophytin anion, or a chlorophyll monomer linked to a 16 kDa polypeptide in a special environment. A secondary acceptor is found in the form of ESR signals characteristic of 2Fe–2S or, more probably, 4Fe–4S centres of an 18 kDa protein; it is called Fe–S centre B (Fe–S.B); another Fe–S centre (Fe–S.A) may take electrons from Fe–S.B. With oxidation–reduction of P700 an optical spectroscopic signal occurs at 430 nm (hence called P430), possibly from the 4Fe–4S centres. The Fe–S centres delocalize electrons, stabilizing them for long enough to enable chemical reactions to take place. The different Fe–S units associated with PSI may allow e^- to be transported to different processes depending on conditions; if $NADP^+$ is limiting then e^- may pass to the electron chain via another centre.

Donation of electrons to ferredoxin

The very negative mid-point potential of the reduced carriers enables ferredoxin (Fd) to be reduced. As P700 is activated cyt f is oxidized and

Fd reduced, indicating its role as electron acceptor. Ferredoxins are electron-carrying iron—sulphur redox proteins, found in animals and plants, with Fe at the active centre (but not as haem), usually as 2Fe—2S (or sometimes 4Fe—4S). Plant ferredoxins are of 10—12 kDa mass and reddish brown in colour and transfer one electron at each step. They have a redox potential of -0.43 V (soluble ferredoxin), characteristic absorption spectra below 500 nm and ESR spectra which are important for their identification and quantitative estimation. Ferredoxins in the thylakoid membrane accept electrons from the secondary acceptors of PSI but soluble ferredoxin is situated on the stromal surface of the thylakoid and may receive electrons from a ferredoxin-reducing substance (FRS), but there is some doubt about the role of this protein.

Ferredoxins from chloroplasts pass electrons to many biological processes, back to PQ in the electron chain, for example, giving cyclic electron transport (see p. 114), or to $NADPH^+$, reducing it, the normal route in photosynthesis:

$$2 \text{ ferredoxin}^- + NADP^+ + H^+ \rightarrow NADPH + \text{ferredoxin} \qquad [5.5]$$

Ferredoxin reduces $NADP^+$ via the flavoprotein enzyme ferredoxin-$NADP^+$ oxidoreductase. The higher plant enzyme is of 40 kDa mass, contains one molecule of FAD, an —S—S— bridge and 4 —SH groups, one of which, buried in the molecule and essential for catalytic activity, is located in the stromal side of the thylakoid. It forms a complex with ferredoxin and $NADP^+$ which binds to a lysine amino acid. The reductase also catalyses other reactions with ferredoxin or NADPH; for example, NADPH reduces cyt f or NAD^+, and NADPH may be oxidized by transferring an electron to ferricyanide, dyes, etc. (diaphorase activity). It is therefore an important control point in the electron chain and may redistribute electrons to other substrates if the normal acceptor, $NADP^+$, is in short supply. Ferredoxin also provides electrons to reduce sulphate and nitrate (Ch. 7) and activates some enzymes of the photosynthetic carbon reduction cycle (see p. 135). Reduced ferredoxin reacts with molecular oxygen forming the superoxide radical O_2^- and H_2O_2 in the Mehler reaction. Ferredoxin links electron transport with the chemical reactions and provides flexibility in metabolism.

$NADP^+$ and NADPH, the oxidized and reduced forms of the photosynthetic reductant with characteristic spectra, are water soluble molecules. The structure (which resembles that of ATP) is complex and maintains a stable redox state under cellular conditions and allows specific binding to enzymes, etc. Reduction involves a 2 e^- transfer and one proton 'adds on'. The simplest model of electron flow from PSI to $NADP^+$ is:

$$P700 \rightarrow A_1 \rightarrow (A_2) \rightarrow (A + B) \rightarrow Fd_{sol} \rightarrow NADP^+ \qquad [5.6]$$

but models with parallel branches have been suggested (see Clayton 1980) allowing more flexible dispersal of electrons if $NADP^+$ limits electron flow.

Energetics of electron transport

To lift an electron from water ($+0.8$ V potential) to ferredoxin (-0.42 V) requires two photons acting in series (Ch. 3). Figure 5.8 shows the Hill and Bendall 'Z' scheme (so-called from its appearance) of electron transport. P680 ejects e^- to Q at about 0 V potential and P680$^+$ (potential of $+1.2$ V) removes e^- from H_2O via S and Z. Electrons from Q pass to P700 at $+0.4$ V; the coupling of electron transport to ATP synthesis is considered in Chapter 6. P700 ejects an electron to ferredoxin at -0.43 V, an accumulated energy of 1.2 V. However, the total energy in the two photoacts is $+0.8$ to -0.2 V in PSII and $+0.4$ to -0.8 V in PSI, a total of 2.2 V. Of the 1 V lost, part is recovered in ATP synthesis. The efficiency of photoactivation and electron transport is $1.4/2.2 \times 100 = 64$ per cent or greater. With a minimum of two photons per e^- transported, eight quanta of red light are needed per oxygen released, an energy of 23.4×10^{-19} J. The energy accumulated by four electrons passing to NADP$^+$ and synthesizing 2.7 ATP is around 8.8×10^{-19} J so maximum efficiency is 38 per cent. However, if the efficiency with respect to the total solar spectrum, not just red light, is calculated the maximum efficiency of the photochemical processes at the reaction centres is about 20 per cent.

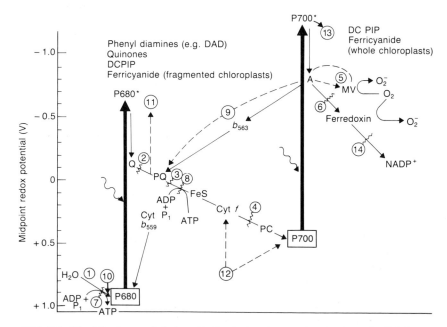

FIG. 5.8 The 'Z' scheme of photosynthetic electron transport and redox potential diagram for the photosystems and components of the electron transport chain. Numbers show the places at which inhibitors act, or at which e^- can be added or removed from the chain. Some compounds acting at these points are listed in Table 5.2.

Table 5.2 Sites of inhibitor activity and electron donation or acceptance identified by numbers on the electron transport scheme (Fig. 5.8), the processes affected and examples of compounds affecting them

Site	Process	Compound
Inhibitors		
1	Water-splitting	NH_2OH; TRIS
2	e^- transport to PQ	DCMU; atrazine
3	e^- and H^+ transport by PQ	DBMIB; DAD
4	e^- transport in Fe–S centres	HCN; mercurichloride; amphotericin B
5	e^- removal from PSI and autocatalytic production of oxygen radicals	bipyridyl herbicides (Paraquat)
6	Reduction of $NADP^+$	DSPD
7&8	Uncoupling ATP synthesis from e^- transport	NH_4^+; FCCP; CCCP; DCCD; Dio-9; DNP; phlorizin
14	$NADP^+$ reduction	2-phosphoadenosine diphosphate ribose
Artificial electron donors and acceptors		
9	e^- carrier from PSI to PQ giving cyclic flow	PMS
10	e^- donor to PSII	Catechol, DPC, ascorbate; phenylenediamine
11	Acceptor from PSII	DCPIP at PQ; KFeCN (fragmented chloroplasts); DAD
12	PSI donors	ascorbate; DPC; PMS; DAD reduced; DCPIP
13	Acceptors from PSI	DCPIP; KFeCN (whole chloroplasts); Paraquat; MV (acceptor from F_A)

Abbreviations

Atrazine	2-Chloro-4-(2-propylamino)-6-ethylamine-5-triazine
NH_2OH	Hydroxylamine
DCMU	3(3,4-Dichlorophenyl)-1,1-dimethylurea, (Diuron)
DBMIB	2,5-Dibromo-3-methyl-6-isopropyl-*p*-benzoquinone, (Dibromothymoquinone)
DCPIP	2,6-Dichlorophenolindophenol
HCN	Hydrogen cyanide
DAD	2,3,5,6-Tetramethyl-*p*-phenylenediamine, (Diaminodurol)
CCCP	Carbonylcyanide-*m*-chlorophenylhydrazone
Dio-9	Antibiotic
DSPD	Disalicylidenepropanediamine
DPC	Diphenylcarbazide
PMS	Phenazinemethosulphate, (5-methylphenazonium-methylsulphate)
FCCP	Carbonylcyanide-*p*-trifluoromethoxyphenylhydrazone
DCCP	Dicyclohexylcarbodiimide
DNP	Dinitrophenol
TRIS	Tris (hydroxymethyl) aminomethane
MV	Methylviologen

Rates of processes in electron transport

Physical reactions within the pigment bed are much faster than electron transport processes. Photon capture and excitation migration to the reaction centre are fast (10^{-15} s and 5×10^{-12} s, respectively). Fluorescence from chlorophyll occurs in 10^{-9} s when PSII is reduced but is slower when it is oxidized. Fluorescence from PSI is very limited as e^- transport from the reaction centres occurs in picoseconds. Electron transfer from reaction centres to acceptors requires $\sim 20 \times 10^{-9}$ s. What happens between 10^{-9} and 10^{-12} s is unknown. An electron is transferred from water to $NADP^+$ in 0.02 s. Water-splitting (0.2 ms) and electron movement from the PSII acceptor to plastoquinone (400 μs) are fast in comparison with e^- movement from plastoquinone to cyt f which requires about 20 ms and is the rate limiting step in the process; the large pool of PQ minimizes this limiting step in e^- transport. Electron transfer from plastocyanin to PSI takes 0.2 ms and from PSI to $NADP^+$ 10 μs. The plastoquinone 'pump' of H^+ across the thylakoid controls the rate of H^+ gradient development, a six-fold difference in rate (16–90 ms) occurring when changing from conditions of high to low back pressure of H^+; proton transport is probably slower (60 ms) than PQ to cyt f transfer and the development of ΔpH requires 30 s or so. Protons diffuse across the thylakoid in 5 s if CF_0-CF_1 (see Ch. 7) is not functioning. Proton transport rates across the membrane may be controlled by the herbicide-binding protein of PSII.

Artificial electron donors and acceptors

Many artificial redox compounds donate electrons to or accept them from the electron transport chain. They have been important in analysis of photosynthetic processes since Hill's discovery in 1937 that ferricyanide accepts electrons from water with oxygen evolution (now called the Hill reaction, and the substances Hill reagents):

$$4\,Fe(CN)_6^{3-} + 2\,H_2O \xrightarrow[\text{chloroplasts}]{\text{light}} 4\,Fe(CN)_6^{4-} + 4\,H^+ + O_2 \qquad [5.7]$$

Changes in the absorbance spectra of ferricyanide or a dye like DCPIP indicates the oxidation–reduction state of the electron acceptors and donors and electron flow. As their redox potentials are known, the potentials of the electron chain components may be determined. Differences in lipid solubility, molecular size, etc., can also be exploited to indicate where a reaction is taking place in the thylakoid or chloroplast.

Electron transport is blocked by compounds which remove electrons from different parts of the chain or are non-functional analogues of compounds in the chain. Hydroxylamine, as mentioned on p. 86, may occupy the position of H_2O in the water-splitting complex and this inhibits water splitting. The viologens, for example, methyl viologen (called commerically

'Paraquat') transfer electrons to O_2, forming singlet oxygen, which destroys lipids, etc. Viologen recycles, so the process is autocatalytic and destroys tissues rapidly in the light. DCMU is a quinone analogue and blocks the acceptor of PSII or between Q and PQ, thus separating PSII from PSI. One molecule of DCMU per 100 chlorophylls completely stops electron transport. Atrazine stops e^- flow at PQ by inhibiting the binding of Q to the protein complex. DBMIB is a structural analogue of PQ and prevents electron transfer to cyt *f*. An analogue of $NADP^+$, phosphoadenosine diphosphate ribose, blocks ferredoxin $NADP^+$ reductase. The antibiotic DIO-9 inhibits ATP synthesis at CF_1.

Electrons are donated to PSI by phenyldiamine and to cyt *f* by DCPIP plus ascorbate. Using DBMIB to block electron flow, DCPIP (plus ascorbate for reduction) donates electrons to PSI and methyl viologen is the acceptor; these reagents provide a test system for PSI and the effects of conditions on it. PSII is measured by electron flow to DCPIP with transport blocked by DBMIB. Open chain electron transport is measured with water as donor and viologen or $NADP^+$ as acceptor. Cyclic electron transport is measured in broken chloroplasts which have lost soluble ferredoxin, by adding PMS which carries electrons from PSI back to PC or cyt *f*. The sites of action of some compounds are shown with approximate redox potentials in Fig. 5.8 and Table 5.2. Some are commercial herbicides such as DCMU, the viologens and atrazine.

Formation of reactive forms of oxygen by photosynthesis

Oxygen in its diatomic form (O_2) which contains two unpaired electrons with parallel spins ('triplet') is not only a product of photosynthesis but is 'assimilated' into different forms as a consequence of it. Light energy (and electrons from other biochemical reactions not considered here) causes the formation of reactive O_2 species (see Elstner 1982):

$$O_2 \xrightarrow{e^-} O_2^{\cdot-} \xrightarrow[H^+]{e^-} HO_2^{\cdot} \xrightarrow[H^+]{e^-} H_2O_2 \xrightarrow[H^+]{e^-} OH^{\cdot} \xrightarrow[H^+]{e^-} 2H_2O \qquad [5.8]$$

superoxide   perhydroxyl radical   hydrogen peroxide   hydroxyl radical

In photosynthesis, excited pigments (e.g. triplet chlorophyll, ^{3}Chl) donate energy to O_2 giving $^1O_2$ (singlet oxygen):

$$\text{Chl} \xrightarrow{h\nu} {}^3\text{Chl}; \ {}^3\text{Chl} + O_2 \rightarrow \text{Chl} + {}^1O \qquad [5.9]$$

or by electron transfer, producing superoxide:

$$\text{Chl} \xrightarrow{h\nu} {}^3\text{Chl} + O_2 \rightarrow \text{Chl}^+ + O_2^{\cdot-} \qquad [5.10]$$

Photosystems may pass e^- to O_2 directly or via intermediates, paraquat, for example, which reduces O_2 to superoxide and recycles to carry more

e^-, or from physiological intermediates such as ferredoxin which produces superoxide or, in the Mehler reaction, H_2O_2:

$$Fd_{red} + O_2 \rightarrow Fd_{ox} + O_2^{\cdot -}$$

$$Fd_{red} + O_2^- + 2H^+ \rightarrow H_2O_2 + Fd_{ox} \qquad [5.11]$$

Hydrogen peroxide and O_2^- react to give $\cdot OH$ in a reaction catalysed by metal salts:

$$H_2O_2 + O_2^{\cdot -} \xrightarrow{Fe_{salt}} O_2 + \cdot OH + OH^- \qquad [5.12]$$

Superoxide reacts with H^+ in the presence of the enzyme superoxide dismutase (SOD), in alkaline conditions, to give H_2O_2 which is destroyed by catalase:

$$O_2^{\cdot -} + O_2^{\cdot -} + 2 H^+ \xrightarrow{SOD} H_2O_2 + O_2$$

$$2 H_2O_2 \xrightarrow{catalase} 2 H_2O + O_2 \qquad [5.13]$$

SOD is found throughout the plant cell including chloroplasts, and some catalase is in the chloroplast, probably bound to PSI and is not soluble. These two enzymes (together with the carotenoid quenching of reactive forms of oxygen) are an important defence against highly reactive oxygen states. Superoxide reacts with unsaturated fatty acids causing lipid peroxidation and thereby destroying membranes, and with chlorophyll causing photobleaching. Accumulation of malondialdehyde indicates destruction of lipids. Superoxide also oxidizes sulphur compounds, NADPH and ascorbic acid, or reduces cyt c and metal ions; such reactive O_2 species must be rapidly removed. SOD is a fast enzyme which, together with catalase, protects the thylakoid. By coupling oxidation–reduction of the thiol (—SH) groups of the tripeptide glutathione (which is at high concentration in chloroplasts) with the reduction state of ascorbic acid and NADP, H_2O_2 and superoxide content is regulated:

$$H_2O_2 + ascorbate \xrightarrow[peroxidase]{ascorbate} dehydroascorbate + H_2O$$

$$dehydroascorbate + glutathione_{red} \rightarrow glutathione_{ox} + ascorbate$$

$$glutathione_{ox} + NADPH \xrightarrow[reductase]{glutathione} glutathione_{red} + NADP^+ \qquad [5.14]$$

Bright light damages mutant plants lacking SOD, catalase or other enzymes which control the oxidation–reduction state of the chloroplast, or when adverse conditions cause stomatal closure and resist the consumption of NADPH allowing high energy states and reducing compounds to accumulate in the light-harvesting apparatus.

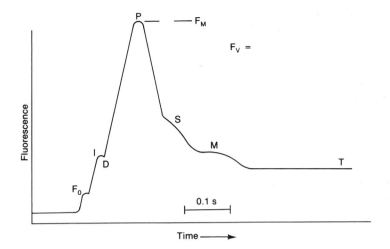

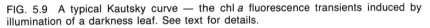

FIG. 5.9　A typical Kautsky curve — the chl *a* fluorescence transients induced by illumination of a darkness leaf. See text for details.

Chlorophyll fluorescence

Up to 90 per cent of the photons incident on the light-harvesting antenna of the photosystems may be captured by chlorophyll *a* and other pigments and the energy is transferred via excitation in the pigment bed to the reaction centres and thence to the chemical reactions. However, not all of the energy is used for photochemistry; thermal deactivation and excitation transfer to non-fluorescent pigments, e.g. to the PSI antenna, contribute to the use of energy. One of the energy dissipation routes is by fluorescence which is emitted at 685 nm and 740 nm at room temperatures. Chlorophyll *a* in solution emits some 30 per cent of the light absorbed as fluorescence but under physiological conditions the maximum is about 3 per cent and the minimum about 0.6 per cent. Most emission at normal temperatures is from chl *a* of PSII, with five major components at 680 nm, produced by LHCII, 685 nm, from CP43, 695 nm, from CP47, 720 nm, from the PSI core and 735 nm, from LCHI. PSI only produces significant fluorescence at low temperature (77 K) because P700 is relatively stable and can still trap excitation energy and dissipate it as heat but $P680^+$ rapidly returns to the ground state. The rate of fluorescence emission, F, is proportional to the light absorbed, I, to the fraction, β, of the energy reaching the PSII with chl *a* concentration [chl PSII] and quantum yield, φ_f (fluorescence relative to radiationless dissipation, d, transfer to other molecules (t) and photochemistry (p)). The relation is:

$$F = I \times \beta \times [\text{chl PSII}] \times \varphi_f \tag{5.15}$$

The rate constant K (units s^{-1}) for each process expresses the rate of decay; the greater the constant the shorter the intrinsic lifetime of a pigment molecule, $\tau = 1/K_f + K_d + K_p + K_t$. With β and [chl PSII] constant during fluorescence changes the quantum yield of fluorescence, $\varphi_f =$ intensity of fluorescence/intensity of absorbed light or $K_f/(K_f + K_d + K_p + K_t \cdot P)$, where P is the fraction of PSII reaction centres that can accept energy from chlorophyll.

At F_0 on the induction curve the PSII reaction centres are fully oxidized, $P = 1$ and the quantum yield of fluorescence is given by $K_f/(K_f + K_d + K_p + K_t)$. At maximum fluorescence, F_m, in the presence of DCMU or after a saturating light pulse, Q_A is blocked so the maximum yield of fluorescence, φ_{fm}, is obtained. The variable fluorescence, F_v, is $F_m - F_0$, which is also the difference between the respective quantum yields and shows the yield of photochemistry. The maximum yield of photochemistry is given by the ratio of K_p to the sum of the other rate constants.

Variable fluorescence, F_v, is a very important component derived from the fluorescence signals for it reflects the balance of energy between that in the antenna and reaction centre and that used by the photochemical and other processes. It is related to the efficiency with which energy is used and therefore to the pathways of chemical processes downstream of the photosystems. As fluorescence provides a signal that can be measured easily with suitable photodetectors via relatively simple amplifiers and filters to remove interfering light, it provides a signal to detect events at the heart of photosynthesis continuously and thus dynamically and so has become a very important technique for understanding the links between the light reactions and metabolism (see Fig. 5.4 and Ch. 12).

In the sequence of reduction of electron chain intermediates each step has a different time constant; the initial charge separation of P680 and reduction of pheo is very fast (3 ps) whereas pheo $\rightarrow$ Q_A is much slower (100–200 μs) with open RCs and 300 μs when closed. Overall reduction of PQ requires 1–2 ms. Consequently the fluorescence emission from a population of photosystems after a period in darkness is very dynamic, the so-called fluorescence transient or Kautsky curve (Fig. 5.9). In darkness all the photosystem components and acceptors are uncharged but when light is absorbed by the antenna P680 is oxidized to P680$^+$ and fluorescence increases rapidly from F_0 to an 'inflection' point called I, caused probably by reduction of Q_A molecules linked to PSII but not connected to Q_B so they are rapidly reduced; in spinach up to 30 per cent of PSII may be in this state. As Q_A is progressively reduced in the other, attached, PSII particles, so the fluorescence rises very rapidly. The dip 'D' reflects the imbalance between fluorescence from attached and unattached Q_A. When Q_A is fully reduced, and point P is reached fluorescence is at the maximum F_m. Further reduction of Q_B and PQ uses electrons from Q_A and so the fluorescence signal decreases. During the decrease in fluorescence from P, transients S and M often appear before steady-state fluorescence is reached,

T. These reflect a number of processes that are poorly understood, involving feedback control of energy transfer by such processes as build up of the proton gradient in the thylakoids and rearrangement of light-harvesting complexes and the induction of carbon and nitrate metabolism.

If the electron transport from Q_A to Q_B is blocked by the herbicide DCMU, which binds to the Q_B sites on the D1 protein on the PSII reaction centre, then no further transfer of electrons is possible and fluorescence is maximal. Also with a very brief pulse (2 s) of light which saturates all photosystems, Q_A is fully reduced and F_m is achieved. The rate of Q_A reduction is proportional to the photon flux, co-operation between PSII units and their heterogeneity, the size of the PQ pool and its rate of re-oxidation and the 'downstream' electron transport. However, during this phase F_v is smaller than expected from the state of Q_A reduction because excitation is transferred from closed to open centres. Also there may be different types of PSII units, PSIIα and PSIIβ, the former in appressed grana and linked to the electron transport chain, the latter in non-appressed grana and probably without direct connection to the electron transport chain. PSIIβ may be damaged units undergoing repair or breakdown and have smaller antenna size but many features of this model are still controversial.

Fluorescence is quenched, i.e. decreased, by a number of factors, as we have seen, and quenching coefficients with values of $0-1$ may be assigned to the processes. Thus quenching, q, of variable fluorescence is described by:

$$q = (F_v - F_v')/F_v \quad \text{or} \quad F_v'/F_v = 1 - q \qquad [5.16]$$

where F_v' is in the quenched state and provides valuable information on the causes of the changed fluorescence. The main physiological mechanism of quenching is by photochemical quenching q_P (or q_Q), which depends on oxidized Q_A and shows the proportion of excitons that are converted to chemical energy in PSII. Other mechanisms are non-photochemical quenching, q_N (or q_{NP}), resulting from energy-dependent quenching (q_E) or state transition quenching (q_T) or photoinhibitory quenching (q_I). Energy-dependent quenching derives from the development of the pH gradient in the thylakoids and has a half-time of about 1 min. As pH builds up it affects the membrane's capacity to allow electron flow, so there is a strong dependence of q_E on pH. The mechanism is not well understood but there is evidence for a role of zeaxanthin in the changes; possibly the carotenoid interacts with the antenna, increasing the rate constant and k_D although other models of the mechanism, e.g. alternative cyclic photochemical processes or quenching in the antenna, have been suggested.

The state transition quenching comes from the energy transfer from PSII to PSI during state 1 to state 2 transitions, which is a shift from high to low fluorescence. This mechanism is thought of as a change in the exciton distribution in LHCII from PSII to PSI. In the light LHCII becomes phosphorylated; light activates (via electron flow and reduced PQ) a protein kinase located on the stromal surface. This kinase phosphorylates threonine residues on the exposed LHCII, using ATP (plus Mg^{2+}) and requires about 10 min, a time-scale similar to the state 1 to state 2 changes. It was once considered that this modification enabled the protein to migrate from granal to stromal regions and donate energy to PSI. However, a number of alternative possibilities have been suggested, e.g. that a complex of PSIIβ units, LHCII in the phosphorylated form and PSI is formed, allowing energy to 'spill over'. The photoinhibitory quenching, q_I, is related to damage caused by excessive light energy and has a half-time of about 40 min. This quenching results from increased non-photochemical processes and is shown as a decrease in F_v/F_m following exposure of leaves to bright light and correlates with a loss of quantum yield of photosynthesis. It is thought that q_I is related to damage to the D1 protein of PSII caused when the energy load on the system is excessive. This may occur when CO_2 and O_2 are not available to act as substrates for electrons. Photorespiration in C3 plants may function to reduce the energy load and protect thylakoids. Recovery from photoinhibition may involve transfer of damaged PSII to the stromal membranes, resynthesis and insertion of D1 protein; these units may be PSIIβ.

References and Further Reading

Amesz, J. (1977) Plastoquinone, pp. 238–46 in Trebst, A. and Avron, M. (eds), *Encyclopedia of Plant Physiology* (N.S.), Vol. 5, *Photosynthesis I*, Springer-Verlag, Berlin.

Amesz, J. (1983) The role of manganese in photosynthetic oxygen evolution, *Biochim. Biophys. Acta*, **726**, 1–12.

Anderson, J.M. and **Andersson, B.** (1988) The dynamic photosynthetic membrane and regulation of solar energy conversion, *TIBS*, **13**, 351–55.

Andersson, B., Jansson, C., Ljungberg, U. and **Åkerlund, H.-E.** (1985) Polypeptides on the oxidizing side of photosystem III, pp. 21–31 in Steinback *et al.* (eds), *Molecular Biology of the Photosynthetic Apparatus*, Cold Spring Harbor Laboratory.

Babcock, G.T. (1987) The photosynthetic oxygen-evolving process, pp. 125–58 in Amesz, J. (ed.), *New Comprehensive Biochemistry*, Vol. 15, Elsevier, Amsterdam.

Barber, J. (1987) Photosynthetic reaction centres: A common link, *TIBS*, **12**, 321–26.

Barber, J. (ed.) (1992) *Topics in photosynthesis, vol II, The Photosystems: Structure, Function and Molecular Biology*. Elsevier Science Publishers, Amsterdam.

Barber, J., Chapman, D.J., Gounaris, K., Marder, J.B. and **Telfer, A.** (1989) Structural and functional properties of the isolated photosystem two reaction centre, pp. 81–9 in Barber, B. and Malkin, R. (eds), *Techniques and New Developments in Photosynthesis Research*, Plenum Press, New York/London.

Bennett, J. (1991) Protein phosphorylation in green plant chloroplasts, *A. Rev. Plant Physiol. Plant Mol. Biol.*, **42**, 281–311.

Blankenship, R.E. and **Parson, W.W.** (1978) The photochemical electron transfer reactions of photosynthetic bacteria and plants, *A. Rev. Biochem.*, **47**, 635–53.

Bricker, T.M. (1990) The structure and function of CP_{a-1} and CP_{a-2} in photosystem II, *Photosynthesis Res.*, **24**, 1−13.

Bruce, B.D., Malkin, R., Wynn, R.M. and **Zilber, A.** (1989) Structural organization and function of polypeptide subunits in photosystem I, pp. 61−80, in Barber, J. and Malkin, R. (eds), *Techniques and New Developments in Photosynthesis Research*, Plenum Press, New York/London.

Brundvig, G.W., Beck, W.F. and **Paula, J.C. de** (1989) Mechanism of photosynthetic water oxidation, *A. Rev. Biophys. Chem.*, **18**, 25−46.

Clayton, R.K. (1980) *Photosynthesis: Physical Mechanisms and Chemical Patterns*, I.U.P.A.B. Biophysics Series, Cambridge University Press.

Cramer, W.A. and **Crofts, A.R.** (1982) Electron and proton transport, pp. 387−467 in Govindjee (ed.), *Photosynthesis*, Vol. 1, *Energy Conversion by Plants and Bacteria*, Academic Press, New York.

Deisenhofer, J., Epp, O., Miki, K., Huber, R. and **Michel, H.** (1985) Structure of the protein subunits in the photosynthetic reaction centre of *Rhodopseudomonas viridis* at 3Å resolution, *Nature*, **318**, 618−24.

Demmig-Adams, B. (1990) Carotenoids and photoprotection in plants: A role for the xanthophyll zeaxanthin, *Biochim. Biophys. Acta.*, **1020**, 1−24.

Dilley, R.A., Theg, S.M. and **Beard, W.A.** (1987) Membrane-proton interaction in chloroplast bioenergetics: localised proton domains. *A. Rev. Plant Physiol.*, **38**, 347−89.

Diner, B.A. and **Joliot, P.** (1977) Oxygen evolution and manganese, pp. 187−205 in Trebst, A. and Avron, M. (eds), *Encyclopedia of Plant Physiology* (N.S.), Vol. 5, *Photosynthesis I*, Springer-Verlag, Berlin.

Dutton, P.L. (1986) Enegy transduction in anoxygenic photosynthesis, in Staehelin, L.A. and Arntzen, C.J. (eds), *Encyclopedia of Plant Physiology* (N.S.), Vol. 19, *Photosynthesis III. Photosynthetic Membranes and Light Harvesting Systems*, Springer-Verlag, Berlin.

Elstner, E.F. (1982) Oxygen activation and oxygen toxicity, *A. Rev. Plant Physiol.*, **33**, 73−96.

Ferguson, S.J. and **Sorgato, M.C.** (1982) Proton electrochemical gradients and energy transduction processes, *A. Rev. Biochem.*, **51**, 185−217.

Govindjee and **Coleman, W.J.** (1990) How plants make oxygen, *Sci. Amer.*, **262**, 42−51.

Govindjee and **Whitmarsh, J.** (1982) Introduction to photosynthesis: Energy conversion by plants and bacteria, pp. 1−18 in Govindjee (ed.), *Photosynthesis*, Vol. 1, *Energy Conversion by Plants and Bacteria*, Academic Press, New York.

Govindjee, Amesz, J. and **Fork, D.J.** (eds) (1986) *Light Emission by Plants and Bacteria*, Academic Press, New York.

Hall, D.O. and **Rao, K.K.** (1977) Ferredoxin, pp. 206−16 in Trebst, A. and Avron, M. (eds), *Encyclopedia of Plant Physiology* (N.S.), Vol. 5, *Photosynthesis I*, Springer-Verlag, Berlin.

Hansson, Ö. and **Wydrzynski, T.** (1990) Current perceptions of photosystem II, *Photosynthesis Res.*, **23**, 131−62.

Hauska, G. (1977) Artificial acceptors and donors, pp. 253−65 in Trebst, A. and Avron, M. (eds), *Encyclopedia of Plant Physiology* (N.S.), Vol. 5, *Photosynthesis I*, Springer-Verlag, Berlin.

Izawa, S. (1980) Acceptors and donors for chloroplast electron transport, pp. 413−675 in San Pietro, A. (ed.), *Methods in Enzymology*, Vol. 69, *Photosynthesis and Nitrogen Fixation*, Part C, Academic Press, London.

Joliot, P. and **Kok, B.** (1975) Oxygen evolution in photosynthesis, pp. 388−412 in Govindjee (ed.), *Bioenergetics of Photosynthesis*, Academic Press, New York.

Kok, B., Forbush, B. and **McGloin, M.** (1970) Co-operation of charges in photosynthetic oxygen evolution. 1. A linear four step mecanhism, *Photochem. Photobiol.*, **11**, 457−75.

Krause, G.H. and **Weis, E.** (1991) Chlorophyll fluorescence and photosynthesis: The basics, *A. Rev. Plant Physiol. Plant Mol. Biol.*, **42**, 313−49.

Kushner, D.J. (1985) The *Halobacteriaceae*, pp. 171−214 in Woose, C.R. and Wolfe, R.S. (eds), *The Bacteria*, Vol. VIII, *The Archaebacteria*, Academic Press, Orlando.

Lichtenthaler, H.K. (ed.) (1988) *Application of Chlorophyll Fluorescence*, 366 pp., Kluwer, Dordrecht.

Malkin, R. (1982) Redox properties and functional aspects of electron carriers in chloroplast photosynthesis, pp. 1−47 in Barber, J. (ed.), *Electron Transport and Photophosphorylation, Photosynthesis 4*, Elsevier Biomedical Press, Amsterdam.

Michel, H. and **Deisenhofer, J.** (1986) X-ray diffraction studies on a crystalline bacterial photosynthetic reaction center. A progress report and conclusions on the structure of photosystem II reaction centers, pp. 371−87, in Staehelin, L.A. and Arntzen, C.J. (eds), *Encyclopedia of Plant Physiology* (N.S.), Vol. 19, *Photosynthesis III. Photosynthetic Membranes and Light Harvesting Systems*, Springer-Verlag, Berlin.

Michel, H. and **Deisenhofer, J.** (1988) Relevance of the photosynthetic reaction centre from purple bacteria to the structure of photosystem II, *Biochemistry*, **27**, 1−7.

Mitchell, P. (1975) The protonmotive Q cycle: A general formulation, *FEBS Lett.*, **59**, 137−9.

Oakamura, M.V. and **Feher, G.** (1992) Proton transfer in reaction centres from photosynthetic bacteria, *A. Rev. Biochemistry*, **61**, 861−96.

Oettmeier, W. (1992) Herbicides of photosystem II, pp. 349−408 in Barber, J. (ed.), *Topics in Photosynthesis Vol. II, The Photosystems: Structure, Function and Molecular Biology*, Elsevier, Amsterdam.

Ort, D.R. (1986) Energy transduction in oxygenic photosynthesis, pp. 143−96, in Staehelin, L.A. and Arntzen, C.J. (eds), *Encyclopedia of Plant Physiology* (N.S.), Vol. 19, *Photosynthesis III. Photosynthetic Membranes and Light Harvesting Systems*, Springer-Verlag, Berlin.

Rees, D.C., Komiya, A., Yeates, T.O., Allen, J.P. and **Feher, G.** (1989) The bacterial photosynthetic reaction centre as a model for membrane proteins, *A. Rev. Biochem.*, **58**, 607−33.

Renger, G. (1983) Photosynthesis, pp. 515−42 in Hoppe, W. *et al.* (eds), *Biophysics*, Springer-Verlag, Berlin.

Rochaix, J.-D. and **Erickson, J.** (1988) Function and assembly of photosystem II genetic and molecular analysis, *TIBS*, **13**, 56−9.

Rutherford, A.W. (1989) Photosystem II, the water-splitting enzyme, *TIBS*, **14**, 227−32.

Skulachev, V.P. (1988) *Membrane Bioenergetics*, Springer-Verlag, Berlin.

Stevens, Jr. S. and **Bryant, D.A.** (eds) (1989) *Light-energy Transduction in Photosynthesis: Higher Plant and Bacterial Models*, The Pennsylvania State University, University Park.

Stoeckenius, W. (1979) A model for the function of bacteriorhodopsin, pp. 39−47, in Cone, R.A. and Dowlin, J. (eds), *Membrane Transduction Mechanisms*, Society of General Physiologists Series 23, Raven, New York.

Trebst, A. (1980) Inhibitors in electron flow: Tools for the functional and structural localization of carriers and energy conservation sites, pp. 675−715 in San Pietro, A. (ed.), *Methods in Enzymology*, Vol. 69, *Photosynthesis and Nitrogen Fixation*, Part C, Academic Press, London.

Trumpower, B.L. (ed.) (1982) *Function of Quinones in Energy Conserving Systems*, Academic Press, New York.

Velthuys, B.R. (1980) Mechanisms of electron flow in photosystem II and towards photosystem I, *A. Rev. Plant Physiol.*, **31**, 545–67.

Walker, D.A. and **Osmond, C.B.** (eds) (1989) *New Vistas in Measurement of Photosynthesis*, The Royal Society, London.

Yocum, C.F. (1984) Photosynthetic oxygen evolution: An overview, pp. 239–42 in Sybesma, C. (ed.), *Advances in Photosynthesis Research*, Vol. 1, Martinus Nijhoff/Dr W. Junk Publishers, The Hague.

Youvon, P.C. and **Marrs, B.L.** (1987) Molecular mechanisms of photosynthesis, *Sci. Amer.*, **256**, 44–9.

CHAPTER 6

Synthesis of ATP: photophosphorylation

Metabolic processes, such as the assimilation of carbon dioxide, protein synthesis and ion pumping, require ATP. Mitochondria and aerobic bacteria synthesize ATP by respiration of preformed substrates. Despite the different sources of energy for photo- and respiratory ('oxidative') phosphorylation, both involve electron flow along a chain of redox components coupled to ATP synthesis; this chapter considers the mechanism of photophosphorylation in chloroplasts.

The metabolic role of ATP

ATP has two anhydride (pyrophosphate) bonds. Hydrolysis to ADP yields -31 kJ mol^{-1} of energy. Under cellular conditions the energy values of the ATP/ADP + P_i reaction may be $50-60$ kJ mol^{-1}. The energy of the reaction comes from the electrostatic repulsion of negative charges on the phosphate groups, and to resonance stabilization and the large enthalpies of solvation of the reaction products. The Gibbs free energy released in the reactions is 'coupled' by biochemical mechanisms to do work, for example, the ion ATPases couple the hydrolysis of ATP with the transport of ions. In many metabolic reactions hydrolysis of ATP is stoichiometrically linked to the chemical transformations. ATP is used in most cellular reactions, directly or via other phosphorylated nucleotides, for example, guanosine and uridine nucleotides (e.g. GTP and UTP), to which ATP transfers P_i groups.

$$\text{ATP} + \text{GDP} \xrightarrow{\text{nucleotide diphosphokinase}} \text{ADP} + \text{GTP} \qquad [6.1]$$

ATP is required in metabolic pathways where phosphorylated intermediates are interconverted, for example, the photosynthetic carbon reduction cycle. Cleavage of ATP at position I allows the P_i group to be transferred to water (the enzymes are called ATPases) or to other compounds in the presence of suitable enzymes. There are three different mechanisms: (1) the enzyme may itself be phosphorylated (e.g. ion ATPase), (2) no covalent bond forms between enzyme and ATP (e.g. adenylate kinase, eqn 6.2) and (3) the enzyme forms phosphorylated intermediates in the course of interchanging groups (e.g. glutamine synthetase, which uses ATP in transferring NH_3 to glutamate to form glutamine). When ATP donates phosphate groups to compounds, it increases their reactivity, for example, glucose is phosphorylated to glucose-6-phosphate by ATP and hexokinase before consumption in glycolysis and respiration. The synthesis of ATP is central to any discussion of photosynthetic processes. The anhydride bonds of ATP are not, as often said, 'richer in energy' or of 'higher energy' than those of many other compounds and do not therefore 'drive' metabolism simply by providing energy. The ability of a chemical reaction to do work is related to the state of equilibrium of the reaction, the further from equilibrium the more energy available. The ATP reaction is important because under the conditions in the cell it is displaced from equilibrium and the Gibbs free energy change, ΔG, is favourable for doing metabolic work. ATP has an intermediate phosphate group transfer potential, as defined by the free energy of hydrolysis, and can function as a phosphate group carrier. In an analogous way NADP and NAD acts as an electron and H^+ carrier in metabolism. ATP is, however, very stable at normal temperatures and near neutral pH in the cell, unless involved in enzyme reactions, when it is an almost universal donor and acceptor of phosphate groups to other molecules, activating them in biochemical reactions.

Different cell compartments contain ATP, which because of differences in rates of reactions, will be in different equilibrium states. Exchange of phosphorylated compounds (often not ATP directly) takes place between compartments where reactions consume or produce ATP, so regulating the ATP pools in different parts of the cell. Cell metabolism is balanced with respect to the energy available for ATP synthesis and supply of substrates. Turnover of ATP in cells is rapid: the total 'pool' in leaf cells may be broken down and resynthesized within 500 ms; therefore metabolism responds quickly to the supply of — and demand for — ATP. If synthesis slows, metabolism also slows and as different pathways require different amounts of ATP or are differentially regulated by the concentration of ATP or ADP, so the response of the system is modified. The proportions of ATP, ADP and AMP in cellular compartments, under different metabolic conditions which change the requirement for nucleotide, is controlled by adenylate kinase:

$$ATP + AMP \xrightleftharpoons{\text{adenylate kinase}} 2\ ADP \tag{6.2}$$

This means that the forms of adenylate are regulated close to an optimum for the many processes involved in metabolism. The phosphorylation state in tissues is expressed by the energy charge (EC) (Atkinson 1977):

$$EC = \frac{[ATP] + \frac{1}{2}[ADP]}{[AMP] + [ADP] + [ATP]} \qquad [6.3]$$

An EC of 1 would be a condition of all ATP, and an EC of 0, all AMP. Atkinson suggests that the ATP-regenerating enzymes have minimum velocity at large EC and maximum at small, whereas enzymes consuming ATP act in reverse. So in cells with rapid synthesis of ATP compared to demand for ATP, EC is large and ATP synthesis is slow. Conversely with little ATP synthesis and large demand, EC is small and ATP synthesis is rapid, given the required conditions. Equilibrium is attained between supply and demand because of the response of enzyme systems to EC. However, the rate of reactions is not linear with EC but changes most rapidly as EC decreases from approximately 0.9 to 0.6 and only slowly with a further decrease. Metabolism is therefore very sensitive to small changes in EC and is closely regulated by the supply of, and demand for, ATP.

Control of enzyme reactions is not only by EC or the availability of ATP, ADP or AMP as substrate but by these molecules acting as allosteric effectors of enzyme reactions. The effectors bind away from the reaction site and change the catalytic behaviour of the enzyme. An important photosynthetic example is the effect of ATP on 3-phosphoglycerate kinase of chloroplasts (see p. 127); the enzyme uses ATP in formation of 1,3-diphosphoglycerate and is stimulated at high EC. It also generates ATP in the reverse reaction, which is inhibited by ATP and high EC (0.9–1.0) and stimulated by low EC (0.7). Both forward and reverse reactions are inhibited by AMP. Such complex control based on phosphorylated adenylates provides for a very subtle balance between processes and is an essential feature of cellular metabolism and maintenance of cellular homeostasis.

Another measure of the role played by adenylates in metabolism is the phosphorylation potential, $P = (ATP)/(ADP)$ (P_i) which takes account of the influence of inorganic phosphate but does not involve the adenylate kinase system as does EC.

Measurement of ATP

Because of the rapid turnover of ATP, tissues or cells must be killed quickly (milliseconds) if the state of the system is not to change during extraction of ATP; this is done by plunging tissues into very cold solvents, for example, pentane at $-20\ °C$, or by clamping between the jaws of metal tongs at liquid nitrogen temperature ($-196\ °C$). Adenylates are extracted by solvents, which also denature enzymes. The concentration of ATP may be measured by one of several methods, for example, by chromatographic separation from other nucleotides on ion exchange columns and detection of ATP with ultraviolet light after elution. Enzymatic methods are frequently used to measure the ATP in extracts, for example, hexokinase converts glucose to glucose-6-phosphate using ATP, the glucose-6-P is oxidized by $NADP^+$

and the resultant NADPH is detected spectrophotometrically, as 1 mol ATP consumed produces 1 mol NADPH. Another sensitive method for ATP, measures the bioluminescence produced when an extract from the light organs (lanterns) of fireflies, containing luciferin and the enzyme luciferase, reacts with ATP. The emitted photons are measured with a sensitive photometer.

Energy transducing mechanisms in phosphorylation — the chemiosmotic theory

ATP is synthesized from ADP and P_i; an enzyme, ATP synthase and an energy source are required. The enzyme and the ATP production system are similar in all organisms: the requirements are an intact membrane enclosing a space or opening to the exterior of the cell (see Ch. 1, p. 10) and the ATP synthase which has a basal part, called F_0, in the membrane and a head part, F_1, attached to F_0 and protruding into the cytosol (Fig. 6.1). As discussed (p. 119) the flux of H^+ from the lumen into the cytosol through the proton pore (F_0), regulated by the enzymatic 'plug' (F_1), generates ATP. In mitochondria the enzyme is called MF_0F_1 and is on the inner (matrix) surface of the inner membrane; in photosynthetic bacteria

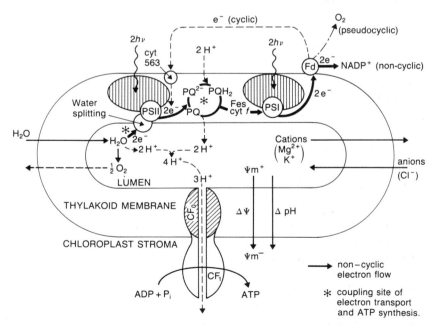

FIG. 6.1 Schematic relationship between electron transport driven by light, proton accumulation in the thylakoid lumen and ATP synthesis by the enzyme complex CF_0-CF_1. $\Delta\psi$ and ΔpH are, respectively, the electrical potential difference and H^+ gradient providing the proton motive force for ATP synthesis.

it is BF_0F_1 and is on the inside of the membrane. The enzyme of chloroplasts is called coupling factor (CF_0F_1), with CF_1 projecting from the thylakoid membrane into the stroma (Fig. 6.1). In chloroplasts the 57 kJ of energy required for synthesis of a mole of ATP is provided by the energy lost by photoenergized electrons as they are transported in the thylakoid membrane along a chain of redox carriers, as discussed in Chapter 5.

The mechanism by which the redox potential energy of the electron in the membrane is coupled to the synthesis of the anhydride bond of ATP was a matter for heated debate for many years. Several hypotheses were advanced to account for the observations that electron transport was coupled at three sites (in mitochondria) or two (in chloroplasts) to ATP synthesis and that the demand for ATP and the supply of ADP and P_i could regulate electron transport. The relationship between ATP synthesis and electron transport was measured by determining ATP production or P_i consumption in relation to O_2 consumption in mitochondria or O_2 evolution in chloroplasts. Any hypothesis also had to account for ion accumulation (e.g. calcium in mitochondria) which was known to be related to the electron transport processes and could be driven by ATP hydrolysis in the absence of electron transport. Another phenomenon requiring explanation was how artificial compounds of very diverse type, applied to mitochondria or chloroplasts, could uncouple ATP synthesis from electron transport.

Electron transport was thought to produce a high energy form of energy-transducing intermediate (often called $\sim$ 'squiggle') which could provide energy for ATP synthesis and related processes. Uncouplers, it was suggested, prevented the formation of $\sim$ or destroyed it. If a component in the membrane, electron carrier or protein for example, altered its configuration as the redox potential changed, allowing ADP and P_i to bind and form the anhydride bond, it should have been detectable spectroscopically or by other means. Despite much experimentation no 'high energy' chemical intermediate was identified and there is no direct coupling between the electron transport chain and ATP synthesis.

The chemiosmotic hypothesis, developed by Mitchell, has provided a general mechanism for coupling the energy in electron transport to phosphorylation. It combines into a coherent scheme many experimental observations related to ATP synthesis, for example, the need for intact organelles (e.g. mitochondria) or membrane vesicles (e.g. the thylakoids in chloroplasts), the requirement for coupling factor attached to the membrane and the need for electron transport. It also accounts for the observed increase in alkalinity of the chloroplast stroma in the light, in addition to the other features mentioned previously. The hypothesis is that photophosphorylation in chloroplasts and in photosynthetic bacteria and respiratory chain phosphorylation in mitochondria employ the same basic energy transducing mechanism, driven by a proton flux through the ATP synthetase enzyme. Energy from electron transport is conserved as a high concentration of protons on one side of the vesicle membrane and a low concentration on the other. The high concentration is outside the cell membrane in the photosynthetic bacteria, between the inner and outer mitochondrial membranes, and inside the thylakoid lumen in chloroplasts. In photosynthesis, the energy of electrons activated by light reactions is conserved by coupling electron transport in thylakoid membranes (Fig. 6.1) to H^+ transport across the membrane into the thylakoid. Capture of four photons leads to accumulation of 2 H^+ from oxidation of one molecule of water and 2 H^+ are transported from the stroma (which becomes alkaline) by the plastoquinone pump (p. 87).

The difference in H^+ concentration across the membrane is equivalent to a gradient in pH, ΔpH. Electron transport and proton pumping also cause an electrical potential difference, $\Delta\psi$, to develop across the membrane, which in chloroplasts is positive inside the thylakoid lumen and in mitochondria, negative in the matrix. Under the influence of $\Delta\psi$ and of ΔpH acting together, protons in the thylakoid lumen will tend to move from the lumen, across the membrane, to the stroma to preserve electrical neutrality and decrease the gradient of H^+. This force is called the proton motive force, pmf (also called the proton electrochemical potential, $\Delta\mu_H^+$). Thus:

$$pmf = \Delta\mu_{H^+} = \Delta\psi + \Delta pH \qquad [6.4]$$

The Gibbs energy change for the transfer of 1 mol of H^+ down a gradient of H^+ between the inside, i, and outside, o, of the membrane vesicle in the absence of an electrical potential, is:

$$\Delta G = 2.3\,RT\log_{10}\frac{[H^+]_o}{[H^+]_i} \qquad [6.5]$$

The Gibbs energy change for the transfer of 1 mol of ions down the electrical potential gradient $\Delta\psi$ (in millivolts) is:

$$\Delta G = -mF\Delta\psi \quad \text{where } m \text{ is the number of charges (for protons,} \qquad [6.6]$$
$$m = 1), \text{ and } F \text{ is the Faraday constant } (9.65 \times 10^4$$
$$C\ mol^{-1}).$$

As Gibbs energy differences are additive, the total Gibbs energy difference or pmf resulting from the transfer of 1 mol of ions down an electrical potential gradient of $\Delta\psi$ (mV) and an H^+ concentration gradient is:

$$\Delta G = -mF\Delta\psi + 2.3\,RT\log_{10}\frac{[H^+]_o}{[H^+]_i} \qquad [6.7]$$

In units of electrical potential the combined energy for the two sources of proton flow, with $m = 1$, is:

$$\Delta\mu_{H^+} = pmf = \Delta\psi - \frac{2.3\,RT}{F}\Delta pH \qquad [6.8]$$

which at 30 °C becomes, with units of millivolts:

$$\Delta\mu_{H^+} = \Delta\psi - 60\Delta pH \qquad [6.9]$$

Lipid membranes have very low permeability to H^+. The protons, however, can flow in a controlled manner through the CF_0-CF_1 enzyme complex, producing the change in enzyme configuration required for ATP synthesis. Possibly a certain 'pressure' of H^+ is needed to change the enzyme configuration. The flow of protons is called protonicity, by analogy with electricity and it is a property of the flow of H^+ in CF_1 which synthesizes ATP. The chemiosmotic hypothesis does not provide a description of the molecular mechanism of energy transduction at the level of enzyme sites (enzymological studies may provide that), but rather a mechanism of coupling ATP production with electron transport.

The magnitudes of $\Delta\psi$ and ΔpH have been measured by several techniques. Synthetic lipophilic ions (e.g. tetraphenyl-phosphonium) called Skulachev ions after the discoverer, penetrate lipid membranes, even though charged, due to the extensive π orbital system (p. 22). Their absorption spectra change with conditions so that, with careful calibration, an estimate of $\Delta\psi$ can be made optically. The ΔpH has been measured from fluorescence quenching of 9-aminoacridine in chloroplasts. Microelectrodes have also been employed to measure pH and ion concentrations. Carotenoids, bound in the thylakoid (and other membranes) are important indicators of $\Delta\psi$. They respond to the large electrical field (3×10^5 V cm^{-1}) which develops, for example, on illumination, by a rapid (nanosecond) change in absorption towards longer wavelengths. This electrochromic shift is readily measured without disturbing the conditions and may be calibrated to provide a measure of $\Delta\psi$; however it reflects the $\Delta\psi$ only in the immediate vicinity of the carotenoid in the membrane, not necessarily at the membrane surface.

Under steady illumination $\Delta\psi$ is 10–50 mV across the thylakoid membrane, whereas ΔpH is 3 units, equivalent to 180 mV, and therefore the most important component of pmf in thylakoids; in mitochondria $\Delta\psi$ is the most important component. However, in suddenly illuminated chloroplasts $\Delta\psi$ may also be more important than ΔpH in developing pmf; within 10^{-8} s $\Delta\psi$ develops due to electron transport but ion transport is much slower. The positive charge in the thylakoid lumen causes anions to move in from the stroma, to balance the electrical charge. Chloride ions (particularly *in vitro*) enter to join the H^+ in the lumen; in Jagendorf's instructive phrase 'thylakoids pump hydrochloric acid into themselves in the light'. Cations, Mg^{2+} and K^+, move into the stroma and with time $\Delta\psi$ decreases as H^+ accumulates. Up to 1 mmol of H^+ accumulates per mg of chlorophyll and would increase the membrane potential and, as the lumen is small (10^{-21} m^3, p. 55), a small change in H^+ (1 mmol) would greatly alter pH and could damage the membrane. However, most of the protons (99 per cent) are buffered by proteins in the lumen. Large $\Delta\psi$ is, possibly, important with the onset of illumination and during fluctuations in intensity, allowing a pmf to develop before H^+ accumulates appreciably, so that ATP synthesis may start quickly.

ATP synthesis can be completely separated from electron transport if the ΔpH is caused by subjecting intact thylakoids to an acid–base transition. Thylakoids are incubated in the dark, with a buffer solution, for example, succinic acid at pH 4, which diffuses into the lumen providing a controlled internal concentration of H^+; then the external pH is raised quickly. With a gradient of four pH units ATP is synthesized, but a gradient smaller than two pH units is ineffective. Even with electron transport inhibited by DCMU ATP is synthesized, showing the pH gradient to be the driving force, not electron flow. A general relationship between ATP synthesis and the ΔpH is shown in Fig. 6.2a. ATP synthesis is related to the cube of the H^+ concentration (Fig. 6.2b), evidence that synthesis of one ATP requires 3 H^+ to flow through coupling factor. However, there is still uncertainty about the 'true' value, because there is a basal rate (non-phosphorylating) of H^+ flux, so that the phosphorylation per H^+ depends on ΔpH; values of 2.4 H^+ per ATP have been suggested. For chloroplasts the accepted $H^+/2\ e^-$ ratio is 4, for non-cyclic electron transport. With 3 H^+ required per ATP synthesized, the ATP/2 e^- ratio is 1.33; with 2.4 H^+/ATP the ATP/2 e^- ratio is 1.7.

Paths of electron flow and coupling to phosphorylation

Three paths of photosynthetic electron flow linked to ATP synthesis have been recognized and are shown in Fig. 6.1. ATP synthesis coupled to a 'linear' flow of electrons from water to $NADP^+$ is called non-cyclic photophosphorylation because the electrons pass on to an acceptor, which passes them to metablic reactions, and do not return to the electron transport chain. In cyclic photophosphorylation, electrons are cycled from PSI back to the electron transport chain. Pseudocyclic photophosphorylation involves electron flow to O_2, and to H^+ and water rather than to $NADP^+$ and is

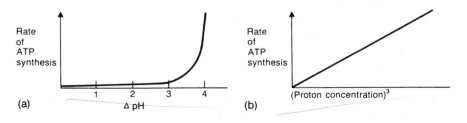

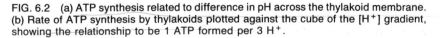

FIG. 6.2   (a) ATP synthesis related to difference in pH across the thylakoid membrane. (b) Rate of ATP synthesis by thylakoids plotted against the cube of the $[H^+]$ gradient, showing the relationship to be 1 ATP formed per 3 H^+.

a variant of non-cyclic electron transport, but with a different electron acceptor. It is, of course, the electron movement which is cyclic or non-cyclic not phosphorylation. However, the inexact but historical term remains.

Non-cyclic photophosphorylation

With adequate substrates to oxidize NADPH, primarily CO_2, linear (i.e. 'non-cyclic') flow of electrons is coupled to ATP synthesis. Both PSII and PSI are needed. The action spectra for both CO_2 assimilation and ATP synthesis are very similar and they saturate at similar light intensities. Measured ATP/2 e$^-$ ratios for non-cyclic photophosphorylation are between 1.5 and 2; higher ratios may be obtained if allowance for the basal rate of electron transport is made. However, it is not established if the basal rate occurs when the processes are coupled or only when uncoupled. It is important to know the *in vivo* rates of ATP formation per H$^+$ because at least 1.5 ATP must be synthesized for each 2 e$^-$ if most of the ATP required for CO_2 fixation is produced non-cyclically, as is probable.

Cyclic photophosphorylation

When non-cyclic electron flow is prevented, electrons from PSI or more probably ferredoxin (shown by the greater sensitivity to antimycin, an Fe−S inhibitor), pass back to plastoquinone via cytochrome 563, a *b* type cytochrome also called cyt b_6 (Fig. 6.1). Cytochrome involvement is shown by spectral changes. The ratio of cytochromes to PSI is 2:1, so two cytochromes take one electron each from ferredoxin fed by two different PSI centres and reduce plastoquinone. Cyt b_{563} has a potential of -0.18 V and donates e$^-$ to plastoquinone at about zero potential. Coupling with ATP synthesis is associated with the transfer of H$^+$ from plastoquinone to the thylakoid lumen. Only energy of PSI is used, that is, it is driven by light above 680 nm. With no net transport of e$^-$, water is not split and no O_2 evolved. However, when the acceptors of PSI are fully reduced the e$^-$ flow cannot start; a slow flux of e$^-$ from PSII to PSI maintains the correct redox potentials ('poises' the system) so that electrons can move. DCMU, which stops electron flow from PSII to PSI, inhibits cyclic photophosphorylation by interfering with 'poising'.

The importance of cyclic photophosphorylation *in vivo* is not clear; it may be most important in physiologically intact tissues when non-cyclic electron transport is slowed by lack of CO_2 and O_2, and in dim light. Cyclic and non-cyclic photophosphorylation may co-operate; the former poises the system and the latter generates ATP. Reduced NADP$^+$ and ferredoxin (where nearly all reduced) also regulate electron flow in cyclic photophosphorylation.

Pseudocyclic photophosphorylation

This requires both photosystems, like non-cyclic photophosphorylation but with ferredoxin reducing an 'oxygen reducing factor' which passes electrons to molecular oxygen as the terminal electron acceptor. The two-step O_2 reduction forms the superoxide radical O_2^- and then, by the action of superoxide dismutase (p. 98), hydrogen peroxide. Electrons can also be donated from the reduced acceptors of PSI to H^+ giving H_2, the reaction being catalysed by hydrogenase. Pseudocyclic photophosphorylation is measured by the uptake of O_2 caused by light and by the effect of ADP and P_i on it. However, in photosynthesizing tissues H_2O_2 is destroyed by catalase, O_2 is released and there is no net exchange of O_2. Electrons go from water to O_2 back to water so the process is not cyclic as the same electrons are not recycled. The rate is greater at high O_2 concentration than low and when CO_2 fixation is slow; it saturates at higher light intensity than cyclic photophosphorylation. It has the same $P/2\ e^-$ ratio as non-cyclic photophosphorylation as the same sites of coupling are employed.

Enzyme mechanism of phosphorylation

In chloroplasts CF_0F_1 links the thylakoid lumen to the stroma (Fig. 6.3). CF_1 was first detected by Avron in 1963, as a protein which when removed from the thylakoid, uncoupled photophosphorylation but when replaced, restored ATP synthesis. CF_1 is only loosely attached and is readily removed if the cation concentration (particularly Mg^{2+}) is low, for example, after treatment with EDTA, a cation chelating chemical. CF_1 is thought to move around on the membrane under the influence of ionic and electrical charges.

The CF_1 is composed of five types of subunits α, β, γ, δ and ϵ in the ratio 3:3:1:1:1 and has a molecular mass of 325 kDa. The five subunits have different masses as determined by treating CF_1 with urea, which breaks hydrogen bonds, followed by polyacrylamide gel electrophoresis in the presence of the detergent sodium dodecyl sulphate (SDS). There are three elongated (approximately 10 nm long axis) α subunits of $55-56$ kDa mass and three elliptical ($5-8$ nm) β subunits ($52-54$ kDa) symmetrically arranged alternately around a central cavity (Fig. 6.3); the catalytic (nucleotide binding) sites are either on β subunits or at the interface between the α and β subunits. The γ polypeptide (37 kDa) is probably asymmetrically arranged in the cavity formed by the hexamer and this is thought likely to allow different linkages to α and β and to be important in catalysis. Oxidation and reduction of cysteine residues on γ (probably 4 per polypeptide) is important in its activity. In darkness ATPase is inactive but is activated by dithiothreitol (DTT); it is also activated in the light. Probably this involves the disulphide bonds on the subunits which changes affinity

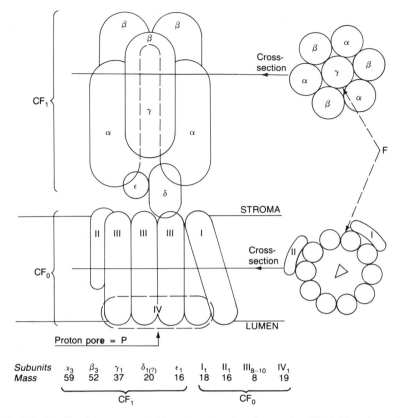

Subunits	α_3	β_3	γ_1	$\delta_{1(?)}$	ϵ_1	I_1	II_1	III_{8-10}	IV_1
Mass	59	52	37	20	16	18	16	8	19

FIG. 6.3 Idealized structure of chloroplast coupling factor, the CF_1 (with polypeptide subunits α to ϵ) CF_0 (with subunits, I, II and III) enzyme complex responsible for synthesis of ATP.

for ϵ, releasing ADP from the β subunits. Subunit δ (21–25 kDa) is located in the 'stalk' holding CF_1 to CF_0; removing δ has little effect on ATPase activity but it is required to block H^+ leakage and thus to regulate photophosphorylation. Subunit ϵ (14 kDa) is also important in regulation of H^+ leakage; removal and addition stimulates and inhibits ATPase activity, respectively.

Although apparently identical the α subunits bind the indicator lucifer yellow vinyl sulfone differently. It binds covalently to a lysine residue (Lys 378) on only one of the three α subunits. This unit may have a role in binding CF_1 to CF_0. Subunit β binds the nucleotides and a number of inhibitory analogues of ATP; there are important amino acid sequences responsible for the binding sites. There is some uncertainty about the number of binding sites with weak and strong binding; possibly there are two strong binding sites.

This hexamer of CF_1 is water filled. Upon activation there are changes in conformation and exchange of water with the medium, as was elegantly shown by incubating CF_1 with tritiated water, removing the enzyme, denaturing it in urea and measuring incorporation of radioactivity. When the complex was energized by illumination or a pH gradient there was exchange of tritium from the medium to hydrogen on the protein. Under de-energizing conditions approximately 100 H atoms per CF_1 remained hidden in the proteins, possibly associated with conformational changes. Proton exchange was essential for ATP synthesis.

ADP + P_i bound to CF_1 and altered the tritium exchange, demonstrating that molecular arrangement of subunits changes with conditions and is involved in catalysis. Sulphydryl groups on the CF_1 polypeptides (measured with N-ethyl maleimide, which reacts with —SH groups on the γ subunit) are hidden within the complex in the dark but become exposed in the light. Possibly four —SH groups on γ subunits control H^+ flux, like a 'plug' in the 'flow' system. Reagents which bind to —SH groups (e.g. dithiothreitol) increase ATPase, suggesting that sulphydryl groups regulate ATP synthesis. Light activates CF_0–CF_1, not only via ΔpH but by the thioredoxin system which regulates other chloroplast enzymes (p. 136) changing sulphydryl groups on subunits. In the unactivated state a much greater ΔpH is needed to drive photophosphorylation.

CF_0 is composed of polypeptides called I (18–19 kDa), II (16 kDa), III (8 kDa) and IV (27 kDa) in the ratio of 1:1:12:1. Together they form the proton pore with hydrophobic portions imbedded in the membranes and hydrophilic sections emerging into the lumen and stroma, the latter attaching to CF_1. I and II may form part of the stalk. Subunit III, which is also called the 'proteolipid' forms the H^+ channel. It has two membrane spanning helices one of which binds dicyclohexylcarbodiimide (DCCD) to a glutamyl residue blocking proton translocation. Subunit IV has five helices spanning the membrane; possibly the proton pore is formed by one of these helices, which has many charged amino acid residues, linking to a polypeptide of III. Thylakoid lipids, e.g. MGDG are required for effective function of coupling factor. CF_0 allows very rapid and specific H^+ transport: 2×10^5 H^+ per CF_0 per second with μ_{H^+} equivalent to 30 mV. This is 10^3 times faster than required for ATP synthesis. The proton pore subunits interact and also with CF_1 via δ, ϵ and α and possibly III with β, judging from studies of MF_0F_1 using fluorescent binding dyes. It is believed that protons alter the conformation of CF_0 and CF_1 so that both pmf and proton flow are required to effect ATP synthesis.

The structure of the site for ATP synthesis on CF_1 and the mechanism by which the terminal anhydride bond is formed are not understood and most models are speculative. Adenylates bind to three sites tightly and to three sites rather weakly but the significance of this is unknown. Enzyme conformation may change with proton concentration or energy state of the membrane; light increases subunit reactivity and may alter the position of subunits, opening up a channel for protons. ADP and P_i bind tightly to CF_1 at two sites, about 90 μm of binding sites per chloroplast. A CF_1-ADP

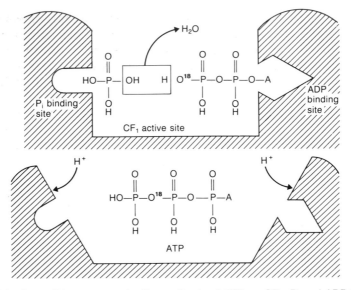

FIG. 6.4   A possible sequence for the synthesis of ATP on CF_1. P_i and ADP bind to the enzyme and in a spontaneous reaction H^+ is removed from ADP terminal phosphate (shown by incorporation of ^{18}O from ADP into ATP) and the OH from P_i. ATP is bound to the complex. When H^+ passes through CF_1, conformational changes release ATP and allow P_i and ADP to bind again, thus 'resetting' the complex. A = adenosyl residue.

and P_i complex may be formed, which is inhibited by arsenate. Studies with ^{18}O-labelled ATP, P_i and H_2O show that as protons pass through CF_0 they remove an oxygen atom on the phosphate group, forming water (Fig. 6.4). The electron on the phosphate group moves to the ADP terminal phosphate forming the anhydride bond. This is thought to require little change in energy. Protons in CF_1 alter the conformation of the peptide chains and change the binding energy between ATP and enzyme complex, allowing ATP to escape into the stroma. ADP + P_i then bind again to the enzyme reaction site in its energized conformation, which then returns to its original form, producing the anhydride bond. One theoretical mechanism is the obligate alternating site hypothesis: two or three sites interact on the enzyme and alternate rapidly for ATP synthesis. The tightly bound products at one site are released due to conformational changes perhaps in the subunit caused by the passage of protons through the complex. Release of ATP allows ADP to bind and phosphorylation occurs spontaneously, the release of ATP requiring energy. Inhibition of ATPase by ADP may reflect 'clogging' of the sites. Removal of subunits regulating the conformational change then prevents phosphorylation but leaves the enzyme in a state capable of hydrolysing ATP. Such interesting ideas remain to be established.

   CF_1 activity is regulated by several factors, including H^+ and the concentration of substrates, ADP and P_i. ATP inhibits ATP synthesis and ADP and P_i inhibit ATP hydrolysis so that the complex is allosterically controlled. Nucleotides bind to multiple sites, not active in ATP synthesis, depending on the energized state of CF_1 (indicating conformational changes in CF_1 proteins) but their role in catalysis is not understood. Energy from the proton gradient is essential for releasing ATP by changing conformation of CF_1; this stage may be the most energy demanding.

In illuminated thylakoids, ATP synthesis is the main function of CF_0–CF_1 but this complex also catalyses hydrolysis of ATP. Treatment of isolated CF_1 with the protein-digesting enzyme trypsin, heat, or sulphydryl reagents (e.g. dithiothreitol) stimulates ATPase activity which requires Ca^{2+}; the ϵ subunit controls ATPase activity. This may have a physiological significance, allowing ATP to drive proton accumulation, providing control over the ionic balance of thylakoids, for example, in darkness, when regulation of the state of the membranes and ionic concentration creates the conditions needed for rapid synthesis of ATP on illumination.

Relationship between ATP formed and e^- and H^+ flow

Mitchell's hypothesis was that 2 H^+ were needed for 1 ATP but, in chloroplasts, the stoichiometry is close to 3, from several lines of evidence. With 2 coupling sites in the electron chain, 8 photons give 8 H^+ for 4 e^- transported to $NADP^+$. Thus, the ATP/2 e^- ratio is 1.33. The energy available to drive synthesis is calculated from eqn 6.9 with ΔpH of 3 or 180 mV and $\Delta\psi$ of 20 mV, a total pmf of 200 mV.

A mole of ATP requires about 30 kJ for synthesis under equilibrium conditions which when converted to redox potential difference, $\Delta E'$, by

$$\Delta G = -nF \Delta E'$$

where n is the number of reducing equivalents and F the Faraday constant, gives about 150 mV. Under cellular conditions the low ATP concentration may require more energy, 57 kJ mol^{-1}, about 230 mV, more than the system can provide. Possibly the ΔpH is 3.5, which would suffice to generate ATP with 3 H^+, as the pmf is 210 mV and $\Delta\psi$ of 30 mV would give about -70 kJ mol^{-1}, sufficient to synthesize ATP under unfavourable equilibrium conditions. Energetically the chemiosmotic hypothesis is feasible and is supported by experimental evidence of many types. It is now the accepted model for energy transduction in phosphorylation and CF_1 requires 3 H^+ per ATP. As drawn in Fig. 6.1 the thylakoid lumen appears as a large open compartment, presumably full of aqueous solution with 'free' protons. However, the luminal space is very small, of molecular dimensions, and full of extrinsic proteins. It is possible that protons bind to these and that only local areas of the membrane provide protons to the coupling factor. The buffering capacity is probably large and may aid regulation of proton flow.

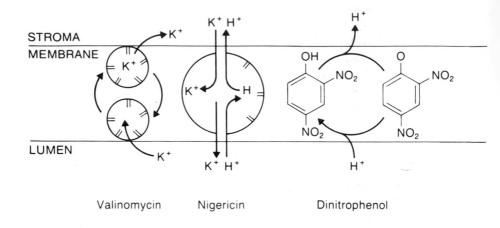

Valinomycin   Nigericin   Dinitrophenol

Uncouplers of ATP synthesis

An intact enclosed vesicle, as in the thylakoid system, is needed for H^+ accumulation with a bounding membrane impermeable to H^+; the lipid membrane is an effective barrier. Illumination of intact thylakoids without ADP or P_i causes H^+ accumulation from water-splitting and a 'back pressure' on the PQ pump, slowing electron transport to a basal rate, corresponding to leakage. With the substrates ATP and P_i, H^+ flows through CF_1, ATP is synthesized, the back pressure drops and electron transport rate increases until equilibrium is attained. Treatments causing loss of ΔpH or $\Delta\psi$ uncouple electron transport and ATP synthesis.

Different uncoupling mechanisms are known. If only some CF_1 complex is removed by EDTA, which chelates Mg^{2+}, then H^+ leaks through CF_0 and no ATP is made. Inhibition of ATP synthesis by arsenate or thiophosphate allows H^+ flow through CF_1 and destroys the ΔpH. Phlorizin (a glucoside from roots) blocks ATP synthesis but prevents H^+ flow. Detergents disrupt the membrane preventing ΔpH formation. Membranes are made 'leaky' by lipid-soluble proton ionophores — H^+ transporting compounds, such as carbonylcyanide p-trifluoromethoxyphenyl hydrazone (FCCP). The negatively charged molecule moves along the $\Delta\psi$ gradient into the lumen, where it is protonated. It moves back to the medium, is deprotonated and recycles, collapsing ΔpH. Dinitrophenol (DNP) (Fig. 6.5) carries anions and cations (H^+) in response to $\Delta\psi$, and is a very effective uncoupler in mitochondria but not in thylakoids. Other ionophores carry ions; for example, valinomycin, a depsipeptide (with alternating hydroxy and amino acids) antibiotic from bacteria, dissolves in the membrane and carries K^+ out of the lumen, changes $\Delta\psi$ and

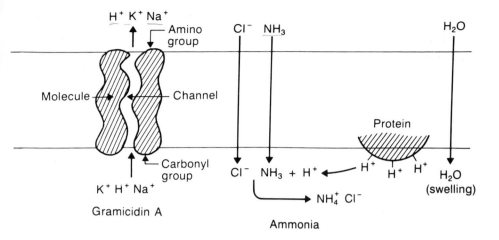

FIG. 6.5 Substances which uncouple ATP synthesis from electron transport by dissipating the H^+ gradient across the thylakoid membrane or consuming H^+ inside the lumen.

uncouples. Together valinomycin and DNP are very effective uncouplers in chloroplasts, carrying H^+ and K^+ from the lumen and collapsing ΔpH and $\Delta\psi$. Nigericin (Fig. 6.5) binds H^+ and K^+ reversibly and transports them across the membrane and changes the pmf. Gramicidin is a peptide which forms a very efficient channel (carrying 10^7 ions s^{-1}) for monovalent ions, including H^+; only one molecule per thylakoid uncouples ATP synthesis completely.

Ammonia is an uncoupler. On entering the lumen (Fig. 6.5) it is protonated to NH_4^+ and destroys ΔpH. Anions enter to restore neutrality, osmotic potential increases and water enters, causing swelling. Ammonia only uncouples above 10^{-3} M concentration as H^+ is an effective buffer. As ammonia is formed under normal physiological conditions in chloroplasts it must be rapidly metabolized to prevent damage.

Photophosphorylation and physiological control

Photosynthesis is regulated, in ways not well understood, by the balance between ATP and NADPH formation and the supply of CO_2, P_i, etc. In dim light non-cyclic electron transport is slow and ATP synthesis limits the rate of metabolism if $NADP^+$ is available as electron acceptor. With increased light intensity electron flow will be non-cyclic if $NADP^+$ is available (with adequate CO_2 or other substrates) and ATP synthesis is adequate for metabolism. However, if the rate of ATP synthesis is limiting, $NADP^+$ is reduced and therefore non-cyclic e$^-$ flow is slowed and cyclic electron flow may increase the rate of ATP synthesis. The same may apply when light is abundant but CO_2 limiting. Pseudocyclic electron flow may also permit ATP synthesis when $NADP^+$ is not available, perhaps in very intense light. Low demand for ATP, which slows the rate of supply of ADP to CF_1 or inadequate P_i supply, inhibits phosphorylation. Photosynthesis by C3 plants (p. 126) requires an ATP/2 e^- ratio of at least 1.5 for CO_2 reduction; as other processes also consume ATP, particularly in the light, either the ratio *in vivo* is greater than *in vitro* or additional ATP is synthesized in other ways. Regulation of the ATP/NADPH ratio is important for photosynthesis but poorly understood. Affinity of components of the e^- transport chain for substrates may be expected to determine the relative flux of e^- into parts of metabolism. Ferredoxin-$NADP^+$ reductase has a much greater affinity for $NADP^+$ than NAD^+ (a K_m of 10^{-5} M compared to 3×10^{-3} M) but under unphysiological conditions it catalyses H^+ transfer between NADPH and NAD^+ and also oxidation of NADPH by several electron acceptors. Also, the e^- flux will depend on environmental conditions (e.g. light) which determine the saturation of the e^- transport chain, relative to CO_2, O_2 and NO_3^- supply. Conditions during growth may also affect the relative amounts of different components. C3 and shade plants, for example, may have many grana and grow in dim light but are relatively inefficient in bright light. Some C4 plants (e.g. *Zea mays*) are agranal and inefficient in dim light, but are very efficient in bright light. Rate of ATP synthesis may limit CO_2 assimilation in bright light in C3 plants and in dim light in C4. Granal number and size may regulate the area of membrane not only for light harvesting and electron flow between photosystems (p. 70) but H^+ flux through CF_1 under dim light conditions.

Such concepts of regulation depend on the environment and assume that the enzyme is unregulated; this is probably not so. CF_1 is almost certainly controlled by feedback mechanisms because of its extremely important position in metabolism and because of the very dynamic nature of thylakoid energetics. In darkness or very dim light ATP is hydrolysed by CF_0–CF_1 and this must be prevented by inactivation of the enzyme so that synthesis and breakdown is balanced to achieve efficient metabolism in relation to energy supply, etc. It may also be possible to regulate the pmf and proton gradient via coupling factor. Dark adapted enzyme is inactive and ATP synthesis stops within milliseconds of cessation of illumination but ATP breakdown requires much longer; regulation is related to the redox state of disulphide links on the subunit which is determined by thioredoxin f.

References and Further Reading

Althoff, G., Lill, H. and **Junge, W.** (1989) The single channel conductance of CF_0, pp. 271–73 in Barber, J. and Malkin, R. (eds), *Techniques and New Developments in Photosynthesis Research*, Plenum Press, New York/London.

Atkinson, D.E. (1977) *Cellular Energy Metabolism and its Regulation*, Academic Press, New York.

Cross, R.L. (1981) The mechanism and regulation of ATP synthesis by F_1-ATPases, *A. Rev. Biochem.*, **50**, 681−714.

Dietz, K.-J. and **Heber, U.** (1989) Assimilatory force and regulation of photosynthetic carbon reduction in leaves, pp. 341−63 in Barber, J. and Malkin, R. (eds), *Techniques and New Developments in Photosynthetic Research*, Plenum Press, New York.

Hinkle, P.C. and **McCarty, R.E.** (1978) How cells make ATP, *Sci. Amer.*, **238**, 104−23.

Jagendorf, A.T. and **Anthon, G.E.** (1985) Unsolved problems in photophosphorylation by higher-plant chloroplasts, pp. 121−40 in Steinback *et al.* (eds), *Molecular Biology of the Photosynthetic Apparatus*, Cold Spring Harbor Laboratory.

Jagendorf, A.T., **McCarty, R.E.** and **Robertson, D.** (1991) Coupling factor components: Structure and function, in Bogorad, L. and Vasil, I.K. (eds.), *The Photosynthetic Apparatus: Molecular Biology and Operation*, Academic Press Inc., San Diego.

Junge, W. and **Jackson, J.B.** (1982) The development of electrochemical potential gradient across photosynthetic membranes, pp. 589−646 in Govindjee (ed.), *Photosynthesis*, Vol. 1, *Energy Conversion by Plants and Bacteria*, Academic Press, New York.

McCarty, R.E., **Shapiro, A.B.** and **Feng, Y.** (1988) Regulation and mechanisms of the chloroplast ATP synthase in relation to function, pp. 290−304 in Stevens, Jr. S.E. and Bryant, D.A. (eds), *Light-energy Transduction in Photosynthesis: Higher Plant and Bacterial Models*, American Society Plant Physiology, Rockville.

Mitchell, P. (1966) Chemiosmotic coupling in oxidative and photosynthetic phosphorylation, *Biol. Rev.*, **41**, 445−502.

Nelson, N. (1982) Structure and function of the higher plant coupling factor, pp. 81−104 in Barber, J. (ed.), *Electron Transport and Photophosphorylation (Topics in Photosynthesis 4)*, Elsevier Biomedical Press, Amsterdam.

Nicholls, D.G. (1982) *Bioenergetics. An Introduction to the Chemiosmotic Theory*, Academic Press, London.

Ort, D.R. and **Melandri, B.A.** (1982) Mechanism of ATP synthesis, pp. 537−87 in Govindjee (ed.), *Photosynthesis*, Vol. 1, *Energy Conversion by Plants and Bacteria*, Academic Press, New York.

Ort, D.R., **Grandoni, P.**, **Oretiz-Lopez, A.** and **Hangarter, R.P.** (1990) Control of photophosphorylation by regulation of the coupling factor, pp. 159−73 in Zelitch, I. (ed.), *Perspectives in Biochemical and Genetic Regulation of Photosynthesis*, Allen R. Liss, Inc., New York.

Pradet, A. and **Raymond, P.** (1983) Adenine nucleotide ratios and adenylate energy charge in energy metabolism, *A. Rev. Plant Physiol.*, **34**, 199−224.

Roth, R. and **Nelson, N.** (1984) Conservation and organisation of subunits on the chloroplast proton ATPase complex, pp. 501−10 in Sybesma, C. (ed.), *Advances in Photosynthesis Research*, Vol. II, Martinus Nijhoff/Dr W. Junk Publishers, The Hague.

Shavit, N. (1980) Energy transduction in chloroplasts: structure and function of the ATPase complex, *A. Rev. Biochem.*, **49**, 111−38.

Westheimer, F.H. (1987) Why nature chose phosphates, *Science*, **235**, 1173−78.

CHAPTER 7

The chemistry of photosynthesis

All dry matter produced by photosynthetic organisms comes from the use of reductant and energy from the light reactions to synthesize chemical products from simple organic starting materials, basically CO_2 and nitrate and sulphate ions. Carbon dioxide reduction is the major energy consuming process in photosynthesis; nitrate and sulphate reduction (p. 4) use less than 5 per cent of total energy. Enzymes of CO_2 assimilation are in the chloroplast stroma, and are affected by conditions there, for example, ATP and NADPH concentrations or products of enzyme reactions, which act as enzyme effectors. Enzyme activity determines the rate of CO_2 fixation and the balance between processes, such as starch synthesis in the chloroplast or carbon export from it, and may regulate respiration and photosynthesis, which have common intermediates and enzymes.

Carbon dioxide assimilation

CO_2 assimilation is a cyclic, autocatalytic, process (Fig. 7.1), variously called the Calvin cycle, reductive pentose phosphate pathway, photo-synthetic cycle or photosynthetic carbon reduction cycle (PCR cycle). PCR cycle emphasizes the carbon, reduction and cyclic aspects. The PCR cycle is the fundamental CO_2 assimilatory process in all photosynthetic organisms, including prokaryotes; it appears to have developed early in evolution and to have retained its characteristics. Additional processes for accumulating CO_2 have arisen which do not replace the PCR cycle, but rather add to it (C4 and CAM mechanisms, see Ch. 9).

Mechanisms of the PCR cycle

There are 11 enzymes involved in this 13 step carboxylation cycle. In the carboxylation step itself, an acceptor molecule, ribulose bisphosphate (RuBP), combines with CO_2 in a carboxylation reaction, in the presence of the enzyme RuBP carboxylase (Rubisco) producing 3-phosphoglyceric

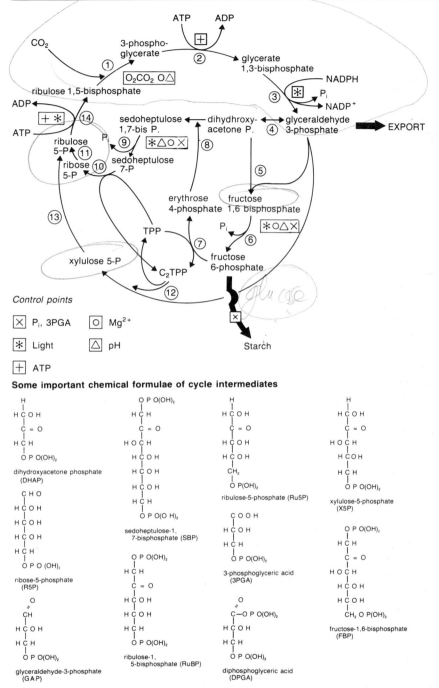

FIG. 7.1 The photosynthetic carbon reduction cycle, with numbered reactions corresponding to the enzymes listed in Table 7.1. The upper part of the figure includes the carboxylation and reduction steps, the lower part shows the regeneration of the CO_2 acceptor ribulose-1,5-bisphosphate (RuBP).

acid (3PGA). The acceptor is regenerated from 3PGA in reactions consuming NADPH and ATP which are produced as a consequence of the light-driven thylakoid reactions. A scheme of this cycle is given in Fig. 7.1 and reactions are identified by numbers which correspond to the enzymes listed in Table 7.1. If more carbon is fixed than is used to regenerate RuBP, carbon is exported from the cycle as triosephosphates or used in starch synthesis in the chloroplast. Control of export is essential to prevent depletion of components of the cycle and to maintain the autocatalytic process. Rate of cycle turnover in the steady state depends on the primary reactions of electron transport and ATP synthesis and on enzyme activity.

Assimilation of CO_2 is described by:

$$3 CO_2 + 9 ATP + 6 NADPH + 5 H^+ \rightarrow C_3H_5O_3P + 9 ADP + 8 P_i$$
$$+ 6 NADP^+ + 3 H_2O + 468 kJ \, mol^{-1} \qquad [7.1]$$

with triosephosphate as the product; for each CO_2 assimilated a minimum of 3 ATP and 2 NADPH + H are needed. The light reactions generate sufficient NADPH but may not produce the required ATP by non-cyclic electron transport (p. 113).

An arbitrary starting point in the PCR cycle is carboxylation of RuBP giving two molecules of 3PGA (reaction (1)), a reaction unique to the PCR cycle and catalysed by Rubisco: details of enzyme and reaction are given later. Reaction (2) uses ATP to phosphorylate 3PGA to a more reactive state in 1,3-diphosphoglyceric acid with two acid anhydride bonds. In reaction (3) NADP glyceraldehyde-3-phosphate dehydrogenase substitutes H^+ for the phosphate group in 1,3-diphosphoglycerate. The enzyme is $NADP^+$ dependent, in contrast to the respiratory enzyme which requires NAD^+. This is the only reduction in the PCR cycle and is of the greatest importance. Glyceraldehyde-3-phosphate (GAP) is converted (4) to dihydroxyacetone phosphate (DHAP); these two compounds are used in (8) and (12) where 3-carbon units are converted to five carbons in the regeneration of RuBP. Triosephosphates are condensed (5) to the 6-carbon compound fructose bisphosphate; the aldolase has maximum activity in alkaline conditions. Dephosphorylation of FBP by fructose bisphosphatase, an enzyme unique to the PCR cycle, gives fructose-6-phosphate (F6P). The reaction has a free energy change of $-25 \, kJ \, mol^{-1}$, so is not reversible and is a control point in the cycle.

Regeneration of RuBP is achieved by interconversion of 3-, 4-, 5- and 6-carbon compounds. Transketolase removes 2-carbon fragments (glycoaldehyde) from F6P and sedoheptulose-7-phosphate (S7P), attached to the thiamine pyrophosphate (TPP) co-factor of the enzyme. Erythrose-4-phosphate (E4P) reacts with DHAP (8) giving sedoheptulose-1,7-bisphosphate. This is dephosphorylated by the sedoheptulose bisphosphatase, unique to the PCR cycle and an important control point. Ribose-5-phosphate (Ru5P) is made from S7P (10), and converted to ribulose-5-phosphate (11). Another 5-carbon sugar, xylulose-5-phosphate,

Table 7.1 Enzymes of the photosynthetic carbon reduction cycle (reactions are numbered as in Fig. 7.1) with their approximate mass, specific activity (SA = $\mu mol\ min^{-1}\ mg\ chlorophyll^{-1}$) and Michaelis constant (K_m). The free energy change ΔG^s, at the steady state physiological (i.e. stromal) concentrations of substrates is for *Chlorella* in 40 Pa CO_2 and 21 kPa O_2. Large negative ΔG^s indicates a probable control reaction

Reaction number	Enzyme	SA	% of total in chloroplast	Total mass (kDa)	Number subunits	ΔG^s	Increased by	Decreased by	K_m
1	Ribulose bisphosphate carboxylase/oxygenase	10	100	550	8	−41	high pH, CO_2 Mg^{2+}, FBP	gluconate, SBP?	CO_2 12 μM, O_2 250 μM, RuBP 40 μM
2	Phosphoglycerate kinase	900	90	47	1	+16	3PGA, ATP Mg^{2+}	DGPA, ADP	3PGA, 0.5 μM, ATP 0.1 μM
3	Triosephosphate dehydrogenase (NADP glyceraldehyde P dehyd.)	100	—	140	4	−6.7	light, NADPH ATP	P$_i$? DHAP	DPGA 1 μM, NADPH 4 μM
4	Triosephosphate isomerase	200	—	53	2	−7.5	alkaline pH	PEP, RuBP, FBP glycolateP, PGA	DHAP 1 mM, GAP 0.4 mM
5,8	Aldolase	100	—	140	4	−1.6	high GAP/DHAP ratio, pH8	RuBP, ADP, PGA	FBP 20 μM, GAP 0.3 mM, DHAP 0.4 mM, SBP 20 μM
6	Fructose bisphosphatase	100	90	140	4	−27	high pH, Mg^{2+}, ATP, reductant, light	P$_i$	FBP 0.2 mM, 800 μM activated
7,10,12	Transketolase	150	—	140	4	−5.9	Mg^{2+}, high pH		Xu5P 100 μM
9	Sedoheptulose bisphosphatase	0.5	—	70	2	−29.7	light, pH, Mg^{2+}	P$_i$	SBP 13 μM (reduced)
11	Ribose-5-phosphate isomerase	10	—	54	—	−0.5	freely reversible, pH 8		R5P 0.2 mM, ATP 0.1 mM
13	Ribulose-5-phosphate 3 epimerase	30	—	46	—	−0.6	freely reversible		X5P 0.5 mM
14	Ribulose-5-phosphate kinase	320	—	83	2	−15.9	reductant, ATP, energy charge?, pH>7	ADP	R5P 2.5–70 μM, ATP 60 μM

is formed (12) from glycoaldehyde and glyceraldehyde-3-phosphate and is converted to Ru5P (13). The 'final step' (14) is the phosphorylation of Ru5P to the more reactive RuBP by Ru5P kinase, another enzyme unique to the PCR cycle, with a free energy change of -15 kJ mol^{-1}, the fourth most negative in the cycle.

Enzymes of the PCR cycle

Many enzymes of the cycle were discovered first in non-photosynthetic tissues; some were not detected in early studies or only at low activity, but improved methods of extraction and assay have substantiated the mechanisms.

Ribulose-1,5-bisphosphate carboxylase/oxygenase (Rubisco)

The carboxylating enzyme (earlier called fraction 1 protein) occurs in all photosynthetic organisms; it comprises 50 per cent of the soluble protein in leaves (6 mg per mg chl) and has been called the most abundant protein on earth. In higher plants the enzyme is made up of 8 large and 8 small subunits (called LSU and SSU of 55 and 15 kDa, respectively) a total mass of 550 kDa; these are arranged as shown in Fig. 7.2. This complex structure

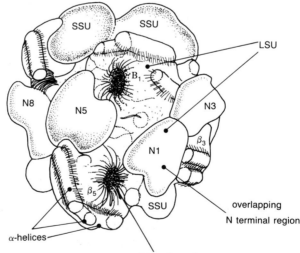

FIG. 7.2   Illustration of the subunit structure of ribulose bisphosphate carboxylase-oxygenase enzyme. Detailed structure based on X-ray crystallography, showing the small subunits (SSU), the N terminal and barrel (B) domains of the large subunits (LSU) numbered 1–8. The α-helices around the B domains are shown and also the 'mouths' of two α/β barrels. N domains of one LSU normally cover the mouths of a barrel on another LSU, e.g. N5 covers the B1 barrel. (After Chapman *et al.* 1988.)

is an L_8S_8 type of Rubisco found in most higher prokaryotes and eukaryotes. There is one catalytic site on each large subunit, i.e. 8 per molecule for the higher plant enzyme. However, this complex multimeric enzyme structure is not essential for catalysis as Rubisco from the photosynthetic bacterium *Rhodospirillum rubrum* has only two LSUs and other bacteria have L_4S_4 and other combinations of units. Rubisco from higher eukaryotes generally has a higher efficiency (specificity factor — see later) than the simple prokaryotic enzymes. LSUs are very similar in amino acid structure in all organisms but the SSUs differ widely. The LSUs are synthesized in the chloroplast stroma and coded in chloroplast DNA whereas the SSUs are coded by nuclear DNA and made in the cytosol (see Ch. 10).

Analysis of Rubisco by X-ray diffraction (Chapman *et al*. 1988) has revealed details of the structure, particularly of the active catalytic site. The large subunit of Rubisco has a NH_2^- terminal domain, called N (of amino acid residues 5–134), a connector region (135–168), a main domain called the barrel (169–432) and the C terminal domains (433–477). The N terminal is made up of four antiparallel β-sheets with four surrounding α-helices; the connector region has a short helix which blocks the end of the barrel. The barrel is made of 8 stranded α-helices and 8 β-sheets, arranged in a complex structure. These β-sheets are in parallel, forming the tube-like barrel; the α-helices surround the β-sheets on the outside and the large structure seems necessary to maintain the active site co-ordinates for effective catalysis. The Rubisco active site is, however, not on one LSU only, it is at the junction of the N terminal of one LSU where its fifth β-strand interacts with the pair of β-strands in loop 6 of the C terminal domain of the other. The active site, at which carboxylation and oxygenation take place, has 3 lysine and 1 glutamate amino acid residues, and is close to the mouth of the barrel. Lysine residues are on one large subunit C domain but the glutamate is on the N domain of the neighbouring large subunit. The phosphate binding site is on loop 5 of the C terminal domain. Thus, the active site of Rubisco is a pocket, formed at the interface between two subunits. Both the α/β barrels and co-operation between subunits in forming an active site are seen in other enzymes, they are not unique to Rubisco; these enzymes may have very little homology between amino acid sequences, etc., yet the active sites are always at the C-end of β-strands. The amino acid structure of the enzyme is most highly conserved in the β-sheet close to the active site at the barrel opening, probably because electronic configuration at the site of catalysis is very critical for carboxylation. Indeed, altering single amino acids at very different positions in the large subunit nearly destroys carboxylation. Such sensitivity to the configuration may make it difficult to reduce the oxygenase activity of the enzyme by manipulation of the site although some progress is being made with modifications to loop 6 of the C-terminal domain.

Small subunits of Rubisco consist of an antiparallel sheet of β-strands with flanking α-helices. A long C terminal extension reaches down to the back of the barrel on an LSU and two SSU extensions contact the barrel of an LSU. This, together with the extensive contact between LSUs, suggests co-operation of LSU and SSU subunits in catalysis but no clear explanation for the structure rôle of the SSU has been advanced although long-range stability in different conditions within the chloroplast may be the reason. Tobacco Rubisco has a specificity factor of about 90 but that of *Rhodospirillum* about 10 so the complexity is associated with greater efficiency (p. 133).

Carboxylation requires substrate CO_2, RuBP and active Rubisco, i.e. the enzyme must be carbamylated (combined with CO_2 in the presence of Mg^{2+}). Early measurements of affinity of extracted enzyme for CO_2, the K_m (Michaelis constant), showed 300 μM of CO_2 was required but the actual concentration in the stroma is probably only 10 μM. Extensive investigation showed that the enzyme must be incubated with CO_2 (in a reaction preceding CO_2 fixation proper) which carbamylates a lysine (No. 201) residue at the active site. This allows the magnesium to co-ordinate (it is not covalently bound) to the site. Six ligands link the magnesium to the protein with aspartate (No. 193), glutamate (No. 194) and histidine (No. 287) all involved in the co-ordination. Carbamylation is slow but the magnesium activation rapid. These processes establish the correct electronic configuration of the active site into which RuBP binds by its phosphate groups (Fig. 7.3) to basic, possibly lysine residues; a tautomeric change in electronic configuration occurs. The C2 of RuBP is normally slightly positive (electron attracting) because electrons are pulled towards the carboxyl O in the molecule. However, loss of a proton from C3 and formation of a keto-enol equilibrium produces a nucleophilic (proton attracting) enediol allowing the CO_2 to react at the negatively charged C2 of RuBP. There is no formal binding site for CO_2 or O_2 at the active site and they are probably co-ordinated in relation to the electronic state of the acceptor. CO_2, Mg^{2+} and the carboxyl group of RuBP are contiguous, as shown by ^{13}C NMR. CO_2 reacts directly with the tautomeric complex. The 6-carbon intermediate form is 2-carboxy-3-ketoarabinitol-1,5-bisphosphate. Water donates OH^- to the C3 of RuBP, cleaving the intermediate to two molecules of 3PGA. Two molecules of the D-stereoisomer of 3PGA are formed in the reaction. However, if hydrolysis of the 6-carbon intermediate proceeds non-enzymatically both the L- and D-isomers are formed. The enzyme controls the stereochemistry of the reaction.

Transition state analogues of RuBP such as 2-carboxyarabinitol-1,5-bisphosphate (CABP) are effective inhibitors because they bind tightly to the catalytic site. This characteristic can be used to measure the number of sites on the protein by using ^{14}C-labelled CABP which sticks to the sites and can be measured. Naturally occurring inhibitors of enzyme activity are known; carboxyarabinitol-1-phosphate (CA1P) is produced by carbohydrate metabolism and binds to sites which are not occupied by RuBP, thereby blocking activity. CA1P appears to be a normal nocturnal inhibitor in many

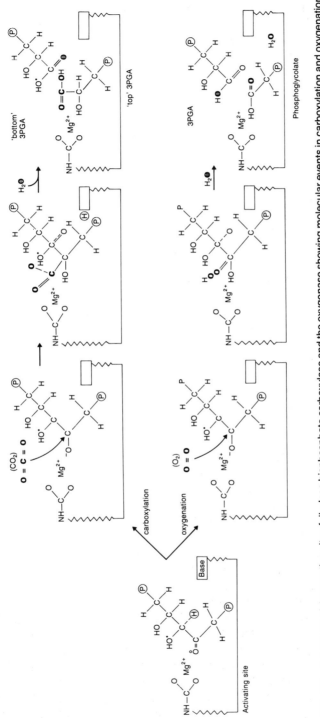

FIG. 7.3 Schematic of the reaction site of ribulose bisphosphate carboxylase and the oxygenase showing molecular events in carboxylation and oxygenation of RuBP to give 3-phosphoglyceric acid (3PGA) or phosphoglycolate and 3PGA, respectively. (After Gutteridge and Keys, 1985.)

higher plants, e.g. bean and tobacco, but not in others, e.g. spinach. Possibly CA1P prevents the consumption of RuBP and intermediates of the PCR cycle in darkness. In the light CA1P is metabolized by specific phosphatases and is not detectable. The inhibitory action of CA1P is reversed by a protein, called Rubisco activase, which was discovered in an *Arabidopsis* mutant lacking active Rubisco. Rubisco activase probably binds to Rubisco and releases CA1P, allowing RuBP to bind to the active site and thus permits photosynthesis to take place. The number of RuBP molecules per active site required to give maximum rates of assimilation is still controversial; probably 1.5−2 RuBP per site is required. It may be necessary to ensure that all sites are filled with RuBP to prevent tight binding inhibitors (which appear to be byproducts of carbohydrate metabolism) from blocking them and inhibiting photosynthesis.

Rubisco is a bifunctional enzyme, catalysing the reaction of molecular oxygen with RuBP at the same catalytic site as the carboxylation. The oxygenation reaction is:

$$RuBP + O_2 \rightarrow 2\text{-phosphoglycolate} + 3PGA \qquad [7.2]$$

Phosphoglycolate is a 2-carbon phosphorylated compound. The mechanism of the oxygenation reaction catalysed by the enzyme is still not understood. Molecular oxygen is relatively stable, is uncharged and is not polarized. It has a size of 1.2×10^{-10} m whereas CO_2 is polarized (although neutral overall) and is 2.32×10^{-10} m. Other types of oxygenase enzyme contain a transition metal (e.g. Fe) or a redox prosthetic group, which transfers electrons; but Rubisco has no such group and there is no comparable mono-oxygenase enzyme. It is not clear how the oxygen is activated. It may be co-ordinated to the Mg^{2+} in the presence of the enediol form of RuBP on the enzyme. This leads to formation of a hydroperoxide intermediate (Fig. 7.3) which is subsequently hydrolysed by water bound to the metal. If the RuBP oxygenase reaction is performed in the presence of $^{18}O_2$, one ^{18}O joins to C2 of RuBP to produce phosphoglycolate labelled in the carboxyl group, and the other is released as $H_2^{18}O$. An O from H_2O reacts with C3 of RuBP to give 3PGA. Possibly the enediol form of RuBP can attract the electrons in C of CO_2 or of O_2; the required electron changes in RuBP are produced by the co-ordinated Mg^{2+} and bound activating CO_2.

That oxygenation and carboxylation occur at the same site is shown by inhibition of both functions by CABP, other intermediates of the PCR cycle and pyridoxal phosphate. Also, both functions are activated to the same degree by CO_2 and Mg^{2+}. However, the two reactions are differentially affected by pH, and by temperature which changes the O_2/CO_2 solubility ratios in solution. If manganese ions replace magnesium ions in the activation of the isolated enzyme, the ratio of oxygenase to carboxylase activity increases.

Competition between RuBP carboxylase and oxygenase reactions

Oxygen and CO_2 compete for RuBP at the catalytic site; they are mutually competitive inhibitors. The rate, V_c, of the carboxylation of RuBP in the presence of competitive inhibition by O_2 with saturating RuBP is:

$$V_c = \frac{V_{cmax} \times C}{C + K_c(1 + O/K_O)} \qquad [7.3]$$

where V_{cmax} is the maximum velocity, C and O are the partial pressures of CO_2 and O_2 in equilibrium with dissolved gases in the chloroplast stroma and K_C and K_O are the Michaelis–Menten constants for CO_2 and O_2, respectively. The rate, V_o, of the oxygenase reaction with saturating RuBP is:

$$V_o = \frac{V_{omax} \times O}{O + K_O(1 + C/K_C)} \qquad [7.4]$$

where V_{omax} is the maximum rate of oxygenation.

The K_m for CO_2 in the oxygenation reaction equals the CO_2 inhibition constant for the oxygenase and O_2 has the same K_m in the oxygenase reactions as K_i in the carboxylase reaction. Oxygen competes inefficiently with CO_2 for RuBP at the catalytic site and only at high molar ratio of O_2 to CO_2 is the oxygenase reaction significant. The specificity factor, $\tau = V_c K_O/V_o K_C$, expresses the intrinsic carboxylation capacity of Rubisco relative to oxygenation.

The two reactions may be expressed as a ratio of oxygenase to carboxylase, α, (Fig. 7.4) which increases as the ratio of O_2 to CO_2 increases:

$$\alpha = \frac{V_o}{V_c} = \frac{V_{omax}}{V_{cmax}} \times \frac{OK_C}{CK_O} \qquad [7.5]$$

The oxygenase activity of RuBP carboxylase is important because the PG formed is oxidized in leaf cells by the glycolate pathway (p. 163), with the release of CO_2. The CO_2 produced is a form of respiration, which occurs in the light as photosynthesis proceeds, and is therefore called photorespiration. It offsets the CO_2 assimilation by the PCR cycle and thus decreases the efficiency of CO_2 assimilation. When the ratio of O_2 to CO_2 in the chloroplast stroma is very small, α approaches zero and photorespiration is negligible. However in air, with the partial pressure of O_2 a thousand-fold greater than CO_2 (p. 231), α is about 0.4; photorespiration is large and 20–30 per cent of the fixed carbon is lost. When α is 2, carbon lost by photorespiration equals gross photosynthesis (see p. 260) and there is no net gain of carbon. Some types of plants have developed mechanisms to avoid the loss of photorespired carbon, viz. C4 and CAM plants; those mechanisms are considered in Chapter 9. Crop plants with large α would be more efficient at CO_2 assimilation but attempts to select this desirable

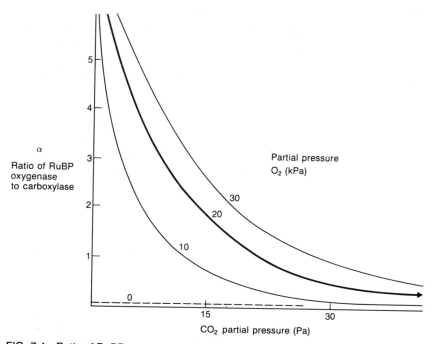

FIG. 7.4   Ratio of RuBP oxygenase to RuBP carboxylase activity, α, in relation to CO_2 and O_2 concentrations for extracted enzyme.

$$\left[\quad \text{nb.} \quad \alpha = \frac{V_o}{V_c} = \frac{V_{omax}}{V_{cmax}} \times \frac{OK_c}{CK_o} \quad \right]$$

trait or to modify the enzyme chemically or by genetic manipulation have so far produced no marked improvement, rather the opposite; selection of more efficient enzyme remains a major challenge.

Control of PCR cycle enzymes

Rubisco is activated by large increases in Mg^{2+} and pH (from 6 to 13 mM and 6 to 8 units, respectively) which occur in the chloroplast stroma with change from darkness to light; also NADPH and metabolite (e.g. F6P, R5P and E4P) concentrations increase the activity of carboxylase and the role of CA1P and Rubisco activase have been discussed. In darkness these compounds decrease, ensuring that the PCR cycle stops, thus avoiding depletion of intermediates and CO_2 acceptor. Fructose bisphosphate inhibits Rubisco activity at higher concentrations which may occur when triosephosphate export from the chloroplast is slowed by phosphate shortage. Rubisco is then slowed, preventing the accumulation of PCR cycle products and depletion of RuBP.

Fructose and sedoheptulose bisphosphatases

Fructose bisphosphatase (FBPase) achieves maximum activity at alkaline pH and with high Mg^{2+}, as expected of a chloroplast enzyme. It is also activated by light and the ferredoxin−thioredoxin mechanism as described on p. 136. FBPase activity is small with low concentrations of FBP but increases greatly above a threshold value of FBP concentration. This characteristic provides control of the PCR cycle and of processes leading to and from FBP. Synthesis of FBP is dependent on the production of triosephosphate; when this is rapid FBP accumulates, stimulating FBPase. If FBP decreases below the threshold the reaction slows, allowing the cycle to attain a new equilibrium. Regulation of SBPase is similar to that of FBPase.

3-Phosphoglycerate kinase and triosephosphate dehydrogenase

3-Phosphoglycerate kinase catalyses a reaction with large positive free energy change and is therefore controlled by the end products, the only PCR cycle enzyme so regulated. ADP and a low ATP/ADP ratio slow the reaction as does accumulation of glyceraldehyde-3-phosphate, regulating the cycle in relation to ATP synthesis and consumption (p. 113). Triosephosphate dehydrogenase is stimulated by reduced ferredoxin and the thioredoxin system. Control of two steps in 3-phosphoglycerate metabolism by products of the light reactions and the PCR cycle provides co-ordination of cycle function in light and dark, preventing large fluctuations in intermediates or, more importantly, their depletion.

Phosphoribulokinase

ATP stimulates this enzyme, which is regulated by energy charge and light via reduced thylakoid proteins, probably by the ferredoxin−thioredoxin system.

Light activation of PCR cycle enzymes

At this point it is useful to consider how light regulates enzyme activity. Several PCR cycle enzymes are controlled by reduced components of the electron transport chain (Fig. 7.5). FBPase, SBPase, $NADP^+$-glyceraldehyde phosphate dehydrogenase and phosphoribulokinase are activated in the light by ferredoxin or proteins linked to it or to reduced acceptors of PSI. The proteins are a water-soluble low mass thioredoxin (12 kDa) with one S−S bond per monomer. There are several forms of the protein *f, mb* and *mc*. The S−S link is very sensitive to reductant status and can be reduced to S−H bonds by both the natural reductant and chemicals, e.g. dithiothreitol. The reduction is catalysed by ferredoxin−thioredoxin reductase, an Fe−S protein of 30 kDa mass with two dissimilar

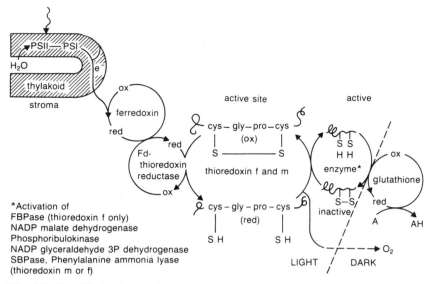

FIG. 7.5 Simplified scheme of the regulation of enzyme structure and activity by the reduction of disulphide bridges with electrons from the thylakoid via thioredoxin and the mechanism of deactivation.

subunits and S—S bridges. A complex of these soluble components acts as a protein modulase and has been called a light effector mediator or LEM. How the thioredoxin system is deactivated in darkness is not well understood; possibly oxidized glutathione or ascorbate slowly deactivate proteins. Probably a continuous flow of e⁻ to O_2 oxidizes the thioredoxin so that inhibition of electron supply is accompanied by enzyme inactivation. Such light effectors provide several points at which light can start and regulate carbon flux in the PCR cycle although their function in intact systems, such as leaves, is still unclear.

Induction and control of the PCR cycle

There is a delay or 'lag' of several minutes between illuminating isolated chloroplasts, after a long period of darkness, and attainment of a rapid, constant rate of photosynthesis. Intact systems such as protoplasts and leaves exhibit a shorter lag, although closed stomata may open slowly and prolong the lag period for photosynthesis in leaves. What limits the rate of photosynthesis during induction and at the steady state? Rapid synthesis of RuBP is required and darkened chloroplasts contain RuBP, which may protect Rubisco and enable the cycle to start quickly. Photochemistry and electron transport are very fast following illumination and NADPH is synthesized within a few seconds but ATP synthesis is slow. Reduction of thioredoxin is fast but activation of enzymes is probably slower, requiring minutes, comparable to the enzyme changes induced by alterations in

Mg^{2+} or pH in the stroma. Over several minutes ATP concentration and the ATP/ADP ratio increase. Synthesis of inhibitors (e.g. 6-phosphogluconate) stops and as their concentration decreases so the PCR cycle increases. As RuBP is synthesized faster than Rubisco activity increases, there is a transient increase in RuBP in tissues in the lag phase followed by a decrease in the steady state. Fructose bisphosphate and sedoheptulose bisphosphate also increase but their conversion to F6P and S7P is slow due to the slow activation (30 s) of the respective bisphosphatases.

If rapid synthesis of 3PGA and the formation of triosephosphates cause depletion of ATP, accumulation of ADP inhibits further activity, thus slowing the rate of R5P and RuBP formation. During induction, as concentration of intermediates rises and falls and enzymes in the cycle are activated at different rates, CO_2 assimilation and the concentration of intermediates oscillate as the system 'hunts' (to use an engineering term) until it approaches dynamic equilibrium, where the rate is determined by a 'limiting factor', either on the rate of supply of light, CO_2, etc., from the environment or in the control mechanisms operating within the system. Oscillation in CO_2 fixation and amounts of intermediates are damped because the several pools act as capacitances in the system, smoothing the demand/supply imbalance and giving only small fluctuations in the generally increasing rate of photosynthesis.

In the induction phase every 5 CO_2 assimilated produces an extra RuBP, if no carbon is removed from the PCR cycle, so that the rate of CO_2 assimilation increases with each turn of the cycle. Starting with 1 μmol RuBP per m^2 leaf turning over in unit time, after five turns of the cycle there will be 2 μmol RuBP and 5 μmol of CO_2 fixed, so the CO_2 fixation rate is 1 μmol per unit time; after two turns there will be 4 μmol RuBP and 15 μmol will have been assimilated, with the rate of CO_2 fixation 2 μmol per unit time. A further turn of the cycle gives 8 μmol RuBP and 4 μmol CO_2 per unit time. This is an exponential increase in activity and there must be a mechanism regulating how excess C is removed from the PCR cycle. If all excess C was immediately exported photosynthesis could not increase in the lag phase but clearly RuBP production rises until it matches the limitations of light reactions or enzyme capacity when the cycle is at steady state and excess C goes to make starch in the stroma or is transported into the cytosol. When the system is limited by CO_2, the amount of light or internal processes such as NADPH and ATP synthesis, so the rate of CO_2 assimilation becomes constant with time. Isolated chloroplasts lose intermediates of the PCR cycle and therefore contain little RuBP and generate it slowly; the lag period is shortened by addition of intermediates, 3PGA, for example, particularly if the ATP/ADP ratio is small, enabling RuBP to be made quickly. However, adding ribulose-5-phosphate increases the lag, because it stimulates the synthesis of RuBP but consumes ATP. Increased ADP then inhibits phosphoglycerate kinase and thereby assimilation.

During rapid photosynthesis the rate of turnover of substrates depends on the size of the pools and the activity and amount of the enzymes. The slowest reaction limits RuBP regeneration, but at present it is not clear which is the limiting factor, although Rubisco is the most likely contender followed by FBPase, SBPase and phosphoribulokinase. Supply of organic phosphate to the chloroplast from the cytosol may limit assimilation slowing export of C from the chloroplast; formation of large pools of phosphorylated intermediates 'locks up' P_i. Also under some conditions, e.g. P deficiency and low temperature, P in the cytosol may be insufficient to supply the chloroplast. Large 3PGA and small P_i concentration stimulates starch synthesis (p. 175), freeing P_i but as it is a slow reaction, the rate of assimilation with deficient P_i is smaller than the rate when P_i is freely available. Rate of ATP synthesis may limit assimilation under some conditions, for example, in leaves grown in dim light. In C3 plants photo-respiration limits assimilation by competitive inhibition of carboxylation by O_2 and because RuBP is consumed in the RuBP oxygenase reaction in addition to the CO_2 release. Phosphoglycolate produced by RuBP oxy-genase is not completely recycled to form RuBP so the amount of acceptor limits assimilation. Without O_2, CO_2 assimilation increases more than expected from preventing photorespiratory CO_2 release, as extra RuBP becomes available for carboxylation.

During rapid photosynthesis, enzymes turn over quickly, possibly at rates approaching the maximum, and therefore control the rate of assimilation, even if light, CO_2 and nutrients are in excess; this is discussed for leaves in Chapter 11. Control of the PCR cycle is complex; interactions between conditions in the stroma and thylakoid membranes, amount of substrates and of control molecules, regulate enzyme activation and reaction rates which determine the overall function of the system and its rate in relation to demand for assimilates.

Metabolite concentrations in the chloroplast stroma

Under steady state conditions the rate of each step in the PCR cycle is constant and the pool size of each metabolite also. In some cases (RuBP) the concentration exceeds the enzyme K_m, in others (DHAP and GAP) K_m and concentration are similar. Concentration of substrates and the K_m of a reaction determine the physiological rate and, because reactions are interconnected, the rate of CO_2 fixation.

When concentrations of CO_2 or O_2 change or light alters or demand for products changes, then the equilibrium conditions and fluxes of material are no longer constant. A finite time (many seconds) is required for transport of material from one pool to another and the rate of reactions is not constant, so fluctuations in the size of metabolite pools are observed. A step decrease in CO_2 concentration to a photosynthesizing leaf (Fig. 7.6) results in an immediate decrease in 3PGA synthesis and in the amount of 3PGA in the

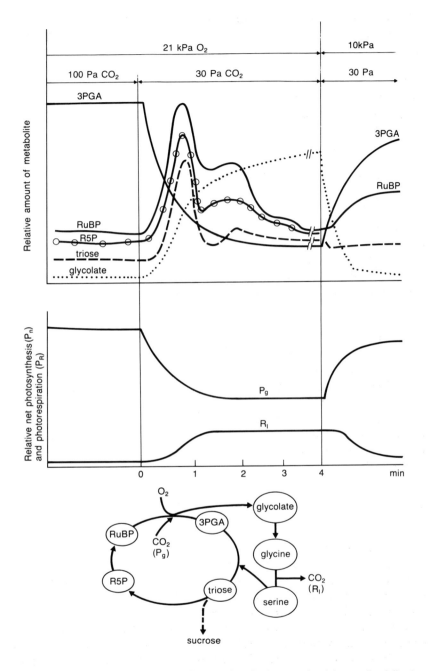

FIG. 7.6 Changes in the metabolite pools of photosynthesizing cells following perturbations in CO_2 and O_2 concentration related to net photosynthesis and photorespiration (semi-schematic). Note the phase shifts between fluctuations in metabolites and CO_2 exchange due to position in the PCR cycle and the different rates at which metabolites are interconverted.

tissue. RuBP is not consumed so the pool increases very shortly after 3PGA falls; triose is not used for RuBP formation so it also rises. As the diminishing pool of 3PGA feeds through to RuBP, transient changes in the amount are observed. Glycolate synthesis increases with the low CO_2/O_2 ratio. In low O_2 concentration, glycolate synthesis slows and the 3PGA increases, permitting more RuBP to be made and faster photosynthesis. In a complex, cyclic system the size of the fluctuations, their rate and phasing between metabolites depends on pool sizes, reaction rates and the magnitude of the change perturbing the system. This was appreciated by Wilson and Calvin (1955), who, after describing such fluctuations, wrote: ' . . . perhaps the most important result of this work is the general insight it gives into the complicated interrelated system of chemical reactions which occur in living systems'.

Exchange of photosynthate between chloroplast and cytosol

In mature leaves products of the PCR cycle and nitrate assimilation do not accumulate in the chloroplast stroma (with the exception of starch), but are transported across the envelope to the cytosol. The fluxes of material and energy, their control and interaction with chloroplast reactions have been studied on isolated chloroplasts, with intact envelopes, free (or relatively so) of the rest of the cell substance and physiologically undamaged as far as can be established. Distribution of assimilates between these chloroplasts and defined media is measured under different conditions of light, CO_2, etc., and with added metabolites, inhibitors, etc. Permeability of the envelope is determined from the distribution of metabolites between the medium and the chloroplasts by, for example, using radioactively labelled substrates. How added compounds affect the rate of photosynthesis and chloroplast reactions is measured by CO_2 and O_2 exchange. The latter is rapidly determined by the polarographic oxygen electrode. A suspension of chloroplasts is placed in a small chamber with an O_2-sensitive electrode, which produces a voltage proportional to O_2 concentration, allowing this to be measured continuously as O_2 is evolved by photosynthesis or removed by respiration. Vigorous stirring and small volume increase the speed of the instrument's response. Inhibitors of photosynthesis or substrates may be added without significantly disturbing conditions.

Methods have been developed for determining the compounds exchanged between cell compartments. Fast, efficient separation of organelles is essential for such metabolic studies. Chloroplasts are rapidly separated from the medium by placing the suspension in a centrifuge tube above a killing solution (e.g. perchloric acid) but separated from it by a layer of inert silicone oil. After the reaction has taken place, the tube is centrifuged rapidly, the chloroplasts penetrate the oil and reactions are stopped by the acid within seconds. Such treatment may alter the behaviour of chloroplasts because conditions differ from those in the intact cell, yet the methods provide much information on processes.

Other techniques have been used to analyse chloroplast and cytosol interaction *in vivo*. Distribution of assimilates, radioactive label, etc. have been measured in separated organelles. Isolation in aqueous solutions allows redistribution of water-soluble substances between parts of the cell and causes 'smearing' of the distribution pattern. Therefore, non-aqueous media are used. Chloroplasts are isolated in organic solvents such as carbon tetrachloride, which prevent water-soluble metabolites mixing during isolation; the method provides information on chloroplast interaction with the cytosol. Techniques and detailed discussion of the results are given by Heber (1974) and by Heldt (1979).

The outer membrane of the chloroplast envelope allows substrates of up to some 10 kDa mass to pass freely into the intermembrane space. In contrast the inner envelope membrane is the permeability barrier between the stroma and cytosol; across this membrane all products of the PCR cycle and all cytosolic requirements for the chloroplast must pass. Transport is generally by means of specific translocators. (For details see Flügge and Heldt (1991) and Douce and Joyard (1990).) Permeability of the membrane has been demonstrated by measuring the volume of chloroplasts which can be penetrated by low molecular mass, neutral substances such as sorbitol or sucrose, or by larger polymeric sugars. The space between the envelope membranes is accessible to sorbitol but the stroma and thylakoids are not (indicating that there is no direct connection between the inner membrane and thylakoid compartments). Changes of osmotic concentration in the medium cause chloroplasts to swell or shrink, altering the proportion of volume accessible to solutes and indicating that the outer membrane is freely permeable but the inner is not.

Movement of compounds across the envelope is not by simple diffusion; CO_2, O_2 and H_2O are major exceptions. CO_2 is soluble in lipid membranes and diffuses rapidly aided by carbonic anhydrase. Membranes have low permeability to cations, many large sugars and phosphorylated intermediates of the PCR cycle (Fig. 7.7). Carbon dioxide, Cl^-, dihydroxyacetone phosphate (DHAP) and P_i penetrate most rapidly; sucrose does not permeate nor does $NADP^+$ or NADPH; ATP and ADP only enter slowly. 3PGA, which might be expected to enter readily, has only limited penetration.

When plants or algae assimilate $^{14}CO_2$ the cytosol contains labelled DHAP, 3PGA, fructose- and glucose-6-phosphates, fructose bisphosphate and the nucleotide sugar uridine diphosphoglucose before labelled sucrose; the delay ('lag') is up to a minute. With longer time intervals, several amino acids are labelled. However, not all are made in the chloroplast and transferred unaltered to the cytosol; most are synthesized from triosephosphate (DHAP), which rapidly passes across the envelope. Fructose bisphosphate,

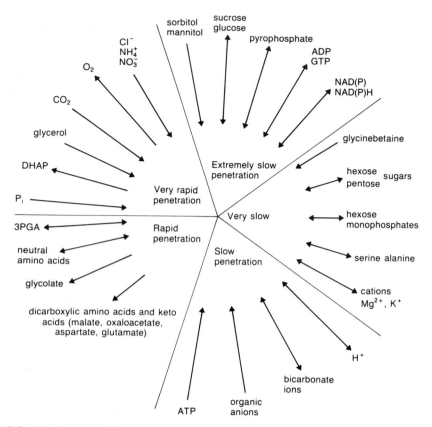

FIG. 7.7   Permeability of the intact chloroplast envelope to substances produced or consumed in photosynthesis by C3 cells, based on information in the references (see particularly Heber 1974).

for example, quickly appears outside the chloroplast but is synthesized from triosephosphate. Sucrose, the major translocated photosynthetic product, is made in the cytosol from DHAP (Ch. 8).

The phosphate and dicarboxylate translocators

DHAP is transported rapidly (16 μmol (mg chlorophyll)$^{-1}$ h^{-1}) and 3PGA at half this rate, by a phosphate-triosephosphate-phosphoglycerate translocator, generally called the phosphate translocator which exchanges P_i from the cytosol.

The phosphate translocator was identified from studies of the permeability of isolated chloroplast by silicon oil centrifugation. It catalyses counter exchange of P_i and triosephosphate (DHAP, GAP and 3PGA). It is the major protein of the inner envelope (which can be isolated by differential centrifugation) and is extracted from the lipid matrix with non-ionic detergents, e.g. Triton X-100. The mature protein (mass 61 kDa) consists of two polypeptide subunits of 29 kDa, is elliptical in shape and about 6.6 nm in length enabling it to span the membrane. It is a very hydrophobic protein with only 36 per cent of polar amino acids and hydropathicity plots indicate that it has seven membrane spanning segments. It is made as a precursor protein which loses a transit peptide when incorporated into the membrane (see Ch. 10). The translocator is a divalent anion exchange protein with very strict specificity, binding to active site P at the end of three carbon chains but not P at the 2-position. So it has large affinity for P_i (K_m 0.2 mM) and for 3PGA (K_m 0.15 mM) but not for 2PGA or glucose-6-phosphate (K_m 65 and 40 mM, respectively). Active sites are probably sulphydryl groups and lysine and arginine residues (since activity is blocked by chloromercuribenzoate and benzene sulphonates, respectively). Possibly DHAP binds to the protein, changing its conformation and allowing P_i to move through its 'channel'. There are differences between translocators in C3 and C4 plants, e.g. the latter is able to transport phosphoenol pyruvate. Characteristics of the translocator are of great importance to metabolite fluxes between compartments and for metabolic regulation. As a consequence of this the DHAP and 3PGA regulation in chloroplasts is achieved. DHAP is divalent, but 3PGA trivalent at neutral or alkaline pH; trivalent 3PGA only slowly equilibrates with the divalent form so in illuminated chloroplasts 3PGA is retained and DHAP exported and the 3PGA/triosephosphate ratio is 10 times greater in the light than in the dark. This enables 3PGA to be reduced at low $NADPH/NADP^+$ ratio and low phosphorylation potential.

Dicarboxylic acids, oxaloacetate, malate and α-ketoglutarate and the amino acids aspartate and glutamate are also translocated by a single molecule dicarboxylate transporter, located on the inner envelope, which does not work by counter-exchange as does the phosphate translocator system, so less rigid metabolic control is obtained.

Translocators in biological membranes are identified by saturation of the rate of transport with increasing concentration of the compound being transported. From the velocity of movement versus concentration, the K_m and V_{max} of the translocator are obtained. Addition of different substances shows the nature of the material transported and if there is competition for the carrier. Analogues of the substrate with related chemical structure compete for the translocator sites and enable the reaction to be analysed. All these characteristics show that an active process is involved, rather than diffusion.

Chloroplast envelopes also regulate the passage of H^+ into the medium. The pH of darkened chloroplast stroma is smaller than that of the medium due to Donnan equilibrium of H^+ with proteins but in the light the pH of the stroma increases as H^+ is pumped into the thylakoid lumen. With an acidic medium or cytosol, H^+ would enter the stroma faster than ions (for example, Mg^{2+}) in the light and decrease the pH gradient required for ATP synthesis. However, the envelope has an ATPase and H^+ is pumped (slowly compared to the thylakoid pump) out of the stroma into the medium in exchange for K^+. Anything which penetrates the envelope and transports H^+ into the stroma inhibits photosynthesis, for example, the nitrite anion (NO_2^-) at pH 7 gives nitrous acid (HNO_2) which enters and dissociates; H^+ remains in the stroma destroying the pH gradient and NO_2^- diffuses back to the medium and recycles H^+. Other weak acid anions, such as glycolate, decrease photosynthesis in the same way.

Shuttles of assimilate, reducing power and energy across the chloroplast envelope

Combination of the phosphate and dicarboxylate translocators and the H^+ pump provides the mechanism for distributing assimilates and energy between compartments. Pyridine nucleotides ($NADP^+$ and $NADPH$ in the chloroplast, NAD^+ and $NADH$ in the cytosol) cannot pass the intact envelope. ATP and other adenylates move by a translocator, but very slowly ($2 \ \mu$mol ATP mg chlorophyll^{-1} h^{-1}) in older leaves although faster in younger. Synthesis of sucrose from triosephosphate requires a rapid ATP supply but respiration, which may be inhibited in the light, is not a likely

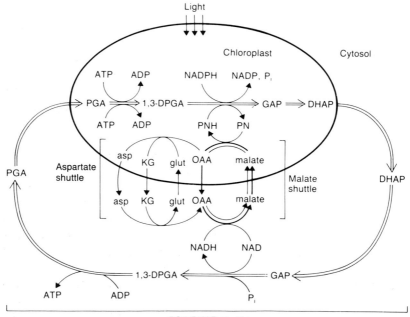

FIG. 7.8   Shuttles of dihydroxyacetone phosphate and 3-phosphoglyceric acid and of dicarboxylic acids between chloroplast and cytosol to generate ATP and NAD(P)H outside the chloroplast in the light or to provide ATP and NAD(P)H to the chloroplast in darkness.

source and ATP is probably synthesized in the cytosol via a shuttle of DHAP (Fig. 7.8). A large cytosolic DHAP/PGA ratio drives the synthesis of ATP and reducing power in the cytosol by triosephosphate oxidation. In the stroma, 3PGA is metabolized to DHAP and in the cytosol DHAP is converted back to 3PGA generating ATP and NADH. The reaction is the reverse of that in the stroma:

$$DHAP + ADP + P_i + NAD(P)^+ \rightleftharpoons 3PGA + ATP + NAD(P)H + H^+$$
[7.6]

Exchange of ATP and pyridine nucleotides (PN) is thus directly linked to assimilate transport. Glyceraldehyde phosphate formed in the cytosol may be oxidized to 1,3-diphosphoglyceric acid without synthesis of ATP, and reduce $NADP^+$ in the cytosol before NAD^+, thus favouring reactions using NADPH. Glyceraldehyde phosphate oxidation by this mechanism proceeds before the glycolytic enzyme is activated because the latter requires much larger concentration of substrate. These mechanisms may provide flexibility in the supply of ATP and NADPH, and a mechanism to balance demands in the cytosol, for example, in protein synthesis, ion exchange, etc., particularly if mitochondria are inhibited. In darkness the DHAP shuttle reverses and supplies the chloroplast with ATP and NADH; starch is metabolized to DHAP which enters the chloroplast and is converted to 3PGA thus forming ATP and NADPH.

Dicarboxylate exchange is also regulated by the proton gradient. Malate dehydrogenase (which catalyses malate oxidation and oxaloacetate (OAA) formation) occurs on both sides of the envelope and the redox potential is determined by OAA, malate and oxidized and reduced pyridine nucleotides. Coupling the H^+ gradient with the distribution of substrate and pyridine nucleotides enables electrons to flow between compartments against a gradient of reduction potential. Malate concentration is 10^3 times greater than oxaloacetate in tissues, and this inhibits transport of OAA. However transaminations with glutamate give aspartate and α-ketoglutarate so that the stromal NADP and cytosolic NAD systems are linked. The malate/aspartate ratio is larger in the light than in the dark in tissues. Together the DHAP and dicarboxylate shuttles regulate the pyridine nucleotides in both compartments. Chloroplast stroma is more reduced than the cytosol in the light and a gradient of phosphorylation potential develops from cytosol to stroma, that is, there is more ATP outside the envelope than inside, as found experimentally by non-aqueous fractionation. Of course in the dark ATP in the stroma is at low concentration but even in the light there is relatively more ATP in the cytosol than in the chloroplast, due to the action of the shuttles, despite ATP synthesis in the stroma.

C4 plants (Ch. 9) have massive flux of assimilate between the cell types during CO_2 fixation, as well as in translocation. Bundle sheath chloroplasts form 3PGA and DHAP; probably the latter is transported in exchange for

P_i as in C3 plants. Large flux of malate or aspartate occurs across the cell membrane but does not involve the chloroplast envelope. Mesophyll cell chloroplasts import pyruvate and 3PGA from oxaloacetate, malate and phosphoenol pyruvate (PEP) which are exported across the envelope to the cytosol. Thus, the type of compounds formed and transported are similar to C3 plants but the proportion and size of the fluxes probably differ greatly.

The translocators provide a 'valve' for regulating the flows of assimilate and energy. By coupling reactions in one compartment with those in another acting in reverse, net transport of energy and reductant is achieved without physical movement of ATP or NADPH and fine control of metabolism is possible. Also specific translocators control the fluxes of triosephosphate and P_i, and organic acids, enabling close coupling of all cellular factors. Thus transitions between dark and light are controlled; for example, starch synthesis occurs in the stroma if P_i is in short supply, or in darkness starch may supply the stroma with ATP. Over-reduction of one compartment is balanced by exchange with another. However, conditions which drive metabolism too far from equilibrium cannot then be balanced and metabolism is damaged. Many interactions and transport processes in the cell are poorly understood and remain to be quantitatively evaluated.

Photosynthetic assimilation of nitrogen and sulphur

An adequate supply of reduced nitrogen compounds is required for plant growth. Light energy is used for reducing nitrate to NH_3 which is used in synthesis of amino acids and proteins. Nitrate reduction in roots, and in leaves in darkness, uses energy and reductant derived via respiration from carbohydrates and is therefore only indirectly dependent upon photosynthesis. Higher plants absorb nitrogen from their environment as the nitrate ion (NO_3^-), or to a limited extent as ammonia (NH_3) or ammonium (NH_4^+), but before nitrate is utilized for amino acid synthesis it is first reduced to nitrite (NO_2^-) and then NH_3 (Fig. 7.9):

$$NO_3^- + 2\,e^- + 2\,H^+ \xrightarrow[\text{reductase}]{\text{nitrate}} NO_2^- + H_2O \qquad [7.10]$$

$$NO_2^- + 6\,e^- + 7\,H^+ \xrightarrow[\text{reductase}]{\text{nitrite}} NH_3 + 2\,H_2O \qquad [7.11]$$

Reduction of NO_3^- to NH_3 requires 8 e^-. Nitrate reductase is a cytoplasmic enzyme and uses electrons from NADH (rather than NADPH) passing via FAD, cytochrome b_{557} and molybdenum. The electrons are probably supplied to NAD^+ in the cytosol from NADPH in the chloroplast by shuttle systems.

Nitrate reductase is a 200 kDa molybdo-flavoprotein enzyme loosely attached to the chloroplast envelope. It is a rapidly turned-over enzyme with a half-life of some four hours. Nitrate reductase activity is controlled

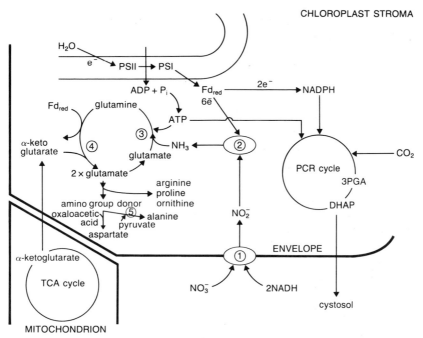

FIG. 7.9 Nitrate reduction in chloroplasts shown schematically with the glutamine/glutamate cycle of ammonia assimilation, but omitting the photorespiratory nitrogen cycle. Enzymes identified in the figure are: (1) nitrate reductase; (2) nitrite reductase; (3) glutamine synthetase; (4) GOGAT; (5) pyruvate aminotransferase.

by the concentration of nitrate and by light. Concentration of ammonia in leaves is usually so small that it does not control activity.

Nitrite is a toxic compound and therefore must be rapidly removed when formed. Reaction 7.11 proceeds with electrons supplied from ferredoxin, not NAD(P)H, and all the intermediates are bound to the enzyme. The nitrite reductase enzyme (60 kDa) is in the stroma and contains sirohaem and two additional atoms of iron and two labile sulphides per molecule. The location of nitrite reductase close to ferredoxin would ensure that nitrite reduction obtains available electrons in preference to nitrate reductase, so that toxic concentrations of NO_2^- would not be reached. In darkness the supply of electrons from respiration probably controls conversion of NO_3^- to NO_2^- and NH_3 reduction is slower than in the light. Assimilation of NO_3^- by isolated intact chloroplasts increases the OH^- ions and changes the pH of the cell; exchange of organic acids balances this.

Ammonia, the product of the nitrite reductase reaction, is toxic uncoupling ATP synthesis from electron transport in the thylakoids; it is therefore rapidly assimilated and the concentration in the leaves is at micromolar values. Assimilation of ammonia involves formation of the amide glutamine from glutamate:

$$\text{glutamate} + NH_3 + ATP \rightarrow \text{glutamine} + ADP + P_i + H_2O \qquad [7.12]$$

The reaction is catalysed by glutamine synthetase, a protein of 360 kDa probably with eight subunits, a pH optimum of 8 and requiring Mg^{2+} for activation; approximately half the enzyme activity in higher plant leaves is found in the chloroplast. Glutamate dehydrogenase (located in the mitochondria and chloroplasts) may operate at very high ammonia concentrations:

$$\alpha\text{-ketoglutarate} + NH_3 + NAD(P)H + H^+ \rightarrow glutamate + NAD(P)^+ + H_2O$$
$$[7.13]$$

Much evidence points to the glutamine synthetase reaction as the primary route for ammonia assimilation; for example, isotopic nitrogen (^{15}N or ^{13}N) from NO_3^- accumulates first in glutamine then glutamate. When $^{14}CO_2$ is given to photosynthesizing leaves containing an analogue of glutamate to block further glutamate metabolism, ^{14}C accumulates in glutamine. Glutamine synthetase is found in many plants; it is a very efficient enzyme with a high affinity (low K_m) for NH_3 (10^{-5} M), the concentration of which is therefore kept at rather low physiological levels.

Glutamate is synthesized from glutamine, regenerating glutamate as an acceptor of NH_3, by glutamate synthase (also called GOGAT, the acronym for the earlier name of the enzyme (glutamine (amide):2-oxoglutarate aminotransferase (oxido reductase NADP)) found in chloroplasts:

$$glutamine + \alpha\text{-ketoglutarate} + Fd_{red} + H^+ \rightarrow 2 \text{ glutamate} + Fd_{ox} \quad [7.14]$$

The dicarboxylic acid α-ketoglutarate is the 'carbon skeleton' for amino acid synthesis. Probably dicarboxylic acids are not formed within the chloroplast but are produced in the mitochondria, from carbon exported as triosephosphate (DHAP) and reimported into the chloroplast by shuttle systems. The glutamine synthetase system links not only the chloroplast carbon and nitrate metabolism but also that of the peroxisomal and mitochondrial photorespiratory nitrogen 'cycles'. Figure 8.4 (see Ch. 8) shows the nitrogen cycle associated with photorespiration. Energy for net nitrate reduction is probably less than 5 per cent of the energy used in CO_2 assimilation. The photorespiratory nitrogen cycle is, however, very active in C3 plants; the flux is some ten-fold greater than the net rate of nitrate reduction. Considerable reductant and ATP is consumed and the photorespiratory N and C cycles together consume over 30 per cent of total available energy. Photorespiratory nitrogen metabolism is important because it decreases the efficiency with which light is used without increasing the net assimilation of nitrate. However, it may be important in regulating the energy supply in metabolism and may protect metabolism from high NH_3 and reductant levels.

Control of nitrate and ammonia assimilation

Nitrate reduction is controlled by nitrate reductase which is a rapidly synthesized and inactivated enzyme so nitrate metabolism is potentially sensitive to factors which affect protein synthesis. Should conditions become unsuitable, then enzyme synthesis slows and also ammonia production. A connection between NO_3^- reduction and the production of carbon skeletons would be provided by ATP or energy charge. As ATP is required for glutamine and protein synthesis, shortage of ATP would inhibit both nitrate assimilation and protein synthesis as well as the PCR cycle. Change in the amount of enzyme may provide longer term control. Nitrate reductase synthesis is stimulated by NO_3^-, enabling the plant to respond rapidly to NO_3^- availability. Another control point in NO_3^- reduction is at the transport of nitrate to the assimilatory sites. Concentration of Mg^{2+} and pH and energy charge may regulate glutamine synthetase; light may stimulate enzyme activity by mechanisms similar to those of the thioredoxin system (see p 136). Glutamate synthase does not appear to be inhibited by end products of the reaction. Light activation and conditions in the chloroplast would provide co-ordination between PCR cycle, NO_3^- reduction, light reactions and demand for products from secondary metabolism.

Synthesis of amino acids in photosynthesis

Some amino acids are rapidly labelled with ^{14}C when leaves assimilate $^{14}CO_2$. Alanine is formed from glutamate by pyruvate aminotransferase with pyruvate derived from 3PGA via phosphoenol pyruvate, possibly in the chloroplasts. The enzyme for transamination of oxaloacetic acid to aspartate is in the chloroplast. Glycine and serine are synthesized by three routes in photosynthesis. In the phosphorylated pathway 3PGA is dephosphorylated to glycerate and this is converted to hydroxypyruvate which is transaminated to give serine. Another route is via the glycolate pathway (p. 163), glycolate giving glycine; one glycine is decarboxylated and its β-carbon added to a second glycine giving serine; the reaction is catalysed by serine hydroxymethyl transferase in the mitochondria. A third route is via phosphoserine. Multiple routes for serine formation probably keep the supply of amino acid balanced when conditions, such as CO_2 supply, vary.

Chloroplasts contain enzymes for synthesis of most protein amino acids; glutamate is an essential precursor for many others, e.g. arginine. However, the last step in the arginine synthetic pathway (arginosuccinate lyase) is cytoplasmic. Aspartate is metabolized in chloroplasts to lysine, threonine and homocysteine but the final methylation of this with methionine occurs outside the chloroplast. Chloroplasts are also dependent on mitochondria for their supply of organic acids for amino acid synthesis. Thus, there are close links between the organelles in metabolism.

Amino acid synthesis requires relatively higher reductant-to-ATP ratio than triose formation. One molecule of aspartate requires 11 NAD(P)H and 10 ATP whereas triose requires 6 NADPH and 9 ATP. Nitrate reduction increases electron flow and accumulation of H^+ in the thylakoids, and may increase ATP synthesis which would benefit CO_2 fixation if the PCR cycle is limited by ATP (p. 113). Thus, both CO_2 and NO_3^- reduction would increase together and stimulate assimilation.

Dinitrogen fixation and photosynthesis

Higher plants cannot photosynthetically assimilate gaseous dinitrogen (N_2) because they lack a nitrogenase enzyme. However, photosynthetic bacteria and blue–green algae (e.g. *Nostoc*) reduce N_2 to ammonia. The nitrogenase is inhibited or destroyed by very small O_2 concentrations so that in the blue–green algae N_2 assimilation proceeds in heterocysts, cells with thick walls which are impermeable to oxygen and only have PSI activity so no O_2 is produced. The electrons for reduction of N_2 (eqn 7.15) come from ferredoxin and are derived from sugars transported into the heterocysts and metabolized to give electrons. Cyclic photophosphorylation probably provides ATP, the requirement for which may be as high as 15 molecules per N_2 because the N_2 molecule is particularly inert.

$$N_2 + 6\,e^- + 12\,ATP + 6\,H^+ \rightarrow 2\,NH_3 + 12\,ADP + 12\,P_i \qquad [7.15]$$

Hydrogen production during photosynthesis

Gaseous hydrogen (H_2) is produced by heterocysts in blue–green algae, some photosynthetic bacteria and primitive eukaryotic algae. Anaerobic environments are essential for the hydrogenase function. Nitrogenase also possesses hydrogenase activity and H_2 can be used as a source of electrons for N_2 reduction via ferredoxin. Electrons from PSI are supplied to H^+:

$$2\,H^+ + 2\,e^- \xrightarrow[\text{hydrogenase}]{\text{light}} H_2 \qquad [7.16]$$

No ATP is required and electrons come from reduced organic substrates, such as NADH or glucose.

Photosynthetic sulphur metabolism

Sulphur, absorbed as the sulphate ion and reduced in leaves by electrons from electron transport and ATP, is incorporated into the amino acids cysteine and methionine, and into sulphydryl groups of co-enzymes and sulpholipids. Reduction of SO_4^{2-} to sulphite and sulphide and incorporation into cysteine in the chloroplasts (Fig. 7.10) is strongly stimulated by light. Sulphate is activated in two stages both requiring ATP; first SO_4^{2-} reacts with ATP giving adenosine phosphosulphate (APS). The enzyme, ATP: sulphate adenylyltransferase (called ATP sulphurylase) which occurs in all organisms able to reduce sulphate, has several subunits, a broad alkaline pH optimum and requires Mg^{2+} (characteristic of stromal enzymes). The sulpho-group of APS is transferred to a carrier thiol (carSH),

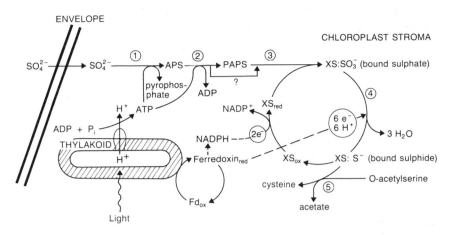

FIG. 7.10 Photosynthetic assimilation of sulphate in chloroplasts, shown schematically, with enzymes: (1) ATP sulphurylase; (2) APS kinase; (3) APS sulphotransferase; (4) thiosulphonate reductase; (5) cysteine synthase. ? denotes uncertainty of PAPS involvement in higher plant SO_4^{2-} reduction.

e.g. glutathione in algae or a larger molecule in higher plants such as phytochelatins (γ-glutamyl-cysteine)$_n$-glycine, by adenosine 5′-phosphosulphate sulphotransferase

$$APS^{2-} + carSH \rightarrow carS\text{-}SO_3^- + AMP^{2-} + H^+ \qquad [7.17]$$

This widely distributed enzyme is probably only in the chloroplast and has a molecular mass of 110 kDa and a K_m for APS of 10 μM; it is probably regulated by changing the amounts and is induced by absence of SO_4^{2-} or large demand for reduced sulphur. Sulphite is reduced by sulphite reductase to free sulphide by reduced ferredoxin in higher plants; energy requirements are similar to those for nitrite reduction. The enzyme is of 2 subunits, 63 and 69 kDa with a sirohaem and 1 Fe_4S_4 centre per subunit and a K_m for sulphite of 20 μM. In the second stage serine reacts with acetyl-S-CoA to give *O*-acetyl-L-serine, catalysed by serine acetyl transferase. Then cysteine synthase reacts the *O*-acetyl-L-serine with sulphide (H_2S) to give L-cysteine. Cysteine synthase is a widely distributed enzyme abundant in chloroplasts with 2 subunits of 34 kDa mass. Interestingly it also catalyses metabolism of selenide (reduced form of selenium) found in serpentine soils and Se enters metabolism, forming compounds toxic to animals.

Methionine is synthesized by a pathway from cysteine, the first step reacting *O*-phosphohomoserine and cysteine giving cystathionine catalysed by a specific synthase; despite its position as the first step in the pathway it is not important in regulation. Cystathionine is cleaved by β-cystathionase to homocysteine which is then methylated to give methionine.

Glutathione (γ-glutamyl-cysteinyl-glycine, GSH) is an abundant low molecular mass thiol in plants where it is very important in detoxifying O_2 species and metals, regulation of redox states of enzymes and in transport of S. Synthesis is by γ-glutamyl-cysteine synthase in the reaction:

$$Glu + Cys + ATP \rightarrow \gamma\text{-Glu-Cys} + ADP + P_i \qquad [7.18]$$

and conversion of the γ-Glu-Cys by adding glycine to the C-terminal of the dipeptide by glutathione synthetase using ATP; the enzyme is active in many different species in cytosol and chloroplast. It has a molecular mass of 85 kDa with an alkaline pH optimum and requires magnesium.

References and Further Reading

Andrews, J.T. and **Lorimer, G.H.** (1987) Rubisco: Structure, mechanisms and prospects for improvement, pp. 132−218 in Hatch, M.D. and Boardman, N.K. (eds), *The Biochemistry of Plants*, Vol. 10, Academic Press, London.

Buchanan, B.B. (1980) Role of light in the regulation of chloroplast enzymes, *A. Rev. Plant Physiol.*, **31**, 341−74.

Calvin, M. and **Bassham, J.A.** (1962) *The Photosynthesis of Carbon Compounds*, Benjamin, New York.

Chapman, M.S., Suh, S.W., Curmi, P.M.G., Cascio, D., Smith, W.W. and **Eisenberg, D.S.** (1988) Tertiary structure of plants Rubisco: Domains and their contacts, *Science*, **241**, 71−4.

Cooper, T.G., Filmer, D., Wishnick, M. and **Lane, M.D.** (1969) The active species of 'CO$_2$' utilised by ribulose diphosphate carboxylase, *J. Biol. Chem.*, **244**, 1081−83.

Delieu, T. and **Walker, D.A.** (1981) Polarographic measurement of photosynthetic oxygen evolution by leaf discs, *New Phytol.*, **89**, 165−78.

Douce, R. and **Joyard, J.** (1990) Biochemistry and function of the plastid envelope, *A. Rev. Cell Biol.*, **6**, 173−216.

Flügge, U.-I. and **Heldt, H.W.** (1991) Metabolic translocators of the chloroplast envelope, *A. Rev. Plant Physiol. Plant Mol. Biol.*, **42**, 129−44.

Gibbs, M., Willeford, K.O., Ahluwelia, K.J.K., Gombos, Z. and **Jun, S.-S.** (1990) Chloroplast respiration, pp. 339−53 in Zelitch, I. (ed.), *Perspectives in Biochemical and Genetic Regulation of Photosynthesis*, Alan R. Liss, Inc., New York.

Gutteridge, S. and **Keys, A.J.** (1985) The significance of ribulose-1,5-bisphosphate carboxylase in determining the effects of environment on photosynthesis and photorespiration, pp. 259−85, in Barber, J. and Baker, N.R. (eds), *Topics in Photosynthesis*, Vol. 6, Elsevier, Amsterdam.

Hanks, J.F., Somerville, C.R. and **McIntosh, L.** (1985) Site-specific mutagenesis of ribulose-1,5-bisphosphate carboxylase/oxygenase, pp. 325−7 in Steinback *et al.* (eds), *Molecular Biology of the Photosynthetic Apparatus*, Cold Spring Harbor Laboratory.

Heber, U. (1974) Metabolite exchange between chloroplasts and cytoplasm, *A. Rev. Plant Physiol.*, **25**, 393−421.

Heber, U., Schreiber, U., Siebke, K. and **Dietz, K.-J.** (1990) Relationship between light-driven electron transport, carbon reduction and carbon oxidation in photosynthesis, pp. 17–37 in Zelitch, I. (ed.), *Perspectives in Biochemical and Genetic Regulation of Photosynthesis*, Alan R. Liss, Inc., New York.

Heldt, H.W. (1979) Light dependent changes of stromal H^+ and Mg^{2+} concentrations controlling CO_2 fixation, pp. 202–7 in Gibbs, M. and Latzko, E. (eds), *Encyclopedia of Plant Physiology* (N.S.), Vol. 6, *Photosynthesis II*, Springer-Verlag, Berlin.

Jensen, R.G. (1980) Biochemistry of the chloroplast, pp. 274–313 in Tolbert, N.E. (ed.), *The Biochemistry of Plants*, Vol. 1, *The Plant Cell*, Academic Press, New York.

Keys, A.J. (1990) Biochemistry of ribulose bisphosphate carboxylase, pp. 207–24 in Zelitch, I. (ed.), *Perspectives in Biochemical and Genetic Regulation of Photosynthesis*, Alan R. Liss, Inc., New York.

Keys, A.J. and **Parry, M.A.J.** (1990) Ribulose bisphosphate carboxylase/oxygenase and carbonic anhydrase, pp. 1–14 in Lea, P.J. (ed.), *Methods in Plant Biochemistry*, Vol. 3, *Enzymes of Primary Metabolism*, Academic Press, London.

Latzko, E. and **Kelly, G.J.** (1979) Enzymes of the reductive pentose phosphate cycle, pp. 239–50 in Gibbs, M. and Latzko, E. (eds), *Encyclopedia of Plant Physiology* (N.S.), Vol. 6, *Photosynthesis II*, Springer-Verlag, Berlin.

Lea, P.J. (ed.) (1990) *Methods in Plant Biochemistry*, Vol. 3, *Enzymes of Primary Metabolism*, Academic Press, London.

Leegood, R.C. (1990) Enzymes of the Calvin cycle, pp. 15–38 in Lea, P.J. (ed.), *Methods in Plant Biochemistry*, Vol. 3, *Enzymes of Primary Metabolism*, Academic Press, London.

Lorimer, G.H. (1981) The carboxylation and oxygenation of ribulose 1,5-bisphosphate: the primary events in photosynthesis and photorespiration, *A. Rev. Plant Physiol.*, **32**, 349–83.

Miziorko, H.M. and **Lorimer, G.H.** (1983) Ribulose-1,5-bisphosphate carboxylase/oxygenase, *A. Rev. Biochem.*, **52**, 507–35.

Ogren, W.I. (1984) Photorespiration: Pathways, regulation and modification, *A. Rev. Plant Physiol.*, **35**, 415–42.

Peterson, R.B. (1990) The RuBP carboxylase/oxygenase model and photorespiration in C_2 leaves, pp. 285–99 in Zelitch, I. (ed.), *Perspectives in Biochemical and Genetic Regulation of Photosynthesis*, Alan R. Liss, Inc., New York.

Pierce, J. (1988) Prospects for manipulating the substrate specificity of ribulose bisphosphate carboxylase/oxygenase, *Physiol. Plantarum*, **72**, 690–8.

Portis, A.R.Jr (1990) Rubisco activase, *Biochim. Biophys. Acta*, **1015**, 15–28.

Ramage, R.T. and **Bohnert, H.J.** (1989) Recent developments in Rubisco research: structure, assembly, activation and genetic engineering, pp. 307–30 in Barber, J. and Malkin, R. (eds), *Techniques and New Developments in Photosynthesis Research*, Plenum Press, New York.

Robinson, S.P. and **Walker, D.A.** (1981) Photosynthetic carbon reduction cycle, pp. 193–236 in Hatch, M.D. and Boardman, N.K. (eds), *The Biochemistry of Photosynthesis*, Vol. 8, *Photosynthesis*, Academic Press, New York.

Roy, H. and **Nierzwicki-Bauer, S.A.** (1991) Rubisco: Genes, structure, assembly and evolution, pp. 347–64, in Bogorad, L. and Vasil, I.K. (eds), *The Photosynthetic Apparatus: Molecular Biology and Operation*, Academic Press Inc., San Diego.

Seeman, J.R., Kobza, J. and **Moore, B.D.** (1990) Metabolism of 2-carboxyarabinitol 1-phosphate and regulation of ribulose-1,5-bisphosphate carboxylase activity, *Photosynthesis Res.*, **23**, 119–30.

Sharkey, T.D. (1989) Evaluating the role of Rubisco regulation in photosynthesis of

C3 plants, pp. 435−48 in Walker, D.A. and Osmond, C.B. (eds), *New Vistas in Measurement of Photosynthesis, Phil. Trans. R. Soc. Lond.*, **B323**, 225−448.

Wilson, A.T. and **Calvin, M.** (1955) The photosynthetic cycle. CO_2 dependent transients, *J. Amer. Chem. Soc.*, **77**, 5948−57.

CHAPTER 8

Metabolism of photosynthetic products

All the carbon for plant metabolism is provided ultimately, by the PCR cycle, so it is difficult to decide where photosynthesis stops and 'secondary metabolism' starts. Purists may argue that the PCR cycle is the limit of photosynthesis and events that consume its products are 'secondary metabolism' and not photosynthesis. The argument is semantic, although for convenience limits must be drawn. Only some secondary processes are discussed because they directly consume products of the light reactions or of the PCR cycle. Photosynthesis is regulated by processes outside the chloroplast involving many aspects of secondary metabolism. At present the way that photosynthetic rate is balanced with demand for assimilates (e.g. for respiration and growth) is poorly understood. Concepts of feedback regulation based on accumulation of assimilates are commonly applied but ill defined. Much better appreciation of the mechanisms is necessary before integration of 'source' and 'sink' processes is fully explained.

In the steady state, carbon assimilated in excess of that needed to regenerate RuBP is used within the chloroplast for synthesis of materials, for example, starch, lipids and proteins, or is exported as dihydroxyacetone phosphate (DHAP) and glycolate which are metabolized to many substances, e.g. sucrose the carbohydrate generally transported in plants, storage polymers (starch, fructans) and amino acids in the cytoplasm of C3 plants (Fig. 8.1); there are large fluxes of organic acids in C4 plants.

Starch synthesis

Starch accumulates in the chloroplast stroma of many species of plants during illumination, forming large granules. During darkness it is remobilized and consumed in respiration. Starch synthesis is restricted to illuminated chloroplasts. Starch is synthesized from fructose-6-phosphate (F6P) of the PCR cycle (Fig. 8.2) which is converted by hexose phosphate isomerase to glucose-6-phosphate (G6P) in a reaction of very small free energy change so that it is not a control step. Production of glucose-1-

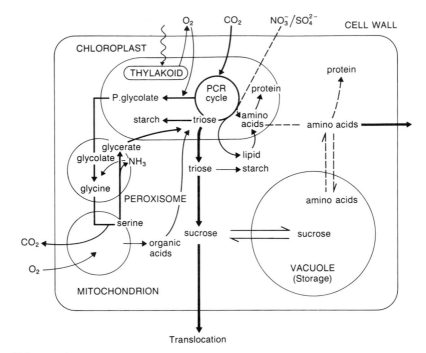

FIG. 8.1   Principal fluxes of carbon and their relation to nitrogen and sulphur assimilation in C3 plant leaf cells in the light.

phosphate (G1P) from G6P by phosphoglucomutase is reversible and gives a ratio of G6P to G1P of about 20. Removal of G1P by subsequent reactions encourages synthesis of more G1P. As with many processes in carbohydrate metabolism, nucleotide sugars are involved in the formation of the starch polymer:

$$ATP + glucose\text{-}1\text{-}phosphate \rightarrow ADP\text{-}glucose + PP_i \qquad \text{[8.1a]}$$

$$ADP\text{-}glucose + (glucose)_n \rightarrow (glucose)_{n+1} + ADP \qquad \text{[8.1b]}$$

where $(glucose)_n$ is a preformed polymer (an α-1,4-glucan primer) to which glucose residues are added. The enzyme responsible is starch synthase (ADP glucose:1,4-α-D-glucan-4-glycosyl transferase). Amylose is converted by a starch branching enzyme (1-4-α-D-glucan:1,4-α-D-glucan-6-glycosyl transferase) to amylopectin with α-1,6-branch points.

Reaction 8.1a is catalysed by the enzyme ADP-glucose pyrophosphorylase which regulates carbon flow; it has a molecular mass of 210 kDa and possibly has four similar subunits. There is a lysine residue at the active site where two molecules of Mg-ATP bind to the four subunits and then four G1P molecules. The enzyme is controlled allosterically by products of the PCR cycle. The rate is increased 10–20 times by increased concentration of 3PGA and less so by F6P. These accumulate if in excess of that needed to regenerate RuBP, and 'signal' the enzyme to proceed with starch synthesis.

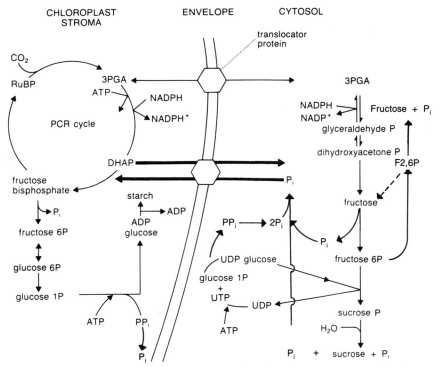

FIG. 8.2 Scheme of carbon metabolism in the chloroplast and cytosol in the light leading to starch and sucrose synthesis.

If 3PGA is depleted, for example, with low rates of photosynthesis or in darkness, the enzyme is inhibited. In addition, ADP-glucose pyrophosphorylase is allosterically inhibited by P_i which interacts with 3PGA; a high ratio of 3PGA/P_i stimulates starch synthesis. When ATP synthesis is insufficient, 3PGA decreases and P_i accumulates, inhibiting starch synthesis and preventing PCR cycle depletion. If consumption of carbon outside the chloroplast is rapid then low 3PGA and increased P_i concentrations slow starch synthesis. Sequestering (binding) P_i into metabolites, for example, by feeding leaves with mannose which is phosphorylated by hexokinase but not remetabolized, decreases P_i concentration and stimulates synthesis of starch. A doubling of the P_i concentration from 1 mM in the medium of isolated intact chloroplasts increases the 3PGA/P_i ratio in the stroma ten fold and the rate of starch synthesis forty fold. Although starch synthesis liberates P_i, rate of the reaction is inadequate to maintain rapid photosynthesis but allows some CO_2 fixation to continue. Plants deprived of phosphate contain much starch which may be a symptom of nutritional inadequacy. Starch synthase may be soluble in the stroma and bound to starch granules, in equal proportion, but their roles are probably similar. Enzyme activity in leaves is barely sufficient to account for rates of starch synthesis. Remobilization of starch may proceed via G1P (catalysed by phosphorylase) and G6P, F6P and fructose bisphosphate to DHAP. ATP is required but may be inadequate in darkness.

Sucrose synthesis

Sucrose is the most abundant oligosaccharide, involved in translocation between organs and storage. It is α-D-glucopyranose + β-D-fructofuranose, linked by an $\alpha(1-2)$glycosidic bond, and is inert in the cell (hence, good for storage), non-reducing and very soluble; 1 g dissolves in 0.5 cm^3 of water at 20 °C. Sucrose from sugar cane and sugar beet is important nutritionally and in international trade. Synthesis is from fructose-1,6-bisphosphate (Fru-1,6-P$_2$) by hydrolysis with the appropriate phosphatase in an irreversible step, forming fructose-6-phosphate (Fru-6-P).

$$\text{Fru-1,6-P}_2 + \text{H}_2\text{O} \rightarrow \text{Fru-6-P} + \text{P}_i \hspace{2cm} [8.2]$$

Fru-6-P is converted to glucose-6-phosphate (Glu-6-P), glucose-1-phosphate (Glu-1-P) and UDP-glucose. Transfer of glucose residues from UDP-glucose to Fru-6-P is catalysed by sucrose phosphate synthase (systematic name: UDP-D-glucose:D-fructose-6-phosphate 2-glucosyl transferase). Sucrose phosphate synthase is not pH-controlled, but is regulated by UDPG and F-6-P and the reaction rate increases rapidly as the substrate concentration increases.

$$\text{UDPG} + \text{Fru-6-P} \xrightarrow[\text{synthase}]{\text{sucrose phosphate}} \text{UDP} + \text{sucrose-6-phosphate} \hspace{1cm} [8.3a]$$

$$\text{sucrose-6-phosphate} + \text{H}_2\text{O} \xrightarrow{\text{sucrose phosphatase}} \text{sucrose} + \text{P}_i \hspace{1cm} [8.3b]$$

Sucrose phosphatase is almost irreversible, so carbon entering the pathway cannot 'flow back'. Free UDP, sucrose and other sugars stimulate the enzyme, whilst sucrose phosphate synthase and the phosphatase may be regulated by Mg^{2+}. Sucrose is also made from UDPG and fructose by sucrose synthase but this enzyme is not believed to be important in leaves.

Sucrose synthesis proceeds in the cytoplasm (possibly the enzymes are loosely bound to the chloroplast envelope), for there is delay in ^{14}C labelling after exposure to ^{14}CO$_2$ and separation of chloroplast from the cytosol shows UDPG pyrophosphorylase (required for UDPG synthesis) to be cytoplasmic.

Carbon for sucrose synthesis comes from DHAP exported from the chloroplast in a strict 1:1 exchange with P$_i$ on the phosphate translocator (see p. 142). Sucrose synthesis consumes energy; 1 UTP is required in the UDP glucose pyrophosphorylase reaction for every 4 molecules of DHAP consumed.

Sucrose synthesis is regulated over long periods by the amounts of enzymes (so-called coarse control) and this is probably the principal way in which plants adapt to conditions. Of course, there are differences between C3 and C4 plants and also within the C4s. In maize, for example, sucrose is probably only synthesized in the mesophyll cells, but in other C4 species it may also be made in the bundle sheath. Short-term regulation (fine control), by metabolic effectors which activate or inhibit enzyme activity, balances sucrose production against demands of growth, respiration, etc., and synthesis of starch and other storage oligosaccharides (e.g. fructans). Storage of sucrose and accumulation of starch, fructans, etc. allows the capacity of photosynthesis to be fully exploited without feedback inhibition of the rate of photosynthesis and permits growth to proceed over long periods without carbon limitation.

Regulation of sucrose synthesis is achieved at three main enzymatic control points in the pathway: cytosolic FBPase, sucrose phosphate synthase and sucrose phosphatase (Fig. 8.2). These enzymes catalyse reactions which are far from equilibrium, as calculated from the free energy changes (see p. 127) derived from estimates of the concentrations of metabolites in the cytosol of leaf cells. The FBPase has a ΔG of -19 kJ mol^{-1}, sucrose phosphatase -11 kJ mol^{-1} and sucrose P synthase -8 kJ mol^{-1}; the concentrations of FBP, UDPG and sucrose P are, respectively, about 0.1, 2 and 0.2 mM in the cytosol. Cytosolic concentration of sucrose is large (40−50 mM) so the reactions must be far from equilibrium to achieve rapid rates of sucrose production and this makes the reactions important sites for regulation. Depending on conditions, between 10 and 85 per cent of the sucrose produced is transported to the vacuole in a few minutes. Cytoplasmic FBPase is the first, and therefore a very important, regulatory enzyme in the pathway. The protein is of 130 kDa molecular mass and differs from the chloroplast enzyme. FBPase has a very low K_m for FBP (2−4 μM), requires Mg^{2+} and is not pH dependent; it shows very strong sigmoidal saturation kinetics with FBP, particularly with AMP present. The enzyme is very strongly inhibited (hundred fold) by micromolar concentrations of the compound fructose-2,6-bisphosphate (F2,6BP); this is not to be confused with FBP, fructose-1,6-bisphosphate.

The role of F2,6BP in regulation of sucrose synthesis in leaves has been demonstrated (see Stitt *et al.* 1987). F2,6BP is found mainly in the cytosol where it is synthesized from F6P and P_i by an enzyme, F6P,2 kinase. The concentration of F2,6BP is also dependent on an enzyme, F2,6BP phosphatase, which degrades the molecule to F6P and P_i. This phosphatase is inhibited by F6P and P_i, in contrast to the kinase. The control mechanism suggested is inhibition of FBPase by F2,6BP when the amounts of F6P and P_i increase, e.g. when flow of C to sucrose slows due to lack of demand. The F6P,2 kinase activity increases, F2,6BP is synthesized and slows the FBPase and, consequently, the conversion of FBP to F6P. At the same time, the F2,6BPase is inhibited by the effectors, so maintaining a large F2,6BP concentration. Conversely, when the F6P and P_i concentrations are low but DHAP and 3PGA concentrations large, the kinase is inhibited and the F2,6BPase is stimulated, thus greatly decreasing the concentration of F2,6BP. This allows FBPase activity to increase and stimulates the flux of C through to sucrose, providing a very rapid switching mechanism which can also link PCR cycle activity and glycolysis.

Sucrose phosphate synthase also regulates C partitioning between starch and sucrose synthesis. Coarse control is obtained by changing activity over periods of minutes to hours, related to the supply and demand for assimilates and to the light—dark transitions; there is marked endogenous rhythm independent of photosynthesis. The amount of enzyme activity observed is usually just sufficient to account for the flux of C to sucrose, so the enzyme may be important in the long term regulation of carbon metabolism. The protein has a molecular mass of 450 kDa, contains an SH— group (which may be required for metabolic regulation rather than catalysis) and is not sensitive to Mg^{2+} or pH. It is, however, activated by G6P, which increases the V_{max} and decreases K_m for F6P. The enzyme is inhibited by P_i but the activity is increased hyperbolically by a large $G6P/P_i$ ratio. Also the products of the reaction, UDP and sucrose P, inhibit sucrose phosphate synthase; there is uncertainty about the control exerted by sucrose with the enzyme from some species more inhibited than that from others. Most importantly sucrose phosphate synthase is activated by light, but probably not by a mechanism of the thioredoxin type (see p. 135). The significance of these characteristics is that the flow of C through the enzyme is increased as the substrate F6P increases (G6P is in equilibrium with it) and as the available P_i decreases, plus the activation by light which may be expected to stimulate the overall carbon flux.

Sucrose phosphate phosphatase rapidly converts sucrose P to sucrose; activity in the leaves is usually ten-fold greater than the activity of the synthase. The phosphatase probably keeps the sucrose P at low concentration, enabling the sucrose P synthetase to function efficiently.

With the characteristics of the enzymes in mind, it is possible to understand how sucrose synthesis is regulated. With the onset of rapid photosynthesis, the concentration of DHAP in the cytosol increases and P_i decreases increasing the concentration of FBP, thus stimulating the FBPase and decreasing the synthesis of F2,6BP. In bright light F6P (and hence G6P) concentrations rise, sucrose P synthetase is activated and consequently the C flux to sucrose increases. However, if demand for sucrose falls, sucrose and sucrose P inhibit the sucrose P phosphatase and synthetase, respectively and F6P increases; the F6P,2 kinase is activated and the F2,6BP phosphatase is inhibited, resulting in an increase of F2,6BP. This slows the FBPase, resulting in a build-up of triosephosphates and fall in P_i. The system then regulates at a different rate of carbon flux. However, if the blockage of sucrose use is severe, e.g. by low temperature, inhibition of growth or deficient P_i, the accumulation of triosephosphate stimulates the carbon flux to starch, which then accumulates in the chloroplast or storage organs. Conversely, slow rates of photosynthesis but large demand for sucrose in growth, respiration, etc. produce plants with very small contents of carbohydrates.

Translocation of sucrose out of the cell is beyond the discussion of photosynthesis although if export is slow, feedback inhibition of photosynthesis may occur, as the storage capacity of cells is reached and intermediates accumulate.

Biosynthesis of chloroplast lipids

One third of the chloroplast dry mass is lipid, mainly synthesized during the development of the leaf, although new synthesis and replacement of lipids continues in mature leaves. Synthesis of fatty acids (Fig. 8.3) such as palmitic, oleic and linoleic acids consumes PCR cycle products, reductant as ferredoxin and NADPH, and ATP. The light reactions, both PSII and PSI, are essential. Synthesis occurs in the stroma where all the required enzymes are found, but involves mitochondria and cytosol. Dihydroxyacetone phosphate moves from the stroma to the cytosol. Pyruvate is formed from the triosephosphate and enters the mitochondria where acetyl-CoA and acetate are produced. Acetate returns to the chloroplasts where acetyl-CoA is again made by a synthetase; it joins to an acyl carrier protein (ACP), a low molecular weight co-factor protein for some 12 enzymes of lipid metabolism, found only in the stroma. ACP is very similar in different organisms (e.g. 40 per cent sequence homology between spinach and *E. coli*).

Malonyl-CoA is also produced in the stroma by carboxylation of acetyl-CoA with bicarbonate by the enzyme acetyl-CoA carboxylase (transcarboxylase) which requires Mg^{2+} (Fig. 8.3). ATP activates the enzyme complex, which is probably a high molecular mass multifunctional enzyme rather than three separate enzymes: a biotin carrier protein (BCCP), biotin carboxylase and BCCP:acetyl-CoA transcarboxylase which initiates the reaction. BCCP is bound to the thylakoid lamellae, although the reason for such close proximity to the light reactions is not known. Malonyl-CoA is transferred to ACP giving malonyl-ACP. Condensation of 1 acetyl-ACP and 7 malonyl-ACP followed by reduction with NADPH and NADH gives palmityl-ACP, which is hydrolysed to palmitic acid or is condensed with more malonyl-ACP and reduced with NADPH to stearyl-ACP. This increases the chain length of the fatty acids from C 16:0 to C 18:0. Stearyl-ACP is reduced (by ferredoxin in the presence of O_2) to oleyl-ACP from which oleic acid is released by hydrolysis. Further modification of the fatty acids occurs in the chloroplast. Palmitoyl- and oleoyl-ACPs may be hydrolysed on the envelope and enter the cytosol for lipid synthesis. Fatty acids may not be formed directly from PCR cycle products and turnover of membrane lipids is slow, for only several hours after exposure to $^{14}CO_2$ is label found in them.

Synthesis of complex lipids occurs in several parts of the cell. Glycolipids, of which mono- (MGDG) and digalactosyl diacylglyceride (DGDG) are the major chloroplast lipids, are synthesized in the chloroplast. MGDG is synthesized by a transferase enzyme bound to the chloroplast envelope which

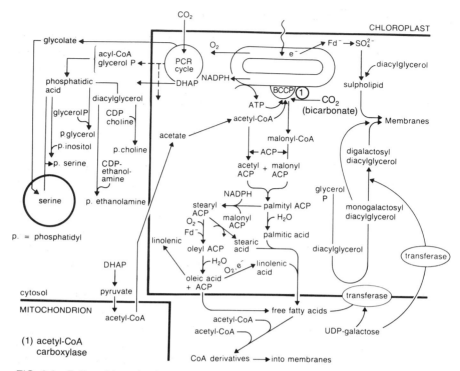

FIG. 8.3 Fatty acid synthesis in leaf cells requires products of the light reactions and the PCR cycle and co-operation between organelles.

attaches a galactose moiety from UDP-galactose to diacylglycerol, whilst DGDG synthesis is by a soluble, stromal enzyme catalysing transfer of two galactosyl residues from UDP-galactose to a monoacyl-glycerol. Sulpholipid synthesis (e.g. sulphoquinovosyl diglyceride) is linked to SO_4^{2-} reduction in the chloroplast.

Phosphatidyl glycerol is a major glycerophosphatide of leaves, constituting over 20 per cent of the total lipid of thylakoids and envelope membranes. Synthesis is from glycerol phosphate, produced by phosphorylation of glycerol, acylated by acyl-CoA to phosphatidic acid. Phosphatidic acid is dephosphorylated to form the diacylglycerols. Membrane glycerophosphatides include phosphatidylcholine and phosphatidylethanolamine, which require choline and ethanolamine derived from glycine and serine, possibly from the glycolate pathway. Phosphatidylinositol is formed from glycerol phosphate in the mitochondria, where phosphatidylglycerol is formed.

This brief summary of fatty acid and lipid synthesis serves to show the complex interaction between the light reactions, carbon assimilation and co-operation between cell organelles. Lipid synthesis is a photosynthetic process requiring much energy; 17 and 8 mol of ATP and NADPH, respectively, are needed for a C18 fatty acid. Regulation of lipid synthesis in relation to other chloroplast functions is not well understood.

Glycolate pathway and photorespiration

As a consequence of the RuBP oxygenase reaction (p. 128) phosphoglycolate is formed in the chloroplast stroma. Alternative synthesis from glycoaldehyde−thiamine pyrophosphate complex from the PCR cycle reacting with hydrogen peroxide is probably negligible. Accumulation of P-glycolate (which occurs when inhibitors are applied) prevents photosynthesis because it inhibits triosephosphate isomerase in the PCR cycle. Thus, P-glycolate must be removed, allowing the carbon and phosphate to be used. P-glycolate is dephosphorylated (Fig. 8.4) by a specific phosphatase, an 84 kDa tetramer with optimum activity under alkaline conditions which occur in illuminated chloroplasts; it is stromal enzyme or attached to the envelope and P_i remains in the stroma and glycolate is excreted. In algae this is by diffusion into the medium and in higher C3 plants into the cytosol. There, metabolism of this 2-carbon compound is by the 'C2 pathway' in which some 11 enzymes recycle part of the carbon back to the chloroplast. Three cell organelles are involved, chloroplast, peroxisome and mitochondrion (Fig. 8.4).

In higher plants glycolate is transported, possibly by a translocator on the chloroplast envelope, into the peroxisome and metabolized to glyoxylate by glycolate oxidase which is at high concentration and induced by light and glycolate. The reaction is almost irreversible and consumes O_2 producing H_2O_2. As the peroxisome is a detoxifying organelle with large catalase activity H_2O_2 is rapidly destroyed; plants lacking catalase are killed in the normal atmosphere where photorespiration accounts for 50 per cent of the total net carbon flux in photosynthesis. Glycolate is transaminated to glycine by glutamate-glyoxylate aminotransferase with glutamate as amino-group donor. This and other aminotransferases (see Fig. 8.4 caption) in the peroxisome use several amino acids as substrates and may link directly to alanine metabolism of the cell. Photorespiratory nitrogen cycling fluxes are most important for they are several times greater than the net NO_3^- reduction in leaves. Glycolate can be converted to formic acid and to oxalic acid which accumulates in many plants as insoluble crystals of calcium oxalate. Glycine is transported to the mitochondria where one molecule is decarboxylated and the 1-C fragment and another glycine are condensed to serine (by glycine decarboxylase and serine hydroxymethyltransferase, respectively) and CO_2 is released and escapes from the mitochondria; this is the CO_2 of photorespiration. Glycine decarboxylase is a complex of four proteins with different linked functions and produces methyltetrahydrofolate which reacts with the other glycine in the serine hydroxymethylhydrofolate reaction.

$$2 \text{ glycine} + H_2O \rightarrow \text{serine} + CO_2 + NH_3 + 2\,H^+ + 2\,e^- \qquad [8.4]$$

Also, NH_3 is released; it reacts in the cytosol and possibly chloroplast with glutamate to give glutamine.

Electrons from the reaction can enter the mitochondrial electron transport chain and, via oxidative phosphorylation, synthesize three molecules of ATP

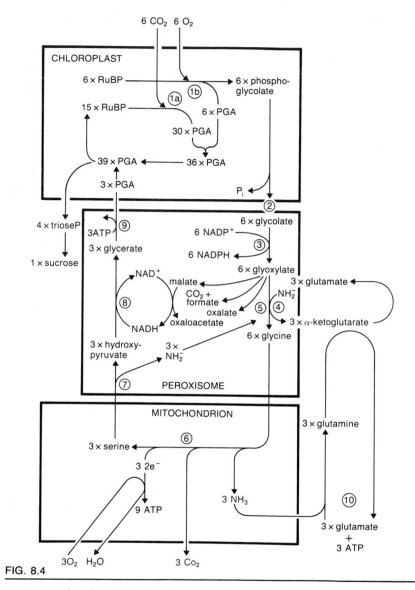

FIG. 8.4

per two molecules of glycine decarboxylated. One ATP is needed to form glutamine in the cytosol and two ATP enter metabolism for sucrose synthesis, etc. Alternatively the electrons may be used in the reduction of oxaloacetate to malate, thereby establishing a shuttle exchange of reducing power with the cytosol.

Serine from the mitochondrial reaction is transferred into the peroxisomes, where it is deaminated by serine-glyoxylate aminotransferase. The resultant hydroxypyruvate is reduced by NADH to glycerate; the reductase enzyme

FIG. 8.4 (left)   Carbon, nitrogen and energy flows in the glycolate pathway and production of CO_2 when photorespiration is one-quarter of the rate of net photosynthesis. Numbers refer to the reactions listed below.

Enzymes of the glycolate pathway:
(1a) ribulose bisphosphate carboxylase; (1b) RuBP oxygenase; (2) phosphoglycolate phosphatase; (3) glycolate oxidase; (4) glutamate-glyoxylate aminotransferase; (5) serine-glyoxylate aminotransferase; (6) serine hydroxymethyl transferase (and glycine decarboxylase); (7) transaminase; (8) hydroxypyruvate reductase; (9) glycerate kinase; (10) glutamine synthetase, GOGAT

Important formulae:

Phosphoglycolate	CH_2OP $\mid$ $COOH$	Glycine	CH_2NH_2 $\mid$ $COOH$
Glycolate	CH_2OH $\mid$ $COOH$	Serine	CH_2OH $\mid$ $CHNH_2$ $\mid$ $COOH$
Glyoxylate	CHO $\mid$ $COOH$	Hydroxypyruvate	CH_2OH $\mid$ CO $\mid$ $COOH$
Glycerate	CH_2OH $\mid$ $CHOH$ $\mid$ $COOH$	Phosphoglycerate	CH_2OP $\mid$ $CHOH$ $\mid$ $COOH$

is a very active peroxisomal dimeric protein of 90 kDa mass. In turn NAD^+ may be reduced by aspartate, malate and oxaloacetate. Glycerate is phosphorylated in the cytosol or more probably chloroplast by glycerate kinase, which returns 75 per cent of the C exported back to the chloroplast. The enzyme is an active monomer of 40 kDa mass which requires Mg-ATP as P-donor. Regulation is unclear; probably light stimulates but there is no effect of pH, Mg^{2+} or AEC (p. 107). Flux of carbon through the pathway consumes NADPH (or NADH) from the light reactions and is an important way of regulating the energy and reductant balances of leaves.

Photorespiration

Photorespiration is important as an oxidative process consuming PCR cycle products and energy. Stoichiometry of the pathway and of glycine decarboxylation is such that for every two molecules (i.e. four carbons (4C)) of glycolate entering the peroxisomes one CO_2 (one carbon atom), or 25 per cent of the carbon is lost when two molecules of glycine (4C) are converted to serine (3C). The carbon flux to glycine depends on the ratio of RuBP oxygenase to RuBP carboxylase reactions (α) (Ch. 7, p. 133) and the O_2/CO_2 ratio. With CO_2 shortage, for example, when stomata are closed because of water stress, the flux of carbon through the glycolate pathway increases relative to that through the PCR cycle (Fig. 8.1), and a greater proportion of newly assimilated CO_2 is lost by photorespiration. A simple model of the system (see Fig. 12.6) uses the characteristics of Rubisco in relation to O_2 and CO_2 concentration. When gross photosynthesis and photorespiration are equal (in leaves at the CO_2 compensation concentration) α is 2 and the flux into the glycolate pathway is four times that in photosynthesis. Photorespiration and glycolate pathway metabolism are a drain on photosynthetic CO_2 fixation, and use extra NADPH and ATP, despite ATP synthesis in the mitochondria. Net carbon fixation decreases because photorespired CO_2 offsets gross assimilation. When O_2 is removed photorespiration stops and gross assimilation increases as the ATP and NADPH consumed by photorespiration become available to synthesize extra RuBP. The amount of NADPH and ATP required for synthesis of a molecule of sucrose depends on the pathway by which the precursors are formed; in an O_2-free atmosphere, 37 ATP and 24 NADPH are required in total for all reactions. However, in air, consumption is 58 and 45 molecules, respectively. Sucrose synthesis with 50 per cent photorespiration demands 72 ATP and 59 NADPH. Competition for PCR cycle carbon between RuBP and sucrose synthesis increases as photorespiration increases and as a consequence carbon flux to sucrose is slowed. At very large α, a flux of carbon from storage into the PCR cycle would be necessary to keep the cycle running.

Biochemically, the glycolate pathway, involving three cellular organelles, may be a method by which C3 plants have partially overcome the effects of a large O_2 and small CO_2 concentration on Rubisco. The pathway scavenges some of the carbon and produces some energy as ATP or reducing power in the cytosol, where it may be needed during assimilation. The oxygenase activity may be 'inevitable' due to the high O_2 concentration of the atmosphere, so that production of phosphoglycolate cannot be prevented. The oxygenation would be the best point for regulation of photorespiration for once phosphoglycolate is produced it must be used as productively as possible to avoid inhibition of photosynthesis and increase efficiency.

Rubisco probably evolved early in photosynthesis of CO_2, when the CO_2/O_2 ratio and CO_2 pressure were very large, and has remained relatively unchanged in mechanism despite the apparent disadvantages of its slow reaction rate (turnover of sites 2 s^{-1}) and the oxygenase activity. As the enzyme has a central role in assimilation, its structure has been strongly conserved; the multiple mutations in amino acid residues thought necessary to increase efficiency must all happen together and this would have a very low probability of occurring; single mutations might be lethal, as lack of photosynthesis would destroy the whole organism and so tend to slow selection of an improved enzyme reaction. Other parts of the photosynthetic system seem to be more 'flexible' and have evolved around Rubisco, for example, in C4 metabolism.

Attempts have been made to find chemicals which block the glycolate pathway and prevent waste of assimilated carbon by photorespiration. Sodium bisulphate reacts with glycolate and forms the α-hydroxy sulphonate, disodium sulphoglycolate, which inhibits glycolate oxidase. Other α-hydroxysulphonates with the formula $R—CHOH—SO_3$ Na, such as HPMS (α-hydroxy-2-pyridine methane sulphonic acid), are effective inhibitors although they may undergo reactions to produce glycolate bisulphate, which is the inhibitor. Isonicotinyl hydrazide (INH) causes accumulation of glycolate by blocking the glycine to serine conversion in the mitochondria. However, these inhibitors that block glycolate metabolism after RuBP oxygenase, stop carbon flow in the pathway and also inhibit photosynthesis because carbon accumulates as an intermediate, preventing the autocatalytic resynthesis of RuBP. Thus, increasing CO_2 assimilation cannot be achieved by blocking the glycolate pathway. It is necessary to stop the oxygenase activity either chemically or genetically by modifying the enzyme, if photosynthesis is to be increased. This goal is yet to be achieved despite much active work.

The glycolate pathway links the metabolism of nitrogen inside and outside the chloroplast. Ammonia released in the conversion of glycine to serine in mitochondria is reassimilated in the cytosol by glutamine synthetase. The glutamine formed is recycled into the chloroplast where it transaminates α-ketoglutarate, using reduced ferredoxin in the presence of glutamate synthase, giving two molecules of glutamate. For each CO_2 released in photorespiration one NH_3 is also produced. With photorespiration 25 per cent of gross photosynthesis, the NH_3 assimilation may be about 5 μmol NH_3 m^{-2} leaf s^{-1}. If reassimilation of ammonia is blocked by methionine sulphoxime (MSO), which inhibits glutamine synthetase, then ammonia accumulates in tissues and photosynthesis stops. Nitrate reduction is only a few per cent of assimilation so the nitrogen turnover during photo-respiration is many times greater than the net nitrogen reduction in the chloroplast. Photorespiration thus involves combined carbon and nitrogen cycles which link the energy and metabolites in chloroplasts, mitochondria and cytosol.

The glycolate pathway may consume excess reduced pyridine nucleotide, 'burning off' reductant and allowing a faster turnover of NADH in the peroxisomes and NADPH in the chloroplast. The decarboxylation of glycine may provide a method of generating ATP in the mitochondria and cytosol in the light, or if the generated electrons can be used to form malate, a regulatory device to balance cellular functions. Consumption of excess reductant may be important when photosynthesis is severely restricted, e.g. in water-stressed plants with closed stomata, when the electron transport chain is fully reduced and chlorophyll absorbs excess energy forming excited states and damaging products (e.g. superoxide, H_2O_2, p. 98). Photo-respiration decreases the energy burden, particularly on PSII which is sensitive to photoinhibition; it may function together with the carotenoids in quenching chlorophyll excited states and singlet O_2 (see Ch. 3, p. 44). C3 plants are predominantly of well watered, often dimly lit environments, where light harvesting may be a limiting factor in growth rather than CO_2 supply. Carbon lost in photorespiration may be relatively unimportant ecologically, compared to the 'safety valve' offered under temporarily adverse conditions. However, long exposure to bright illumination, hot conditions and low CO_2 concentrations cause excessive loss of CO_2 and photochemical damage to C3 plants.

Glycolate metabolism and C4 plants

C4 plants maintain a very large CO_2 and low O_2 concentration at the RuBP carboxylase active site so that RuBP oxygenase activity is small and little phosphoglycolate is formed. Leaves of C4 plants assimilating $^{14}CO_2$ in air produce little ^{14}C-labelled glycine and form serine by non-glycolate routes. This, together with absence of photorespiration, suggests no glycolate metabolism. However, C4 plants make and metabolize glycolate in small quantities. Conditions such as water stress, which may decrease the internal CO_2 concentration, stimulate the formation of glycine. Thus, the glycolate pathway functions in C4 plants, but is only about 10 per cent of that in C3 species under comparable conditions. Any photorespiratory CO_2 is efficiently removed by the PEP carboxylase reaction.

Chloroplast respiration

In higher plants, dark and photorespiration occurs outside the chloroplast, in mitochondria. However, recently a form of respiration has been detected in chloroplasts. This is not entirely unexpected for in photosynthetic bacteria respiratory and photosynthetic electron transport chains are on the same membranes and interact, switching electrons at the cytochrome and quinone steps; membrane potential controls the respiratory component thermo-dynamically. Also in cyanobacteria and green algae thylakoids are involved in respiration with common components, e.g. cytochrome *b/f*, and competition for electrons may control PSII activity. There is also NADH dehydrogenase activity in *Chlamydomonas* and in tobacco chloroplasts, where DNA codes for proteins homologous to mitochondrial NADH dehydrogenase. A pathway may exist for the respiratory oxidation of starch and reducing equivalents in higher plant chloroplasts and regulate the redox state of plastoquinone in thylakoids. Chlororespiration may provide chloroplasts with reductant and substrates and regulate the system under particular physiological conditions.

Interaction of dark respiration and photosynthesis

Transition between light and dark occurs daily for plants and rapid fluctuations between bright and dim light are common in many habitats due to clouds or sunflecks in vegetation. As the light energy incident on the leaf changes so do the fluxes of energy and materials in photosynthetic cells and the ratio of assimilation to 'dark respiration'. How do photosynthesis and respiration interact and what are the controls operating during changes in photon flux and with darkness? Regulation of dark respiration, preventing 'futile cycles', is needed to balance growth with assimilation and prevent depletion of metabolites. This complex area of metabolic interaction has been reviewed by Graham (1980) and little progress has been made in understanding since. Here only some aspects of the problem are considered.

Dark respiration is the consumption of carbohydrates and other compounds to produce energy and carbon substrates (e.g. organic acids) for cellular synthetic processes. The process consumes O_2 and produces CO_2 and H_2O; starch and sugars are degraded to pyruvate by glycolysis, generating ATP. The pyruvate is converted to acetyl-CoA by pyruvate dehydrogenase, situated in the mitochondria. There, acetyl-CoA enters the tricarboxylic acid cycle (TCA cycle or Krebs or citric acid cycle in the mitochondria), where it is metabolized to organic acids (e.g. citrate, fumarate, α-ketoglutarate), releasing CO_2 and forming NADH. NADH is oxidized by the mitochondrial electron transport chain with O_2 as terminal acceptor, giving water. Electron transport is coupled to phosphorylation, and this is the source of ATP in respiration. In addition to glycolysis, the oxidative pentose phosphate pathway consumes glucose via glucose-6-phosphate, which is oxidized by $NADH^+$ to 6-phosphogluconate by glucose-6-phosphate dehydrogenase. The 6-phosphogluconate is oxidized and decarboxylated to ribulose-5-phosphate.

Control of the TCA cycle may be provided by several factors, for example, high ATP/ADP or $NAD^+/NADH$ ratio or low substrate concentration inhibits respiration; the respiratory states are discussed by Wiskich (1980). Also, accumulation of organic acids inhibits cycle activity; oxaloacetate slows succinate dehydrogenase, for example. High ATP/ADP and NADH levels inhibit malate and isocitrate dehydrogenases and these are major control points.

A form of respiration proceeds in the light, for C3 plants photorespire, but it is not clear if photosynthesis or photorespiration or light itself inhibits dark respiration or if it occurs concomitantly. Photorespiration is three to eight times greater than dark respiration (measured in darkness) and involves the mitochondrial electron transport chain, so that dark respiration could proceed during photosynthesis as the chain *per se* is not inhibited by light. NADH produced in the glycolate pathway could also be oxidized by the mitochondria, which do metabolize exogenously applied NADH in darkness. There appears to be no evidence that dark respiration could not take place during photosynthesis; however there are many points at which mitochondrial activity may be regulated.

Respiration of leaves, measured in darkness, is 5–10 per cent of net photosynthesis in plants of brightly lit environments, but from those of shade it may be a much larger proportion. Dark respiration is saturated at low O_2 concentration (1–2 kPa) but photorespiration increases with increasing O_2. Measurements of respiration in the light at different O_2 concentrations extrapolated to zero O_2 suggest that 'dark' respiration is inhibited, but measurements of CO_2 evolution with increasing light at low CO_2 concentration which prevents net photosynthesis, suggests that dark respiration continues. However, at very small photosynthetic carbon flux in the cell, respiration may be stimulated, particularly if the ATP/ADP ratio is low.

Measurements of incorporation of metabolites into TCA cycle intermediates in the light have been made to clarify the role of dark respiration. $^{14}CO_2$ from photosynthesis enters some organic acids and alanine, which is formed via pyruvate, if assimilation is rapid. However, it does not quickly (30 minutes) enter other organic acids or all amino acids. Thus, glutamate becomes radioactive only in some experiments. Long term pulse chase studies suggest that ^{14}C enters TCA cycle intermediates in the light. However, transfer of $^{14}CO_2$ labelled leaves from light to darkness causes an influx of ^{14}C into the TCA cycle, as expected if the flux is inhibited in the light. Leaves fed with ^{14}C-acetyl-CoA, TCA acids, amino acids or pyruvate (from glycolysis) form labelled TCA cycle acids, glutamate and aspartate, consistent with an active TCA cycle. Possibly the metabolic controls of the TCA cycle are bypassed when large concentrations of substrate are added to photosynthesizing tissues, allowing respiration to proceed even if normally blocked or much reduced.

The distribution of $^{14}CO_2$ into assimilates shows continuation of the cycle. Inhibition of photosynthesis by DCMU showed that respiration occurs in the light; however preventing assimilation could stimulate respiration because photosynthetic carbon and energy fluxes are inhibited. Graham (1980) concluded that the TCA cycle continues in the light; with rapid photosynthesis respiration is small and slower photosynthesis encourages TCA cycle activity. Respiration seems to decrease in the first few minutes of illumination but increases later.

The TCA cycle provides carbon skeletons for amino acid synthesis, particularly α-ketoglutarate for glutamate, and this function may be required in the light, for although chloroplasts can synthesize most amino acids, organic acids cannot be made. Mitochondria supply cells in darkness with ATP and NADH but in the light photophosphorylation and photosynthetic electron transport provide ATP and NADPH. Photorespiration gives NADH which may donate electrons to O_2 via the mitochondrial electron transport chain coupled to phosphorylation.

Adenylates in cell compartments

Control of adenylate composition in cellular compartments during dark–light transitions has been examined by rapidly killing and separating chloroplasts, mitochondria and cytosol from leaves or, with greater ease and precision, from protoplasts. Protoplasts are prepared by incubating leaf slices in an enzyme preparation which digests the cellulose walls. The released protoplasts are separated by centrifugation in density gradients of sucrose and sorbitol under controlled conditions. Cell organelles are separated on density gradients and, by measuring marker enzyme complements, the purity of cell fractions is checked. Interconversion of adenylates is inhibited by these rapid separation methods allowing adenylates to be measured (see p 108) in organelles; corrections for cross contamination of fractions are made.

In darkness the ATP/ADP ratio is larger in the cytosol than in the mitochondria and chloroplast stroma; oxidative phosphorylation in the mitochondria maintains a very high ATP/ADP ratio and energy charge (EC) in the cytosol. Shuttles of triosephosphate and 3-phosphoglycerate between chloroplast stroma and cytosol regulate the chloroplast's adenylate levels without direct transfer of ATP. ATP in chloroplasts in darkness is required to maintain ion balance, levels of photosynthetic intermediates and some synthetic processes such as protein synthesis, although the rates may be small in mature chloroplasts. However, protein synthesis continues in darkness in the cytosol, so high EC is necessary. On illumination photophosphorylation in the chloroplast stroma increases rapidly (within 30 s) so that the ATP/ADP ratio increases, but AMP decreases with the result that EC increases. However, the sum of ATP, ADP and AMP remains constant showing that little direct adenylate exchange occurs between chloroplast and cytosol. The cytosolic ATP/ADP ratio and EC increase greatly and those in the mitochondria decrease, suggesting that the cytosol exercises control over respiration. However, after a few minutes illumination, the cytosolic ATP/ADP ratio and EC fall to values greater than in darkness and the mitochondrial values rise but to less than those observed in darkness. On darkening cells there is a decrease in the ATP/ADP ratio and EC values in the chloroplast stroma, and concomitantly a transient decrease in the cytosol and a transient increase in mitochondrial ATP/ADP ratio, before they return to values seen after long periods in darkness. Mitochondrial adenylate would seem to be controlled by cytoplasmic EC which remains high at all times, possibly because there is no AMP in the cytosol and no adenylate kinase. Even in the light the chloroplast EC is lower than the cytosolic, so the photosynthetic carbon cycle and other stromal activities (e.g. protein synthesis) presumably operate at lower EC than those in the cytosol. The reason for and consequences of this are not yet understood.

In the light a high cytosolic ATP/ADP ratio and high EC are generally thought, with little evidence, to depress mitochondrial oxidative phosphorylation and glycolysis. Although photorespiration has been thought to use the mitochondrial electron transport chain in the light, other evidence strongly suggests that the electrons from glycine oxidation may be transferred into the cytosol by malate and aspartate shuttles and not consumed by the mitochondrial electron transport chain. Thus, it may not be essential for mitochondria to function in the light.

A working model of the interaction between dark respiration and photosynthesis and the adenylate systems is that the TCA cycle operates in darkness and transfer of adenylate energy keeps the cytosol EC high and also maintains a minimal EC in the stroma. In light chloroplasts produce ATP, increasing the chloroplast ATP/ADP ratio and EC and transferring energy to the cytosol which experiences greatly increased ATP/ADP ratio and a rise in EC which decreases or stops the activity of the mitochondria allowing the ATP/ADP ratio and EC to re-equilibrate. In dim light a small adenylate supply from the chloroplast might not completely inhibit mitochondria which would provide some energy to the cytosol. ATP is required for many metabolic processes and if demand is large, respiration may continue. Perhaps under conditions where ATP synthesis may be inhibited (e.g. with water stress) respiration increases and maintains cellular adenylate levels. If the demand for energy and substrates between parts of the cell becomes unbalanced then the dynamic equilibria between cell compartments provides for that most important of cellular characteristics, stability. Control of such a complex metabolic network is poorly understood and the above account is speculative.

Chlorophyll synthesis

Synthesis of chlorophyll a and b requires light in higher plants; in darkness such plants lose their pigment and become chlorotic. Chl synthesis is essential for the formation of the mature chloroplast and thylakoid system in higher plants, and enzymes of the biosynthetic pathway of chl are all contained in the chloroplast, either stroma or membrane bound. The processes are closely linked to such fundamental reactions as protein synthesis so that the structure of the mature system is correctly formed. Formation of δ-aminolaevulinic acid, a non-protein amino acid, is a major step in the process of synthesis not only of chl but also haem and other tetrapyrroles. This step is also light regulated; stimulation may occur by the protochlorophyllide reductase complex activity. Different points of entry of carbon into the tetrapyrrole synthetic pathway (5-C pathway) are known, with succinyl-CoA and glycine, 2-oxoglutaric acid or with glutamate providing the C-chain of δ-aminolevulinate. In algae and higher plants glutamate must first be activated at the α-carboxyl by ligating to tRNA^{GLU}. The reaction is catalysed by glutamate tRNA ligase which requires ATP and Mg^{2+}. The single tRNA^{GLU} gene is coded in chloroplast DNA in barley and may also be responsible for activating glutamate during protein synthesis and thus co-ordinates the two pathways in relation to the supply of glutamate and, thence, to nitrogen supply and energy (see p. 146). The ligase is a dimer of c. 51–56 kDa mass and occurs in two forms, one chloroplastic the other cytosolic; both are nuclear encoded and translated on chloroplast ribosomes, the chloroplastic form being imported. The gutamyl-tRNA^{GLU} is reduced by NADPH and a specific dehydrogenase, located in the stroma; it may be a regulatory site. The glutamate 1-semialdehyde produced is converted to the δ-aminolevulinate by a stable transferase in the stroma; the enzyme is an abundant 80 kDa protein encoded by one or two nuclear genes in barley.

Two molecules of δ-aminolevulinate are condensed by the corresponding dehydratase enzyme giving porphobilinogen. The enzyme is a hexamer of 250–300 kDa molecular mass. Four molecules of porphobilinogen are polymerized by porphobilinogen deaminase (also called urogen I synthase) giving the open ring tetrapyrrole hydroxymethylbilane. Hydroxymethylbilane is directly converted to uroporphyrinogen III by the enzyme uroporphyrinogen III synthase, in the stroma. Further steps in chlorophyll synthesis involve the conversion of uroporphyrinogen III to coproporphyrinogen III (enzyme unknown in plants but assumed to be a urogen III decarboxylase) and, thence, to protoporphyrinogen IX. This intermediate is oxidized by protogen oxidase to protoporphyrin IX which, when Mg^{2+} is inserted into the tetrapyrrole, gives Mg-protoporphyrin IX. The latter step is unique to chlorophyll synthesis and is catalysed by magnesium chelatase, a nuclear gene product, which requires Mg^{2+} and ATP.

Further steps in synthesis of chlorophyll are the addition of the propionic acid side chain to ring C of Mg-protoporphyrin IX in a reaction requiring

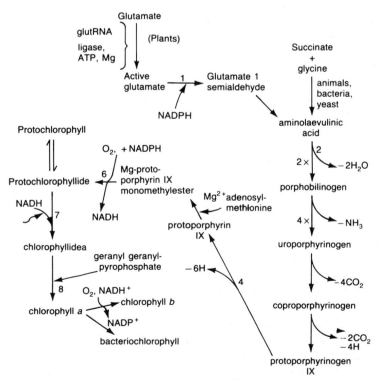

FIG. 8.5 Outline of the biosynthesis of chlorophylls and related molecules.

S-adenosylmethionine bound to the enzyme *S*-adenosylmethionine:Mg-protoporphyrin *O*-methyltransferase. Protochlorophyllide is formed when the Mg-protoporphyrin IX monomethyl ester side chain is cyclized. This requires two enzymes, a soluble and a membrane bound cyclase and requires oxygen and NADH or NADPH. A vinyl reductase reduces the vinyl group at position 8 to an ethyl using NADPH. Protochlorophyllide is converted to chlorophyllide by NADPH:protochlorophyllide oxidoreductase and requires light and NADPH. This enzyme, a single polypeptide of 36−38 kDa using light (acting as a substrate) and NADPH, is a major protein of the prolamellar body in etioplasts; there it is associated with a protochlorophyllide holochrome which converts the pigment to the chlorophyllide using NADPH and undergoes a characteristic spectral shift.

Chlorophyll *a* is produced from chlorophyllide *a* by addition of the phytol group, synthesized via the isoprenoid pathway. Mevalonic acid, a C6 compound, provides the starting point of the pathway. After decarboxylation (loss of a CO_2), dehydration and activation by pyrophosphate a C5 compound is formed (isopentenyldiphosphate or IPP) which contains a double bond. By joining these C5 units into chains the alternating double bond structures of the isoprenoids are formed, e.g. the C15 compound farnesyldiphosphate contains three double bonds and the geranylgeranyldiphosphate (GGPP) is a C20 compound with 4 double bonds. GGPP is synthesized by prenlytransferase in the plastid; it is the precursor of phytol giving the phytyl chain of chlorophyll. Chlorophyllase or chlorophyll synthase may be responsible for the addition of the chain to the tetrapyrrole moiety. Chlorophyll *b* is probably made by a light dependent enzyme which oxidizes the methyl group of ring B; chlorophyll *b* is needed for the formation of the chlorophyll *a/b* binding protein in thylakoids.

Regulation of the synthetic pathway occurs by feedback inhibition by haem and light and oxygen are also very important. Synthesis stops at protochlorophyllide which accumulates in angiosperms in darkness; teleologically it is sensible not to make a light harvesting pigment in darkness. Light stimulates amino acid incorporation into proteins, δ-aminolevulinate, etc. There is close co-ordination and pigment formation, also the tRNA^{GLU} and glutamic acid tRNA ligase are most important in co-operating protein and chlorophyll synthesis. Protochlorophyllide reductase synthesis is regulated by phytochrome and plays a central role in co-ordinating chloroplast development.

Carotenoids

Carotenoids are terpenoids and share the early part of the synthetic pathway with chlorophylls, plastoquinones and hormones such as gibberellins. They are synthesized from acetyl-CoA via mevalonic acid, geranyl pyrophosphate through to phytoene; in the process the length of the chain increases giving a structure of alternate double bonds, the isoprene unit. From phytoene, successive didehydrogenations (catalysed by a desaturase) produce lycopene, and then cyclization results in the formation of carotenoids with the α- and β-ionone rings at the ends of the molecules. Oxygenation of carotenoids produces violoxanthin. Many herbicides block the desaturation steps and thereby carotenoid synthesis; in light, treated plants cannot dissipate excess energy and die. The basic structure of the carotenoid molecule is an isoprene unit and extension of the chain produces monoterpenes, sesquiterpenes (which include farnesol with hormone properties), and redox carriers in the photosynthetic electron transport chain (e.g. plastoquinone), together with abscisic acid, another plant hormone. Diterpenes include the gibberellins, another important group of plant hormones. Steroids are triterpenes whilst carotenoids are tetraterpenes. This important family of compounds was formed early in the evolution of plants. Carotenoids probably functioned as light-harvesting pigments in early photosynthesis and later, after development of the chlorophylls, became more important in regulating the energy state of the chemical reaction centres. When oxygen accumulated in the atmosphere, carotenoids assumed a protective role. The synthesis of carotenoids in bacteria may be prevented by quite low concentrations of oxygen.

References and Further Reading

Bartley, G.E., Coomber, S.A., Bartholomew, D.M. and **Scolnitz, P.A.** (1991) Genes and enzymes for carotenoid biosynthesis, pp. 331–46 in Bogorad, L. and Vasil, I.K. (eds), *The Photosynthetic Apparatus, Molecular Biology and Operation*, Academic Press Inc., San Diego.

Buetow, D.E. (1982) Molecular biology of chloroplasts, pp. 43–88 in Govindjee (ed.), *Photosynthesis*, Vol. II, *Development, Carbon Metabolism, and Plant Productivity*, Academic Press, New York.

Douce, R. and **Joyard, J.** (1980) Plant galactolipids, pp. 321–63 in Stumpf, P.K. (ed.), *The Biochemistry of Plants*, Vol. 3, *The Plant Cell*, Academic Press, New York.

Ellis, R.J. (1984) *Chloroplast Biogenesis*, Cambridge University Press, Cambridge.

Graham, D. (1980) Effects of light on 'dark' respiration, pp. 525–79 in Davies, D.D. (ed.), *The Biochemistry of Plants*, Vol. 2, *Metabolism and Respiration*, Academic Press, New York.

Heber, U. and **Heldt, H.W.** (1981) The chloroplast envelope: structure, function and role in leaf metabolism, *A. Rev. Plant Physiol.*, **32**, 139–68.

Heldt, H.W., Flügge, U.-I., Borchert, S., Brückner, G. and **Ohnishi, J.-I.** (1990) Phosphate translocators in plastids, pp. 39–54 in Zelitch, I. (ed.), *Perspectives in Biochemical and Genetic Regulation of Photosynthesis*, Alan R. Liss, Inc., New York.

Huber, S.C. (1986) Fructose 2,6-bisphosphate as a regulatory metabolite in plants, *A. Rev. Plant Physiol.*, **37**, 233–46.

Huber, S.C., Huber, J.A. and **Hanson, K.R.** (1990) Regulation of the positioning of products of photosynthesis, pp. 85–101 in Zelitch, I. (ed.), *Perspectives in Biochemical and Genetic Regulation of Photosynthesis*, Alan R. Liss, Inc., New York.

Kacser, H. (1987) Control of metabolism, pp. 39–67 in Davies, D.D. (ed.), *The Biochemistry of Plants*, Vol. 11, *Biochemistry of Metabolism*, Academic Press Inc., San Diego.

Kacser, H. and **Porteous, J.W.** (1987) Control of metabolism: What do we have to measure? *TIBS*, **12**, 5–14.

Kannagara, C.G. (1991) Biochemistry and molecular biology of chlorophyll synthesis, pp. 301–29 in Bogorad, L. and Vasil, I.K. (eds), *The Photosynthetic Apparatus, Molecular Biology and Operation*, Academic Press Inc., San Diego.

Keys, A.J. and **Parry, M.A.J.** (1990) Ribulose bisphosphate carboxylase/oxygenase and carbonic anhydrase, pp. 1–14 in Lea, P.J. (ed.), *Methods in Plant Biochemistry*, Vol. 3, *Enzymes of Primary Metabolism*, Academic Press, London.

Keys, A.J., Bird, I.F., Cornelius, M.J., Lea, P.J., Wallsgrove, R.M. and **Miflin, B.J.** (1978) Photorespiratory nitrogen cycle, *Nature*, **275**, 741–43.

Leegood, R.C. (1990) Enzymes of the Calvin cycle, pp. 15–38 in Lea, P.J. (ed.), *Methods in Plant Biochemistry*, Vol. 3, *Enzymes of Primary Metabolism*, Academic Press, London.

Miflin, B.J. and **Lea, P.J.** (1979) Amino acid metabolism, *A. Rev. Plant Physiol.*, **28**, 299–329.

Mudd, J.B. (1980) Phospholipid biosynthesis, pp. 250–82 in Stumpf, P.K. (ed.), *The Biochemistry of Plants*, Vol. 3, *The Plant Cell*, Academic Press, New York.

Oliver, D.J. and **Kim, Y.** (1990) Biochemistry and development biology of the C-2 cycle, pp. 253–69 in Zelitch, I. (ed.), *Perspectives in Biochemical and Genetic Regulation of Photosynthesis*, Alan R. Liss, Inc., New York.

Raymond, P., Gidrol, X., Salon, C. and **Pradet, A.** (1987) Control involving adenine and pyridine nucleotides, pp. 130–75 in Davies, D.D. (ed.), *The Biochemistry of Plants*, Vol. 11, *Biochemistry of Metabolism*, Academic Press, Inc., San Diego.

Robinson, S.P. and **Walker, D.A.** (1981) Photosynthetic carbon reduction cycle, pp. 193–236 in Hatch, H.D. and Boardman, N.K. (eds), *The Biochemistry of Plants*, Vol. 8, *Photosynthesis*, Academic Press, New York.

Roughan, P.G. and **Slack, C.R.** (1982) Cellular organisation of glycerolipid metabolism, *A. Rev. Plant Physiol.*, **33**, 97–132.

Rüdiger, W. and **Schoch, S.** (1988) Chlorophylls, pp. 1–59 in Goodwin, T.W. (ed.), *Plant Pigments*, Academic Press, London.

Stitt, M., Huber, S. and **Kerr, P.** (1987) Control of photosynthetic sucrose formation, pp. 328–409 in Hatch, M.D. and Boardman, N.K. (eds), *The Biochemistry of Plants*, Vol. 10, Academic Press, London

Stitt, M., McC. Lilley, R. and **Heldt, H.W.** (1982) Adenine nucleotide levels in the cytosol, chloroplasts and mitochondria of wheat leaf protoplasts, *Plant Physiol.*, **70**, 971–77.

Thomson, W.W., Mudd, J.B. and **Gibbs, M.** (1983) *Biosynthesis and Function of Plant Lipids*, Proceedings of the 6th Annual Symposium in Botany, American Society of Plant Physiologists, Baltimore.

Tolbert, N.E. (1980) Photorespiration, pp. 488–525 in Davies, D.D. (ed.), *The Biochemistry of Plants*, Vol. 2, *Metabolism and Respiration*, Academic Press, New York.

Wiskich, T. (1980) Control of the Krebs cycle, pp. 243–78 in Davies, D.D. (ed.), *The Biochemistry of Plants*, Vol. 2, *Metabolism and Respiration*, Academic Press, New York.

CHAPTER 9

C4 photosynthesis and crassulacean acid metabolism

The PCR cycle is the only system giving a net increase in chemical energy yet found in living organisms and the mechanism and its control is very similar in all. As 3PGA, a 3-carbon compound, is the first product of the PCR cycle, this mode of carbon assimilation is called C3 photosynthesis (Fig. 9.1a). However, in many families of higher plants, additional metabolic systems have evolved for accumulating CO_2, and passing it to the PCR cycle, increasing the efficiency of photosynthesis, particularly in hot and dry or sometimes saline environments. One group (see Fig. 9.1b), which includes many tropical grasses, assimilates CO_2 into the 3-carbon precursor phosphoenol pyruvic acid (PEP) to produce 4-carbon carboxylic acids as the primary product, and hence are called C4 plants. The acids are formed in mesophyll cells and transferred to bundle sheath cells where they are decarboxylated and the CO_2 assimilated by the PCR cycle. The other major modification (Fig. 9.1c), found in many succulent plants including, but not confined to, the family Crassulaceae is called crassulacean acid metabolism (CAM). Four-carbon organic acids are formed from CO_2 and PEP as in C4 plants, but this synthesis proceeds in darkness using energy accumulated during previous illumination. The acids are decarboxylated in the subsequent light period and the CO_2 is assimilated by the PCR cycle with light energy. CAM plant leaves are not structurally differentiated into tissues with different biochemistry, but CO_2 accumulation and PCR cycle assimilation are separated in time. The changes associated with C4 and CAM photosynthesis are ecologically important as CAM is related to water conservation and C4 to use of intense light under low atmospheric CO_2 concentration. About 3000 species of plant possess C4 metabolism and some 250 exhibit CAM compared with 300 000 with C3 metabolism (putative!). C4 and CAM photosynthesis are probably polyphyletic.

C4 photosynthesis

C4 assimilation occurs in some 18 families of angiosperms (not in gymnosperms) including monocotyledons, particularly the tropical panicoid

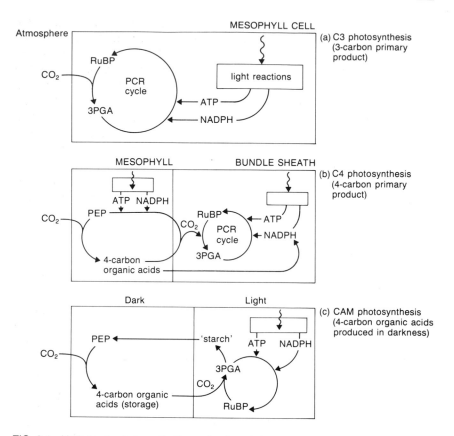

FIG. 9.1   Variation in photosynthetic mechanisms in higher plants: (a) C3 photosynthesis by the photosynthetic carbon reduction cycle (PCR), and synthesis of dicarboxylic acids in (b) C4 photosynthesis and (c) crassulacean acid metabolism related to the PCR cycle.

grasses such as sugar cane (*Saccharum officinarum*), maize (*Zea mays*) and sorghum (*Sorghum vulgare*) and in dicotyledons, for example, in the Chenopodiaceae and Compositae. Even within each genus some species may be C4 whilst others are C3 plants. In *Atriplex* (Chenopodiaceae) some species (e.g. *A. sabulosa*) have C4 and others (e.g. *A. hastata*) C3 photosynthesis. There are nine genera in four families with C3—C4 intermediates. C4 plants generally have higher rates of CO_2 assimilation than C3 plants; photosynthesis does not saturate in bright light and continues at very low concentrations of CO_2. Associated with this are rapid growth rates and higher dry matter and economic yields than C3 plants. Within the C4 plants there is considerable variation in anatomy (discussed in detail for maize in Chapter 4) but the bundle sheath, a ring of large, closely packed cells around the vascular tissue, is distinctive. Arrangement of the chloroplasts in bundle sheath cells is related to the type of C4 metabolism. Large chloroplasts with many thylakoids are arranged (Table 9.1) around the outer wall of the cells (centrifugal) in some aspartate formers (e.g. *Panicum maximum*) and malate formers (e.g. *Zea mays*). Chloroplasts are on the walls nearest to the vascular tissue (centripetal) in NAD-ME types (e.g. *Amaranthus edulis*) which produce aspartate and contain many more large mitochondria close to the vascular tissue than other C4 plants. Bundle sheath cells are larger than mesophyll cells and have a much smaller surface-to-volume ratio and dense, probably impermeable walls which may restrict diffusion of O_2 into them. The peripheral reticulum (p. 67) increases the internal membrane surface area for transport of assimilates via plasmodesmata. In some malate producers (e.g. maize) only stromal thylakoids are present. Some aspartate-producing C4 plants (e.g. *Sporobolis aeroides*) have pronounced grana. In both malate- and aspartate-forming types the mesophyll chloroplasts are granal.

C4 metabolism

There are four main processes in C4 photosynthesis (see Fig. 9.2a). First, carboxylation of PEP produces organic acids in the mesophyll cells without a net gain in energy. Second, the organic acids are transported to the bundle sheath cells. Third, they are decarboxylated there, producing CO_2 which is assimilated by Rubisco and the PCR cycle with a net gain in energy. Fourth, compounds return to the mesophyll to regenerate more PEP. The steps are common to all C4 types. However, the decarboxylation enzymes and the compounds transported between mesophyll and bundle sheath differ. These conclusions are based on the proportions of $^{14}CO_2$ in assimilate products and the time course of labelling (Fig. 9.3) and on the distribution of enzymes between cell organelles and tissue (Table 9.1). When maize photosynthesizes in $^{14}CO_2$, the primary radioactive products are oxaloacetic, malic and aspartic acids. Later, ^{14}C accumulates in PCR cycle compounds (e.g. 3PGA), suggesting flow of carbon from organic acids to the PCR cycle. Extrapolation of the ^{14}C content (measured at steady state, when the fluxes of carbon between pools are constant) back to zero time, shows that more than 95 per cent of ^{14}C enters organic acids and less than 5 per cent enters 3PGA directly. Thus, in the C4 leaf, Rubisco is, effectively, separated from the atmosphere. Rubisco is in C4 bundle sheaths and its activity and content are only one-third of that in C3 leaves (thus decreasing the demand for nitrogen) and has higher specific activity but lower affinity for CO_2 (K_m 30−60 μM compared to 13−30 μM in C3 plants). The specificity factor for C4 Rubisco is c. 70 compared to 107 for wheat. Many other enzymes, common to C3 and C4 plants have similar features but with some differences, e.g. FBPase and SBPase which are similarly light regulated by the redox system.

   Metabolic processes in mesophyll and bundle sheath have been shown by measuring the distribution of enzymes between them using several techniques, including grinding leaves to disrupt the 'softer' mesophyll

Table 9.1 Characteristics of different types of C4 plants, and examples of species with this form of metabolism, including distribution of chloroplasts and enzymes in bundle sheath and mesophyll cells, compared with C3 species

Type of photosynthesis	NADP-ME	C4 PCK	NAD-ME	C3
Examples	*Zea mays*	*Panicum maximum*	*Atriplex spongiosa*	*Triticum aestivum*
	Sorghum sudanense	*Chloris gayana*	*Portulaca oleracea*	*Glycine max*
	Saccharum officinarum	*Sporobolis fimbriatus*	*Amaranthus edulis*	*Pisum sativum*
Bundle sheath	Yes	Yes	Yes	No
Cell size (μm)	113 × 18		40 × 24	
No. chloroplasts/cell	42		39	
bs chloroplast position	Centrifugal	Centrifugal	Centripetal	
Grana	Very reduced	Yes	Yes	
Mitochondria	Few	Few?	Many	
Mesophyll				
Cell size (μm)	56 × 16		38 × 8	60 × 20
No. chloroplasts/cell	31		10	90
Grana	Yes	Yes	Yes	Yes
Major organic acid	Malate	Aspartate	Aspartate	—

Enzyme activity				
NADP malate dehydrogenase	High meso chloropl	Low meso & bs	Low meso & bs chloropl	Very low
NADP malic enzyme	High bs chloropl	Low meso & bs	Low meso & bs	Very low
Aspartate amino-transferase	Low meso chloropl	Very high cytosol	High meso & bs cytosol & mitochondria	Very low
PEP carboxykinase	Low meso & bs	High bs cytosol	Low meso & bs cytosol	Very low
NAD malic enzyme	Low	Low	High bs mitochondria	Very low
NAD malate dehydrogenase	—	Low	bs mitochondria	Very low
Alanine amino-transferase	Low	High bs & meso cytosol	High bs & meso cytosol	Very low
PEP carboxylase	High meso cytosol	High meso cytosol	High meso cytosol	Very low
Pyruvate, P_i dikinase	High meso chloropl	High meso chloropl	High meso chloropl	None
3PGA kinase	Meso & bs chloropl	High meso chloropl	meso & bs chloropl	Chloropl
Rubisco	bs chloropl	bs chloropl	bs chloropl	Chloropl
PCR cycle enzymes	bs chloropl	bs chloropl	bs chloropl	Chloropl
Light reactions	PSI, II meso PSI bs	PSI, II meso PSI, II bs	PSI, II meso PSI, II bs	PSI, II —

bs, bundle sheath; meso, mesophyll; chloropl, chloroplast

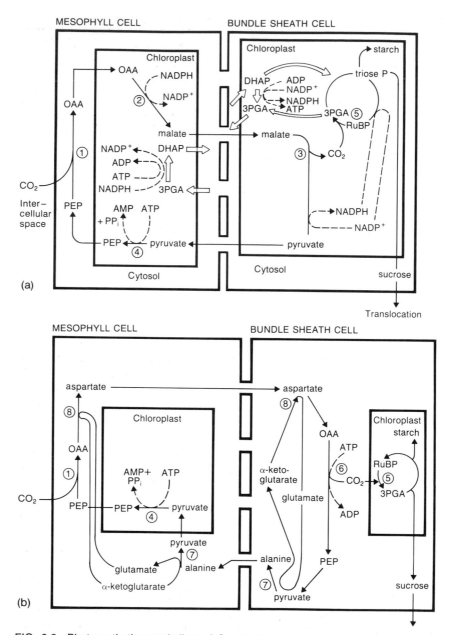

FIG. 9.2 Photosynthetic metabolism of C4 plants, compounds transferred between mesophyll and bundle sheath cells and method of decarboxylation with: (a) NADP requiring malic enzyme or 'NADP-ME' type; (b) aspartate forming and PEP carboxykinase or 'PCK' type of C4 metabolism; (c) aspartate forming and NAD requiring malic enzyme 'NAD-ME' type. Numbers refer to enzymes listed below: (1) PEP carboxylase; (2) NADP malate dehydrogenase; (3) NADP malic enzyme; (4) pyruvate, P$_i$ dikinase; (5) RuBP carboxylase/oxygenase; (6) PEP carboxykinase; (7) alanine aminotransferase; (8) aspartate aminotransferase; (9) NAD malate dehydrogenase; (10) NAD malic enzyme.

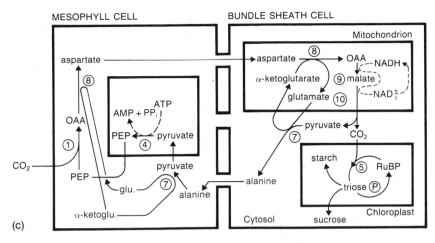

FIG. 9.2 continued.

tissues but not the more 'resistant' bundle sheath; the cells are separated by centrifugation. Also, enzymes are used to disrupt the cell walls of the tissues at different rates. The cells of mesophyll and bundle sheath are distinguished by different enzymes, chlorophyll *a/b* ratios, etc. Chloroplasts and other organelles from the cells can be obtained by normal methods of cell disruption and separation. Distributions of enzymes in tissues and cell organelles clearly show the spatial separation of the carboxylating and decarboxylation processes. PCR cycle and decarboxylating enzymes are in the bundle sheath but not in mesophyll cells, where the only PCR cycle enzymes are for reduction of 3PGA to DHAP.

In all C4 plants carboxylation occurs in the mesophyll cell cytosol, catalysed by PEP carboxylase (PEPc), which consumes bicarbonate ions in reaction with PEP, forming oxaloacetic acid (OAA):

$$\text{PEP} + \text{HCO}_3^- + \text{H}^+ \xrightarrow[\text{carboxylase}]{\text{PEP}} \text{OAA} + \text{P}_i \qquad [9.1]$$

PEP carboxylase (PEPc) is a tetramer of identical subunits of 100 kDa, requires Mg^{2+} and is activated by light ten to twenty-fold. The reaction is irreversible with a large negative free energy change (-30 kJ mol^{-1}) and is therefore a control step in C4 assimilation. It has a pH optimum of about 8 (that of the mesophyll cell cytosol) and an affinity (K_m) of about 300 μM for bicarbonate. Such concentrations are in equilibrium with the internal CO_2 concentration of C4 leaves in air. As PEPc removes HCO_3^-, CO_2 from the atmosphere is rapidly converted to HCO_3^- by carbonic anhydrase even in low CO_2 in which C4 plants can photosynthesize but C3 plants cannot. The apparent K_m for CO_2 is 2 μM and PEPc can remove bicarbonate ions from water in equilibrium with an atmosphere containing 0.5–1 Pa CO_2. PEPc has no oxygenase activity, in contrast to Rubisco, to offset CO_2 fixation. Thus, the enzyme is very efficient in assimilating CO_2, and gives very small compensation concentration, that is, when enclosed in an air-tight chamber in the light C4 plants remove almost all CO_2, in contrast to C3 plants. The enzyme has a large K_m for PEP, the other substrate, which is in considerable concentration in actively photosynthesizing mesophyll cells. The PEP reaction acts as an effective 'CO_2 pump', supplying CO_2 to the PCR cycle, and is efficient at high temperatures, to which many C4 plants are adapted. PEP is allosterically controlled by many photosynthetic assimilates (i.e. they change the rate of reaction without entering into it). If metabolism of the primary products is inhibited, e.g. by slow PCR cycle turnover, malate accumulates and PEP consumption slows thus decreasing the drain of carbon from the PCR cycle. The complex response of PEPc to metabolites probably regulates carbon fluxes in the leaf. Large concentration of glucose-6-phosphate, an important metabolite in starch synthesis (which can be made from PEP), activates PEPc, thus increasing photosynthesis and decreasing the drain to storage.

Oxaloacetate is not translocated to the bundle sheath but is first reduced to malate by NADP-dependent malate dehydrogenase in the mesophyll. Malate dehydrogenase, which maintains the reaction towards malate synthesis, has a high affinity for NADP and OAA, requires light or reduced substances to become active and is specific for NADP. Formation of malate is characteristic of maize and sugar cane, which have NADP malate dehydrogenase; malate is transported to the bundle sheath (Fig. 9.2c) and decarboxylated in the chloroplast by NADP-specific malic enzyme (controlled by Mg^{2+} and pH); hence, this is 'NADP-ME type' of C4 assimilation.

$$\text{malate} + NADP^+ \xrightarrow[\text{enzyme}]{\text{NADP malic}} \text{pyruvate} + CO_2 + NADPH \qquad [9.2]$$

NADP malic enzyme is a polymeric protein of 60 kDa subunits with very complex allosteric regulation with acetyl-CoA, FBP and Mn^{2+} ions.

Pyruvate from decarboxylation of malate is recycled to form PEP, by a reaction with ATP and P_i; the enzyme, pyruvate, P_i dikinase, and the reaction are unique to C4 plants. The enzyme, which is phosphorylated before reacting with pyruvate, is light activated and requires Mg^{2+} and alkaline pH. The enzyme (mass 387 kDa) is a tetramer and undergoes a two-step reaction. An interesting feature is its regulation by a light modulated protein which catalyses removal of P from a threonine residue on the dikinase during activation. The enzyme has a pH optimum of 8, a high specificity and affinity for substrates, e.g. 0.8 mM for OAA and it is inhibited by products of the reaction, thus regulating the activity to the supply of CO_2 or ATP and keeping the PEP cycle and PCR cycle reactions balanced:

$$\text{pyruvate} + ATP + P_i \xrightarrow[\text{dikinase}]{\text{pyruvate, } P_i} PEP + AMP + PP_i \qquad [9.3]$$

Although the PEPc- and pyruvate, P_i dikinase-mediated reactions are common to all C4 species, there are differences in the metabolic routes by which pyruvate is formed, both in the enzymes and their location. A major difference is in the transport of aspartate rather than malate from mesophyll to bundle sheath. Aspartate is rapidly formed in the mesophyll cells by transamination of oxaloacetate by aspartate aminotransferase, with glutamate as the amino group donor.

$$\text{OAA} + \text{glutamate} \xrightarrow[\text{aminotransferase}]{\text{aspartate}} \text{aspartate} + \alpha\text{-ketoglutarate} \qquad [\textbf{9.4}]$$

The enzyme has a pH optimum of 8, a high affinity for substrates, and is inhibited by malate. In bundle sheath cells it deaminates aspartate. Aspartate is transported to the bundle sheath by a dicarboxylate shuttle (see p. 144).

There is distinction between C4 plants in the enzyme decarboxylating organic acids in the bundle sheath (Table 9.1). In malate formers, as already mentioned, decarboxylation is by NADP-requiring malic enzyme in the chloroplast of bundle sheath cells (NADP-ME type). In one group of aspartate formers, called the PCK type because of the role of PEP carboxykinase, aspartate is probably deaminated in the bundle sheath cell cytosol by aspartate aminotransferase, giving OAA. This is phosphorylated by PEP carboxykinase to PEP, releasing CO_2 which is fixed in the PCR cycle (Fig. 9.2b) and PEP is converted to pyruvate, which is transaminated to alanine (by a specific enzyme with glutamate as amino group donor). Alanine returns to the mesophyll cell where it is deaminated to pyruvate and phosphorylated to PEP.

The 'NAD malic enzyme (or NAD-ME) type' converts aspartate to oxaloacetate which is reduced to malate by the bundle sheath cell mitochondria and decarboxylated by an NAD-specific malic enzyme.

$$\text{malate} + \text{NAD}^+ \xrightarrow[\text{enzyme}]{\text{NAD malic}} \text{pyruvate} + CO_2 + \text{NADH} \qquad [\textbf{9.5}]$$

This contrasts with the NADP-ME type located in the chloroplast. Pyruvate is converted to alanine by the specific aminotransferase and returned to the mesophyll cell cytosol for conversion to the acceptor.

Energetics of C4 photosynthesis

Compared with C3 plants, C4 plants require an extra 2 or 3 molecules of ATP for fixation of 1 CO_2, viz. 5 or 6 ATP and 2 NADPH because of the conversion of pyruvate to PEP. The NADP-ME type (Table 9.2) requires 2 ATP and 1 NADPH in the mesophyll cells whereas the NAD-ME and PCK type require only 2 ATP and PCK types use 1 ATP in the bundle sheath. Photochemical competence differs between forms of C4 plants (Table 9.1). Bundle sheath chloroplasts of NADP-ME types have little PSII and may synthesize ATP by cyclic photophosphorylation. This greatly decreases

Table 9.2 Energy requirements of mesophyll (meso) and bundle sheath (bs) cells for ATP and NADPH per CO_2 fixed in C4 species. C4, is requirement in that part of pathway; PCR cycle, requirement in photosynthetic reduction cycle

	NADP-ME		NAD-ME		PCK	
	meso	bs	meso	bs	meso	bs
ATP						
C4	2	0	2	0	2	1
PCR cycle	1	2	0.5	2.5	1	2
Total	3	2	2.5	2.5	3	3
NADPH						
C4	1	0	0	0	0	0
PCR cycle	1	0	0.5	1.5	1	1
Total	2	0	0.5	1.5	1	1

oxygen accumulation in bundle sheaths. NADPH is supplied in NADP-ME types from the mesophyll by a malate shuttle; for each malate decarboxylated 1 NADPH is released. If NADPH supply is insufficient for the PCR cycle, reductant might be provided by a shuttle of DHAP from the mesophyll chloroplast into the bundle sheath where it is oxidized to 3PGA giving NADPH and ATP.

$$DHAP + ADP + NADP^+ \rightarrow 3PGA + ATP + NADPH \qquad [9.6]$$

The 3PGA from the bundle sheath is recycled back to the mesophyll to provide the substrate.

Bundle sheath cells of NAD-ME types have some non-cyclic electron flow to $NADP^+$, those of PCK types have less PSII activity and more cyclic photophosphorylation. Mesophyll chloroplasts of NAD-ME type have larger capacity for cyclic ATP synthesis than the NADP-ME type, which like the PCK type have both PSI and II. Aspartate formers probably have a normal balance of PSI and II activity in the bundle sheath chloroplast and appear not to require transport of reductant.

The advantages of the modifications to the photochemical apparatus and of the complex shuttles of reductant and ATP are not clear. C4 plants are able to absorb incident radiation at high intensity and use it for CO_2 assimilation but they are less efficient at low intensity. Light harvesting and distribution to the different photosystems may be more efficient if the functions are separated. ATP production by cyclic photophosphorylation (minimizing the production of O_2 and reductant) may overcome a limitation to the rate of PCR cycle activity, necessary for the efficient CO_2 accumulating system of PEP carboxylase to be fully exploited. Decreased PSII in bundle sheath chloroplasts decreases O_2 production, so that Rubisco, which has somewhat higher K_m for CO_2 in C4 compared to C3 plants, functions at low O_2 and high CO_2. PEPc is very efficient at assimilating CO_2 at very dilute concentrations allowing large CO_2 accumulation in the sheath. When organic acids from ten mesophyll cells supply one bundle sheath cell, decarboxylation gives a ten-fold increase in CO_2 concentration. In air this may give a partial pressure of 150–300 Pa of CO_2. As bundle sheath cells are impermeable and O_2 production inside them (in the NADP-ME type at least) is small, the CO_2/O_2 ratio is large. Thus, C4 photosynthesis is insensitive to O_2 and photorespiration is very small with smaller light and CO_2 compensation concentration than C3 plants. If C4 and C3 plants (e.g. maize and bean) are illuminated together in the chamber, the C4 plants remove CO_2 from the atmosphere more efficiently than the C3 species, which deplete their carbohydrate reserves and eventually die. Thus, efficient photosynthesis and C4 metabolism which eliminates or minimizes photorespiration, are linked. However, some phosphoglycolate is formed and there is carbon flux through the glycolate pathway. Under conditions such as water stress, where CO_2 supply is limited by stomatal closure, an increase in glycine and serine formation indicates that more carbon enters the glycolate pathway but photo-respiration, measured as CO_2 production and from the difference in assimilation between low (1 kPa) and high (21 kPa) O_2, does not increase. As PEPc is efficient, any CO_2 produced is reassimilated and only under extreme conditions does CO_2 escape from maize leaves in the light. C4 plants have a greater photosynthetic rate than C3 plants at smaller stomatal conductance, so that they have a larger water use efficiency (WUE = photosynthesis/transpiration) and smaller total water loss in drought-prone environments. This slows the onset of water stress and may decrease its severity under conditions of intermittent rainfall.

What is the biological rationale of C4 metabolism? It requires considerable movement of assimilates across cell membranes, imposing an additional energy burden, but probably transport between cells is by simple diffusion. C4 photosynthesis is superior to C3 because it allows efficient CO_2 assimilation, even in dilute CO_2. Stomatal conductance is smaller in C4 than C3 plants, helping to conserve water, but the resulting low internal CO_2 concentration (see Chs 11 and 12) has little effect on photosynthesis as PEPc is so efficient at gathering CO_2. Differences in photosystems, chloroplast structure, etc. in C4 types probably relate to the requirements for balanced energy supply and reductant and the importance of the CO_2/O_2 ratio in their different environments. Most C4 plants are produc-tive under bright sunlight in hot, often dry but not desert, conditions. However, C4 species are also found in temperate and in shaded conditions. The C4 temperate grass *Spartina townsendii* is found in salt marshes, where C4 metabolism may be an advantage under salt and osmotic stress.

The efficiency of C4 species has been achieved by extensive modification of the 'normal' C3 plant. Anatomy is differentiated at subcellular and cellular level and the enzyme complement is modified quantitatively (e.g. more PEP carboxylase), and qualitatively, with a unique enzyme, PEP carboxykinase. Chloroplast structure and light reactions are modified to optimize ATP and NADPH supply. Attempts have been made to transfer C4 features by crossing C3 and C4 species of *Atriplex* (e.g. *A. triangularis* (C3) and *A. rosea* (C4)). Crossing C3 and C4 species of *Atriplex* gives fertile F_1 hybrids, which are intermediate morphologically and anatomically, with a weakly developed bundle sheath. The enzymes are C4, with PEP carboxylase and Rubisco, but the enzymes are not compartmented as in the C4 parent, for example, Rubisco is in all cells and not restricted to the bundle sheath. Also, C4 acids are synthesized but not effectively metabolized. This leads to inefficient photosynthesis and growth. Plants intermediate between C3 and C4 metabolism, with differing degrees of resemblance to the main types are represented in many families. The genera *Panicum* and *Moricandia* have intermediates very C3-like, but with reduced photorespiration and a 'Kranz-type' anatomy. Species of *Flaveria* have very diverse features culminating in the full C4 syndrome. Genetic processes responsible for the development of C4 characteristics and the role of selection pressure from the environment are not understood, but provide valuable material for understanding evolution of photosynthetic characteristics. The genome of C3 predecessors has been, presumably, modified as a result of a long evolutionary process by selection under environmental pressure. If C4 photosynthesis has arisen more than once (as the different metabolic types suggest) then present day C3 species may include individuals with more C4-like characteristics, which might be more productive agricultural crops in hot, dry or saline conditions. Selection has been attempted but no C4 or high efficiency plants have been obtained.

Crassulacean acid metabolism

CAM plants assimilate CO_2 and synthesize organic acids, mainly malate, during darkness but assimilate little CO_2 in the light (Fig. 9.3). In darkness stomata open, offering a low resistance to CO_2 diffusion into leaves. Carboxylation by PEPc, in the β position of PEP, forms oxaloacetate, probably in the cytosol (Fig. 9.4). Oxaloacetate is reduced by NADH to malate by cytoplasmic NAD^+ malate dehydrogenase. Malic acid is transported as hydrogen malate across the tonoplast and accumulates (150 μ equivalents per gram fresh weight) in the vacuole; there may be other pools in leaf cells. The pH of the vacuole becomes very acidic although metabolism is protected by compartmentation. PEP is derived from storage carbohydrate, probably glucans (starch and a dextran, a glucose polymer), mobilized to produce 3PGA and PEP; ATP is consumed and synthesized in these reactions so that the energy consumption is balanced. NADH from oxidation of glyceraldehyde phosphate reduces oxaloacetate; as with ATP, storage carbohydrate supplies the energy for CO_2 fixation in the dark.

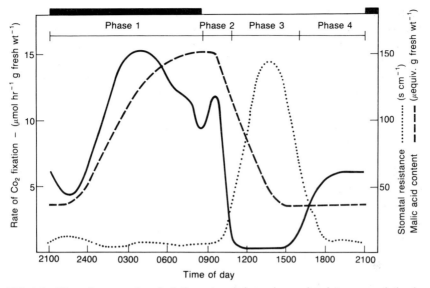

FIG. 9.3 Diurnal course of assimilation, stomatal opening and malate accumulation in a CAM plant.

Early in the dark period malate accumulation and CO_2 fixation are slow, but increase in the middle of the dark period before decreasing. This pattern results from stimulation of CO_2 fixation by a higher PEP concentration which overcomes the inhibition of PEPc by malate. Phosphofructokinase which converts fructose-6-phosphate to fructose bisphosphate is a hundred-fold less sensitive to PEP than that of C3 plants. Control of processes in CAM has been reviewed by Osmond (1978) and Winter (1985).

With illumination of CAM leaves, CO_2 fixation from the atmosphere decreases rapidly to a very low rate and stomata close. Malate is transported out of the storage compartment and is decarboxylated by NADP malic enzyme, and perhaps NAD malic enzyme, which is found in many families with CAM, or by PEP carboxykinase which is more important in others (e.g. Bromeliaceae, Liliaceae). They are designated NADP malic enzyme and PEP carboxykinase types, respectively. The CO_2 released, probably in the cytosol, enters the chloroplast and is assimilated by the PCR cycle with ATP and NADPH from electron transport. Pressure of CO_2 in the tissue may reach 1000 Pa and oxygen rises from 21 to 26 kPa but the RuBP carboxylase reaction is favoured and photorespiration minimized. Malate decarboxylation is controlled by a low $NADP^+/NADPH$ ratio, as $NADP^+$ is the acceptor for reductant. Decarboxylation could occur with $NADP^+$ in the dark but ATP is required for decarboxylation and is available (in the required amount) only in light. Pyruvate is phosphorylated to PEP and recycled to triosephosphate and thence to carbohydrates (mainly glucan)

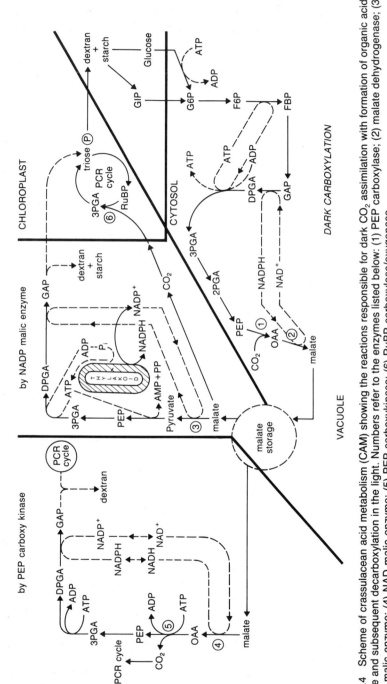

FIG. 9.4 Scheme of crassulacean acid metabolism (CAM) showing the reactions responsible for dark CO₂ assimilation with formation of organic acid, storage and subsequent decarboxylation in the light. Numbers refer to the enzymes listed below: (1) PEP carboxylase; (2) malate dehydrogenase; (3) NADP malic enzyme; (4) NAD malic enzyme; (5) PEP carboxykinase; (6) RuBP carboxylase/oxygenase.

which are stored. PEP carboxykinase CAM plants generate oxaloacetate from malate and NADH by malate dehydrogenase in the cytosol (or possibly chloroplast).

Early in the light period there is little CO_2 fixation because PEP carboxylase is inhibited by malate, but as this is decarboxylated the PEPc fixes more CO_2 in the light. After long periods in the light and extensive decarboxylation, CO_2 from the atmosphere is assimilated directly by Rubisco and the PCR cycle, giving rise to storage glucans. Photorespiration occurs in CAM plants as a result of a low CO_2/O_2 ratio, when decarboxylation has depleted the storage materials.

CAM permits CO_2 accumulation in darkness when stomata are open, without substantial loss of water, for the plants are predominantly of arid habitats with hot, dry and very bright days, which would cause much water loss if stomata were open. A very high ratio of CO_2 fixation to water loss is achieved under conditions lethal to most C3 plants. Energy accumulated in the light is used to generate the substrate for the PEPc reaction at night when low temperatures (characteristic of desert and alpine environments) favour malate accumulation. Large vacuoles enable CAM plants to store both malate and water, obviously important as a buffer against desiccation.

References and Further Reading

Apel, P. (1988) Some aspects of the evolution of C4 photosynthesis, *Kulturplanze*, **36**, 225–36.

Cockburn, W. (1985) Variation in photosynthetic acid metabolism in vascular plants: CAM and related phenomena, *New Phytol.*, **101**, 3–24.

Edwards, G.E. and **Walker, D.** (1983) *C3, C4: Mechanisms, and Cellular and Environmental Regulation of Photosynthesis*, Blackwell Scientific, Oxford/University of California Press, Berkeley.

Lüttge, U. (1987) Carbon dioxide and water demand: Crassulacean acid metabolism (CAM), a versatile ecological adaptation exemplifying the need for integration in ecophysiological work, *New Phytol.*, **106**, 593–629.

Osmond, C.B. (1978) Crassulacean acid metabolism: A curiosity in context, *A. Rev. Plant Physiol.*, **29**, 379–414.

Osmond, C.B. and **Holturn, J.A.M.** (1981) Crassulacean acid metabolism, pp. 283–328 in Hatch, M.D. and Boardman, N.K. (eds), *The Biochemistry of Plants*, Vol. 8, *Photosynthesis*, Academic Press, New York.

Osmond, C.B., Adams, W.W.III and **Smith, S.D.** (1989) Crassulacean acid metabolism, pp. 255–80 in Pearcy, R.W., Ehleringer, J., Mooney, H.A. and Rundel, P.W. (eds), *Plant Physiological Ecology, Field Methods and Instrumentation*, Chapman and Hall, London.

Ting, I.P. and **Gibbs, M.** (eds) (1982) *Crassulacean Acid Metabolism*, American Society of Plant Physiology, Waverly Press, Baltimore, Maryland.

Ting, I.P. (1985) Crassulacean acid metabolism, *A. Rev. Plant Physiol.*, **36**, 595–622.

Winter, K. (1985) Crassulacean acid metabolism, pp. 329–87 in Barber, J. and Baker, N.R. (eds), *Photosynthetic Mechanisms and the Environment*, Elsevier, Amsterdam.

CHAPTER *10*

Molecular biology of the photosynthetic system

The photosynthetic apparatus has been considered at different levels of structure and function which result from expression of genetic information within the organisms: this is a central tenet of biology applicable to plants (Feifelder 1983). It is a considerable intellectual challenge and of great potential practical importance to understand how the structure of the functional photosynthetic system is achieved. How is the genetic information organized? What does it code for, how is it transcribed and translated into the components of the system? How are the components assembled into the final structures and what are the mechanisms? In a mature photosynthetic system how is regulation achieved by genetically determined mechanisms (Vierstra 1989)? Renewal of short-lived or damaged components and regeneration and senescence processes are regulated in photosynthetic tissues; how do genetic and metabolic processes interact to achieve correct function? Such genetic—molecular—physiological questions are being asked and answered.

Plants have unique features of genetic organization, with DNA in the chloroplast as well as the nucleus and mitochondria; aspects are considered in this chapter which provides a brief guide to the molecular biology of photosynthesis and of relevant parts of the plant's genome. Detailed discussion of many of the topics touched upon here is contained in the references (Ellis 1984; Steinback *et al.* 1985; Grierson 1991). Whilst concentrating on the photosynthetic system, it should not be forgotten that photosynthetic organisms depend on the regulated expression of other, non-photosynthetic features of their genome which determine their ability to function in a complex environment.

The nucleotide base sequences of DNA in the genome contain the information required for synthesis of proteins with structural, catalytic and transport functions. These are needed to make the complex structures of photosynthetic organisms, e.g. light-harvesting complexes, the stromal enzyme Rubisco and the phosphate translocator in chloroplasts. Enzymatic proteins synthesize other components of the photosynthetic system (e.g. lipids and energy-gathering pigments of thylakoid membranes). For active transcription and translation of genetic information into proteins, the components of the protein synthesizing machinery (e.g. ribosomal proteins, rRNA) are required.

This chapter considers the higher plant chloroplast and nuclear genomes, and the links between them, and discusses how this genetic information leads to the photosynthetic system. The effects of environment, particularly light, on the mechanisms and regulatory processes in gene expression required for development of the photosynthetic system are considered.

Cellular organization and genetic information

Marked differences exist between photosynthetic prokaryotes and eukaryotes not only in structure and metabolism but in the way their genetic material is organized. The genome of prokaryotes is not contained within a nucleus, neither is the photosynthetic apparatus separated from the body of the cell, as it is in eukaryotes with their membrane-bound nuclei, chloroplasts and mitochondria which all contain DNA. Nuclear, chloroplastic and mitochondrial DNAs are referred to as nDNA, cpDNA and mtDNA, respectively (Dyer 1984). The eukaryotic nuclear genome is separated from the protein synthesizing machinery in the cytosol by membranes but the chloroplast, as in the prokaryotes, contains the necessary components for producing proteins. In eukaryotes there is complex interaction between the organellar genomes and their products. Nuclear genes for proteins found exclusively in chloroplasts, for example, are transcribed in the nucleus and the mRNA species produced are exported to the cytosol and translated there; the proteins are transported into the chloroplast. Transport of chloroplast-coded proteins from the plastid into the cytosol or nucleus has yet to be demonstrated.

Chloroplast genomes and the proteins encoded by the genes, together with development of chloroplasts in photosynthetic eukaryotes will be considered in some detail, but the nuclear genome is only briefly discussed in relation to photosynthesis, although it provides most of the information required for the photosynthetic system and the plant's development. The mitochondrial contribution is ignored. Here, the discussion is restricted, for reasons of space and coherence, largely to photosynthetic components.

Genetics of nuclear and chloroplast information

Many of the characteristics of higher plants are transferred between generations with segregation of nuclear genes and their combination according to Mendel's Laws of Inheritance. However, certain features of higher plant inheritance such as pigmentation associated with leaves, are non-Mendelian (see Tilney-Bassett 1984), derived from the maternal germ-line and cytoplasmically transmitted. They are maternally inherited. Part of the plant's genome is chloroplastic and this DNA is transferred from generation to generation via the plastids of the maternal parent. Plastids may not be transferred through the pollen tube to the zygote or maternal cytoplasm in the egg cell, preventing expression of paternal traits. Inheritance of plastid characteristics may also be biparental. Information from both parents is incorporated through the plastids, although to different degrees, depending on the species of plant. In zonal *Pelargonium* (*Pelargonium* × *Hortorum*) biparental control of chloroplast pigmentation (chlorophyll content, green or white plastids) is under nuclear control via regulation of plastid DNA replication. In approximately two-thirds of some 60 higher plant genera examined some components of the photosynthetic system are maternally inherited; the remaining third have biparental inheritance. There is no simple taxonomic basis for this. Regulation of the mode of inheritance appears to change easily with only a few genes involved (Tilney-Bassett 1984). Different forms of inheritance of specific chloroplast components have now been shown, even within an enzyme complex. For example, the 1–4 different forms of the small subunit (SSU) protein of ribulose bisphosphate carboxylase-oxygenase (Rubisco, see Ch. 7) are nuclear encoded and inherited in a Mendelian fashion. However, the 3 forms of the chloroplast encoded large subunit (LSU) protein of the same enzyme are maternally inherited in *Nicotiana gossei* × *N. tabacum* crosses. This was demonstrated by using differences in behaviour of the polypeptides under isoelectric focusing (see Tilney-Bassett (1984) for detailed discussion and references).

Organization of the chloroplast genome in higher plants

Genetic continuity between generations via the maternal germ line and the plastids was eventually associated with the physical location of DNA in organelles. Dyer (1984) discusses the chloroplast genome and its organization. Chloroplasts were shown, by staining DNA with Feulgen and other indicator dyes and by optical and electron microscopy, to contain DNA. Also, areas of low electron density in chloroplasts were found to be associated with DNA. Isolation of cpDNA from plastids, free of nuclear DNA contamination, was eventually achieved. The large, coiled cpDNA molecule is difficult to isolate; strands sheer during isolation and nucleases cut the molecules.

Information about the chloroplast genome, the molecular biology of photosynthesis and understanding of the way in which the nDNA and cpDNA contribute to the formation of structures in the photosynthetic system has grown dramatically following identification of the separate forms of DNA in 1962 (Palmer 1985). Amounts and characteristics of cpDNA have been analysed in many organisms, including photosynthetic bacteria and algae. Complete genome analysis of cpDNA from a liverwort (*Marchantia polymorpha*) and tobacco (*Nicotiana tabacum*) (Shinozaki *et al.* 1989; Sugiura 1989) has been achieved, together with partial analysis of others, e.g. maize (*Zea mays*), rice (*Oryza sativa*) and some legumes. The

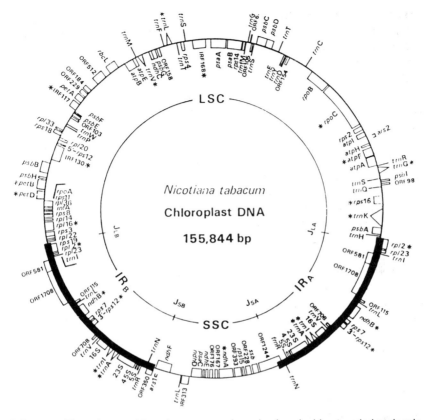

FIG. 10.1 The tobacco chloroplast genome is a circular, double stranded molecule. Genes are indicated, e.g. *rbc*L for the large subunit of Rubisco, *atp* for gene encoding proteins of the coupling factor complex. Genes on the outside of the circle are on the A strand and are transcribed anticlockwise. Genes inside the circle are encoded on the B strand and transcribed clockwise. The inverted repeats (IR$_A$ and IR$_B$) are shown as heavy lines. The position of the small single copy (SSC) region and large single copy (LSC) region is indicated. Some of the genes shown are identified in Table 10.1 and discussed in the text (after Shinozaki *et al.* 1989).

structure of cpDNA and details of specific proteins for which it encodes are now well understood. Despite the chloroplasts ability to make some of its own components, its competence is very limited; no thylakoid complex is made entirely of proteins encoded in cpDNA. Protein products of nDNA are always involved. The large subunit of Rubisco is the only part of any photosynthetic carbon reduction cycle enzyme encoded by cpDNA (Ch 7, p. 128). In addition, protein synthesis in the chloroplast involves nDNA as well as cpDNA products. Chloroplasts depend upon the nucleus for a major part of their protein complement, which is of the order of 400 different proteins (both structural and functional). In theory, with an average genome of 150 kbp, each cpDNA could code for 120–130 products (only 20–30 per cent of the total complement). Some 100 polypeptides are made by isolated chloroplasts with 80 stromal, 20 thylakoid and a few envelope components; the remaining polypeptides derive from genes encoded by nuclear DNA. Cytosolic proteins, encoded by nDNA, are integrated into the chloroplast to provide the effective photosynthetic structures. Components encoded by the cpDNA are synthesized in the chloroplast. Methods by which proteins are transported between different parts of the cell have been described although the detailed mechanisms remain to be analysed (Keegstra *et al.* 1989).

Following early success in isolating chloroplasts which could carry out photosynthetic carbon metabolism (p. 58), the concept of chloroplast autonomy prevailed and attempts were made to maintain chloroplasts *in vitro* by supplying only simple 'nutrients', e.g. inorganic phosphate and CO_2. However, higher plant chloroplasts are not autonomous. A constant supply of many structural components is needed for their function. They are intimately integrated developmentally as well as metabolically into the life of the cell, despite their complexity and separation from the cytosol. Chloroplasts are widely regarded as derived from an oxygen evolving, cyanobacterial form of prokaryote which entered into symbiosis with an early eukaryotic cell. Integration of the two organisms into a very tightly regulated system has, presumably, been achieved by the evolution of complex biochemical control mechanisms and appropriate genetics (Gray 1991).

There are many reasons why chloroplasts are regarded as prokaryotic organelles whereas the nucleus and cytosol are eukaryotic (see Gray 1991). There are, for example, differences at the level of transcription of DNA into mRNA by RNA polymerase, which binds to a region of DNA called a promoter — a sequence of bases recognized by the α-subunit of polymerase. In prokaryotes there are several promoter sequences known, e.g. the Pribnow 'box' sequence of bases, variations on TATAAT, acting as transcription or regulatory sequences. Promoter regions of eukaryotic genes have the TATA or Hogness box and enchancer elements, sequences which regulate the intensity of gene expression. In eukaryotes a number of specific transcription factors are needed plus a larger range of RNA polymerases; the primary mRNA transcript is usually much longer than the mRNA on which the protein chain is made and the intervening non-coding base sequences ('introns') are excised and the remaining coding exons are rejoined or spliced by complex enzymatic systems before the protein is made. RNA polymerases of bacteria and chloroplasts are homologous, suggesting common early ancestry.

In prokaryotes, genes of related functions are often adjacent and polycistronic transcription is more common than monocistronic, enabling the proteins of a metabolic sequence of enzymes to be made in one event from the 'reading frame' of the RNA. Eukaryotic genes are often monocistronic and in different parts of the genome. Their transcription therefore requires more complex regulation with a single environmental or metabolic stimulus acting as a co-ordinating signal.

Translation of mRNA differs in prokaryotes and eukaryotes. In the former, chain initiation requires *N*-formylmethionine transfer ribonucleic acid, tRNA^{fmet}, but the latter use methionine tRNA, tRNA^{met}. Differences are observed in the ribosomes of prokaryotes and eukaryotes, which are 70S (2.5×10^6 molecular mass) and 80S (4.3×10^6), respectively, made up of different sizes of large and small subunits, numbers and types of proteins and rRNA molecules. Ribosomes of prokaryotes are of 50S and 30S subunits whereas those of eukaryotes are composed of 40S and 60S subunits. Ribosomal functions are also affected differently in the prokaryotes and eukaryotes by antibiotic inhibitors of protein synthesis, e.g. prokaryotic ribosomes are inhibited by chloramphenicol but not by the cytoplasmic ribosomal inhibitor cycloheximide. Many aspects of chloroplast gene expression are prokaryotic.

The chloroplast genome

Chloroplasts are partially autonomous genetically containing DNA which replicates independently. They are highly polyploid organelles containing many identical copies of cpDNA per chloroplast (in wheat up to 300). Molecules of cpDNA are often aggregated and linked to the thylakoid membranes forming 'nucleoids' resembling the organization of the bacterial genome. Most mature chloroplasts have 10–20 nucleoids with 2–24 cpDNA molecules in each. During cell and plastid division nucleoids are transferred with thylakoids more or less equally into daughter chloroplasts. Location of cpDNA molcules in chloroplasts is species dependent, e.g. centrally in dicotyledons and peripherally in monocotyledons (Dyer 1984). As leaves expand (as a consequence of cell growth) cpDNA is synthesized faster than the chloroplasts divide and the number of genomes increases per chloroplast and may rise to 12 000 copies per cell. Later, chloroplasts divide without further cpDNA synthesis, decreasing the cpDNA content per chloroplast. In a cell with, say, 20 chloroplasts, each with 20 nucleoids containing 20 cpDNA genomes a total of 8000 copies of each single gene exist. This contrasts with a single nucleus per cell and relatively few copies of nuclear genes even in polyploid plants.

The cpDNA is a double stranded, single covalently closed circular DNA molecule (Fig. 10.1), of density 1.679–1.699, with a molecular mass of 82–96 million Dalton (MDa) and a circumference of 40–50 μm depending on the species of plant (Shinozaki *et al.* 1989; Sugiura 1989). It is tightly coiled in a spiral. There is a large proportion of adenine and thiamine in cpDNA (61–64 per cent), suggesting a high content of non-coding nucleotide base sequences. DNA of chloroplasts is not associated with histones, as is DNA in nuclei.

Differences in size between the largest and smallest cpDNAs are considerable; in algae from 81 to 275 kbp and, in the angiosperms, 111 to 182 kbp in monocotyledonous plants and 117 to 165 kbp in dicotyledons. This is about ten times greater than the mitochondrial genome of mammals, twice that of yeast but only a twenty-fifth of the *E. coli* genome and one-thousandth the size of the very small nuclear genome, of the angiosperm, *Arabidopsis thaliana*.

Chloroplast DNA gene maps

Gene maps for cpDNA have been constructed by using restriction endonucleases to cut the molecule at particular sites followed by restriction fragment length polymorphism analysis. These maps are complete for *Marchantia* (Ohyama *et al.* 1989) and tobacco (Shinozaki *et al.* 1989) or partial (e.g. *Vicia* bean). Structures of the higher plant cpDNA molecules differ in details but they are very similar, with considerable homology in the position and location of the coding sequences and their products (Sugiura 1989).

The tobacco genome

A gene map for *Nicotiana tabacum* is shown in Fig. 10.1. Shinozaki and co-workers (Shinozaki *et al.* 1989) sequenced the entire cpDNA by cloning a set of overlapping restriction fragments of cpDNA which is 155 844 bp in size. One strand of the circular molecule is called the A strand and is transcribed anticlockwise. The gene for the large subunit of Rubisco, *rbcL*, is on the A strand. The complementary strand (B) is transcribed clockwise. This cpDNA shows a characteristic feature of most higher plant cpDNAs examined to date (legumes are the exception) viz. the presence of two large inverted repeat (IR) sequences of 25 339 base pairs, separated by a large and a small single copy region (LSC and SSC, respectively) where the genes are only present once. In *N. tabacum* the LSC and SSC are 86 684 and 18 482 base pairs long, respectively. About 120 gene products have been identified including RNA polymerases and other necessary enzymes for components of protein synthesis (Table 10.1). They include all rRNAs and tRNAs (4 and 30 products, respectively in tobacco), 23 ribosomal proteins (approximately 60— the remainder are from cytosolic synthesis) and about 20 (plus *c*. 8 putative) proteins. Of the 53 proteins, 9 are predicted but not yet identified in the chloroplast. Most of the membrane protein complexes have components coded in the cpDNA, e.g. coupling factor, PSI and PSII complexes and cytochrome b_6-f complex but the LHCP is entirely nuclear encoded. Surprisingly only one soluble stromal polypeptide of the photosynthetic carbon reduction cycle (the LSU of Rubisco) is coded in cpDNA, all others are in nDNA.

Chloroplast genes encoding RNAs and proteins

Chloroplasts contain the means of expressing the information in cpDNA. There are genes for 4 different rRNAs in the IRs, i.e. two copies of each and they resemble genes in *Anacystis*, a cyanobacterium. The 4.5S rRNA is very similar to the 3′ end of the prokaryotic 23S rRNA. The plant 70S ribosomes contain about 60 ribosomal proteins, of which 20 have been identified in the tobacco genome (plus 3 repeats) some of which contain introns. Introns do occur in cpDNA but the number and size is variable (more than 100 of 400–1000 base pairs in size), and they seem to change quickly and are not conserved even where the evolutionary origin of the plants is similar. Introns occur in the ancestors of *Marchantia* and the angiosperms, e.g. two in *trn*I and *trn*A were acquired 450–500 million years ago in the Chlorophyceae, the common ancestor, and suggests strong conservation of cpDNA. Introns of 0.5–2.5 kbp occur in 6 different tRNA genes (greater than in tRNAs of any other organisms; Maréchal-Drouard *et al.*. (1991)) and in 9 protein genes (including *atp*F, *ndh*A, *pet*B, *pet*D) of tobacco. Introns may have regulatory functions in gene expression. When the introns are excised from the message, the coding sequences or exons are spliced together by ligases to give the functional mRNA transcript from which protein is made, but splicing in chloroplasts has not been demonstrated, perhaps because of the need for particular conditions or proteins.

Some 20 *rps*2 genes for 30S ribosomal proteins, 3 in IRs, making a total of 23 (4 with introns), 8 *rpl* genes for 50S ribosomal proteins and probably components of RNA polymerase (the *rpo*A, B and C genes for the α, β and β^1 subunits) plus initiation factor 1 (*inf*A) and DNA binding protein (*ssb*) have been identified in chloroplasts. Ribosomes of the chloroplast are sensitive to chloramphenicol and erythromycin and thus of prokaryotic type, which together with clear homology with bacteria genes supports the prokaryotic ancestry of chloroplasts. An rRNA operon exists in all IRs examined in higher plants and cyanobacteria (probably the symbiotic ancestor of the chloroplast) suggesting its retention from the earliest symbiotic phase. Some 30 tRNA species occur as part of the protein synthesizing machinery, with 7 repeated, a total of 37, sufficient to code for all tRNAs functions. Ribosome binding sites on mRNA of some chloroplast genes have initiator sequences of the Shine–Dalgarno type.

Proteins encoded by cpDNA

Some of the major proteins coded in the chloroplast genome are listed in Table 10.1. Aspects of their coding, synthesis and function are now considered.

rbcL gene for Rubisco

Rubisco is the major enzyme component of the chloroplast, often 40–60 per cent of the total soluble protein of leaves. The LSU is encoded by cpDNA in the *rbc*L gene, in a sequence of 1431 base pairs (bp)

Table 10.1 Components of the chloroplast which are coded by chloroplast DNA genes and their size in amino acids coded for. Those genes which are polycistronically transcribed are shown as PC, those monocistronically transcribed as MC. ⋆ denotes light activation

Chloroplast	Gene designated	Gene makes:	Gene size (codons)	Transcription
ATP synthesizing complex (CF$_1$–CF$_0$)	*atp*A	CF$_1$ α subunit	507	
	*apt*B	CF$_1$ β subunit	498	
	*atp*E	CF$_1$ ε subunit I	133	PC
	*atp*F⋆	CF$_0$ subunit III	184	
	*atp*H	CF$_0$ subunit IV	81	
	*atp*I	CF$_0$ subunit IV	247	
Cytochrome *b*–*f* complex	*pet*A	cytochrome *f*	320	
	*pet*B⋆	cytochrome *b*$_6$	215	PC
	*pet*D⋆	17 kDa polypeptide subunit 4 cytochrome *b*$_6$–*f*	160	
Photosystem I	*psa*A	P700–Chl *a* protein	750	
	*psa*B	P700–Chl *a* protein	734	PC
	*psa*C	8 kDa 2 (4Fe–4S) protein	81	
	*psa*J	4.9 kDa protein		
	*psa*I	4.0 kDa protein		
Photosystem II	*psb*A	D1 (32 kDa, Q$_B$) protein	353	MC
	*psb*B	47 kDa chl *a* protein (CP 47)	508	
	*psb*C	44 kDa chl *a* protein	473	
	*psb*D	34 kDa protein, D2	353	PC
	*psb*E	9 kDa cytochrome *b*$_{559}$	83	
	*psb*F	4 kDa cytochrome *b*$_{559}$	39	
Photosystem II	*psb*G	24 kDa polypeptide	284	
	*psb*H	10 kDa phospho-protein	73	
	*psb*I	4.8 kDa reaction centre	52	PC
	*psb*K	2.4 kDa polypeptide		
	*psb*L	5 kDa polypeptide	38	
Rubisco	*rbc*L	large subunit	477	MC
NADH dehydrogenase	*ndh*A	ND1		
	*ndh*B	ND2	subunits	
	*ndh*C	ND3	of	
	*ndh*D	ND4	respiratory	
	*ndh*E	ND4L	enzyme	
	*ndh*F	ND5		

Table 10.1 Continued

Chloroplast	Gene designated	Gene makes:	Gene size (codons):	Transcription
Ferredoxin	*frx* A	4Fe−4S		
	frx B	type		
	frx C	ferredoxins		
Genes for gene expression and genetic apparatus				
Ribosomal DNA	23S rDNA			
	16S rDNA	components		
	5S rDNA	of the		
	4.5S rDNA	ribosomes		
Transfer RNA	30 different	components		
	trn genes	of protein synthesis		
Ribosomal proteins	20 different	30S + 50S		
	rpl genes	ribosomal proteins		
RNA polymerase	*rpo* A	subunit α		
	rpo B	subunit β^1		
	rpo C	subunit β		
Initiation factor I proteins	*inf* A	component of protein synthesis		

corresponding to 477 amino acids (Dean *et al.* 1989; Manzara and Gruissem 1989). There are no introns and there is considerable homology with the cyanobacterial gene (e.g. *Anacystis nidulans* with 472 amino acids). The LSU is highly conserved, with sequences from different species having more than 80 per cent homology when the silent base changes in the cpDNA are allowed for. This is much greater than the homology for the nuclear encoded SSU (40 per cent) between organisms. In eukaryotes the single copy of *rbc* L is monocistronically transcribed in the chloroplast and is under post-transciptional control, i.e. processing of RNA. The gene has a strong promoter region which is light inducible. There are very many more genes for the LSU than for the SSU and it is still unclear how the necessary stoichiometry of the LSU and SSU components is achieved. In C3 plants, expression of the Rubisco genes occurs only where chloroplasts are found, and the expression is much greater in leaves than stems. C4 plants have Rubisco genes in both mesophyll and bundle sheath cells but expression is restricted to the bundle sheath.

atp *genes for coupling factor*

The ATP synthase (CF_0F_1) complex has 6 of its 9 subunits encoded by the B strand of cpDNA in all species, the rest by nDNA (Hudson and Mason

1989). The chloroplast genes are in two separate loci 40 kbp apart in the large single copy region, with those for β and ϵ (atpBE or β operon) in one locus and IV,III,I and α (atpIHFA or α operon) in the other. These are both conserved in clusters but the distance between them is variable, depending on species. The most abundant transcript of these genes is of atpH, as CF_0III is the most abundant subunit in the complex. A polycistronic transcript of the operon is probably made and then cleaved into a mixture of mono-, di- and polycistronic products. The β operon is *c.* 700 bp upstream from rbcL and, as it is on the opposite strand, transcribed in the opposite direction. The stop codon of atpB (UAG) and the adenine base preceding it form the initiation codon of atpE (AUAG), so they form a single operon. The β operon promoter in spinach lies 454 bp upstream of the 5′ initiator codon for atpB and has two sequences, TTGACA and TGTATA, resembling other chloroplast promotors although there are species differences. Knowledge of the regulation of transcription of these operons is poor; there are complex patterns of mono- and polycistronic transcripts. No evidence exists for close co-ordination of gene expression between the two operons in the chloroplast or with the nuclear genes for coupling factor subunits.

There is considerable similarity in subunit structure and function and in organization of *atp* genes between the chloroplasts of higher plants, cyanobacteria and bacteria. A particular order of genes is probably not essential for function. Genes for α and β subunits are strongly conserved in different organisms, possibly because of their role in catalysis. Such similarity in the very fundamental aspects of coding for this essential enzyme supports the theory of the prokaryotic endosymbiont origin of chloroplasts.

psa *genes for Photosystem I components*

The PSI complex has two subunit apoproteins (A1 and A2) of P700 coded in the *psa*A and *psa*B genes in the middle of the LSC region of cpDNA separated by 25 base pairs (in tobacco) (Herrmann *et al.* 1985). They are without introns, have a single transcriptional start upstream of the *psa*A genes and are co-transcribed (polycistronic). This may aid the maintenance of the correct 1:1 stoichiometry in the complex. Regulation of expression occurs at translational and post-transcriptional levels with complex control by light and developmental stages. The proteins are *c.* 750 and 730 amino acids long (about 83 kDa). Gene *psa*C, the first photosynthetic component found in the SSC region of cpDNA, codes for a 9 kDa polypeptide of 81 amino acids, the apoprotein of Fe−S centres A + B of PSI. This *psa*C gene is co-transcribed with a gene (*ndh*D) for a subunit of NADH dehydrogenase. In addition, *psa*J and *psa*I code for 4.9 and 4.0 kDa proteins of the PSI complex in barley. Other components of PSI are encoded by the nuclear genome and are discussed later.

psb *genes for Photosystem II*

Photosystem II has nine components coded in the cpDNA of tobacco but the 33, 24 and 18 kDa polypeptides of the water-splitting apparatus are encoded by nDNA. The chloroplast genes are located at 6 different sites within the large single copy region. The D1 (also called the 32 kDa or Q_B) protein of the PSII reaction centre is encoded by *psb*A and transcribed monocistronically as a 34 kDa precursor which is then processed to form the 32 kDa polypeptide; the coding sequence is 317 codons. The D1 protein binds herbicides, e.g. atrazine (p. 95) and mutants resistant to the herbicide have single base alterations in the gene code related to specific changes in the amino acid sequence. The mRNA is stable in the dark (more so than the mRNA from *rbc*L) although the product is unstable and rapidly degraded in darkness. Indeed, the protein itself turns over very rapidly (p. 80). Gene *psb*B codes for a 51 kDa polypeptide; it contains an intron and is upstream of genes for cytochrome b_6 (*pet*B) and *pet*D. The 4 genes *psb*B, *psb*H, *pet*B and *pet*D form a single operon, called the *psb*B operon, transcription of which has been described; *pet*B and *pet*D contain introns which are removed from the primary transcript, the exons are spliced and the mRNA processed to give multiple smaller mRNAs. The *psb*C and *psb*D genes overlap by 53 bp; they code for the 44 kDa and D2 proteins of PSII. In addition, the *psb*E and *psb*F genes are coding sequences for the 9 and 4 kDa subunits of cytochrome b_{559} which are co-transcribed. Other PSII genes (*psb*G and I) have been identified as such by Western blotting but their function is not established (for the relevant literature see Buetow *et al.* (1989); Chitnis and Thornber (1989)).

pet *genes of the cytochrome* b−f *complex*

This complex (p. 89) has 6 polypeptide components involved in the electron transport pathway between PSI and PSII. The three genes for cytochrome *f*, cytochrome b_{563} and subunit IV (*pet*A, B, D) are contained in cpDNA. The other 3, coded in the nDNA, include the Rieske Fe−S protein which has a vital role in the assembly of the complex. The precursor protein is made in the cytosol as a 26 kDa polypeptide which is transported into the chloroplast and then processed to the 19 kDa protein. *pet*A is 4 kbp downstream of *rbc*L on the A strand and separated from it by several open reading frames (ORFs). These are base sequences which could be transcribed, but with no identified protein product. *pet*A is co-transcribed with an associated ORF as a large polypeptide and the N-terminal 35 amino acids removed during processing. Cyto *f* has a hydrophobic C-terminal sequence (amino acids 250−271) which allows positioning in the thylakoid membranes. Genes *pet*B and *pet* D are close together and some 15 kbp from *pet*A, so it is likely that they are not polycistronic and therefore are under transcriptional control. The *pet*B and *pet*D genes are co-transcribed with *psb*B and *psb*H. Each gene has a single intron and very small initial exons (6 and 8 bp only) which are very highly conserved in all genomes. The sequences suggest that there are 4 and 3 hydrophobic, transmembrane segments in each protein, respectively.

Genes for NADH dehydrogenase

Possible protein products of open reading frames have been identified by comparing their base sequences with DNA sequences from other organisms (using a DNA library) for which products have been identified. The ORF's of cpDNA contain some 6 genes with strong homology to sequence components of NADH dehydrogenase from human mitochondria and *Chlamydomonas*. This enzyme functions in mitochondrial respiration, taking electrons from NADH and transferring them to ubiquinone. However, the enzyme has not been described in higher plant chloroplasts and may be either a relic of earlier respiratory activity there, or may function in chlororespiration (see p. 168). The genes are *ndh*A, D, E and F in the SSC region, *ndh*C in the centre of the LSC region and *ndh*B in the inverted repeats. Single introns occur in *ndh*A and *ndh*B genes; homology of these genes in several different higher plant and the liverwort genomes is strong.

Other plant chloroplast genomes

Several other genomes have been studied in detail (Bogorad and Vasil 1991a). The liverwort *Marchantia* (Ohyama *et al.* 1989) has, in comparison with tobacco, a somewhat smaller cpDNA genome of 121 024 bp, but in general the cpDNA structure, the gene products, etc., are very highly conserved. As with tobacco *Marchantia* cpDNA has two large inverted repeat (IR) sequences of 10 058 bp with a duplicated complement of genes; each contains genes for rRNAs. The IR sequences are separated by LSC and SSC regions of 81 095 and 19 813 bp, respectively. The cpDNA may code for up to 136 products, including all the rRNAs and tRNAs and proteins associated with the ribosomes of the chloroplast (4 rRNA, 32 tRNAs, 19 ribosomal proteins and 25 RNA polymerase subunits). Twenty genes for polypeptides of photosynthetic membrane electron transport and other membrane components have been identified. Further gene sequences have been related, through sequence homology with other organisms, to $4Fe-4S$ type ferredoxin from bacteria (2 ORFs), and 7 genes have strong homology with NADH dehydrogenase of human mitochondria (7 ORFs) as in tobacco. There is also a component of nitrogenase (1 ORF); other genes resemble those genes coding for the permeases of *E. coli* membranes and an antenna protein of a cyanobacterium. Such similarity beteween these two very different organisms suggests a common origin of the cpDNA and considerable conservatism in the evolution of the genome.

There is rather more variation in cpDNA than the basic pattern of tobacco and *Marchantia* chloroplast DNA suggests (Palmer 1985; Gray 1991). Legumes such as pea (*Pisum sativum*) and broad bean (*Vicia faba*, genome 123 kbp) have no large repeated sequences and are thus unique among land plants so far examined. *Vicia* cpDNA has only one set of rRNA genes, about 30 tRNA genes and encodes the LSU of Rubisco, the α, β, ϵ and subunit III of ATP synthase as well as others described for the tobacco and liverwort genomes. In legumes, the gene *rpl*22 is nuclear but in other angiosperms it is chloroplastic. The reasons for this and the mechanism by which the transfer occurred (probably from cpDNA to nDNA) are obscure. Variation in the IR is the main cause of differences in size of the cpDNA in different species. Reasons for the particular structural features of cpDNA and the differences between species are largely unknown although it is speculated that it may be related to processes of gene copying. Possibly, the IR was present in the cpDNA of the common ancestor of land plants (note *Marchantia*) and the major changes, e.g. loss of the IR in legumes, occurred later.

Nuclear encoded chloroplast proteins

The presence of genes in the nuclear genome for many chloroplast components was shown by genetic analysis, for example, the distribution of defective pigmentation into the offspring of mutant plants (see Tilney-Basset 1984). Table 10.2 lists some of the proteins associated with the photosynthetic apparatus which are encoded by nDNA (Taylor 1989). Virtually all aspects of the chloroplast structure involve nuclear gene products, raising interesting questions about the way proteins are made in one cell compartment and transferred to another and about co-ordination of chloroplast development. Indeed, the evolutionary history of this prokaryotic organelle in a eukaryotic environment is an intriguing subject.

Protein synthesis

The protein synthesis system of plastids has components restricted to the chloroplast (rRNA and tRNA) and components which are nuclear (elongation and termination factors plus enzymes which modify RNAs) but some components appear in both compartments (ribosomal proteins and initiation factors). The locations of some genes for different components are shown in Table 10.3.

rbcS genes

The SSU of Rubisco is nuclear encoded by the *rbc*S multigene family in all eukaryotic chlorophytes (all green algae and land plants) so far examined. The number of genes differs with the organism, e.g. 5 (pea and tomato),

Table 10.2 Some chloroplast proteins which are nuclear encoded and regulated by light (R = red, B = blue, UV = ultraviolet)

Chloroplast component	Gene designated	Gene products	Light response	Photoreceptor
Rubisco	*rbc* S (multigene family)	small subunit	+	R, B, UV
Rubisco activase	*Rca*		+	?
Electron transport	*pet* C	cytochrome *b/f* Rieske Fe–S protein		
	pet E	plastocyanin	+	R, B
	pet F (Fed-1)	ferredoxin NADP oxidoreductase	+	?
Photosystem II	*psb* O	33 kDa water-	+	R
	psb P	24 kDa splitting	+	R
	psb Q	18 kDa polypeptides	+	R
	psb R	10 kDa	+	R
Photosystem I	*psa* D	18 kDa polypeptides	+	R, B
	psa F	17 kDa of the complex		
	psa E	9.2 kDa		
	psa G	10.8 kDa		
	psa H	10.4 kDa		
Light-harvesting chlorophyll protein	*cab* (multigene family)	polypeptides of the chlorophyll protein complex of PSII	+	R, B, UV
ATP synthase (coupling factor)	*atp* C *atp* D *atp* G	γ polypeptides δ of the complex II		
Glyceraldehyde phosphate dehydrogenase	*Gap* A *Gap* B		+ +	? ?
Ferredoxin 1	*Fed*-1		+	
Glutamine synthase 2	GS2		+	
Phytochrome	*Phy*		−	R
Protochloro-phyllide reductase	*Pcr*		−	R
Nitrate reductase			+	(?R)
PEP carboxylase			+	?
Phospho-ribulose kinase			+	?
Flavonoid biosynthesis enzymes				
Chalcone synthase	*Chs*		+	R, B, UV
Phenyl-amoniumlyase	*Pal*		+	UV

Table 10.3 Compartments in which chloroplast protein translation components are coded. + = present; + + = abundant; − = absent

Translation system component	Genes in plastid	Nucleus
tRNA	+	−
rRNA	+	−
r-proteins	+	+ +
elongation factors	−	+
termination factors	−	+
rRNA processing and modification enzymes	−	+
r-protein modifying enzymes	−	+
aminoacyl-tRNA synthases	−	+
mRNA maturation enzymes	−	+

8 (petunia) and 13 (duckweed) but there are only 2 *rbc* S genes at one locus in *Chlamydomonas*. The 5 *Rcb* S genes in tomato are at 3 genetic loci on 2 chromosomes; *rbc* S-1 on chromosome 2 and *rbc* S-2 on chromosome 3, both of which encode a single gene product and *rbc* S-3 on chromosome 2 consisting of 3 genes arranged in tandem over a 10 kb region, *rbc* S-3A, -3B and -3C; they are monocistronically transcribed and differ in number and size of introns. Although these *rbc* S genes differ between and within species, the coding regions are highly conserved, e.g. 91−100 per cent in the tomato genes when silent nucleotide substitutions are allowed for. Probably the evolutionary pressure to conserve the functional protein was greater than the pressure to conserve the nucleotide sequences. Despite their similarity, there is substantial difference in the ability of these nuclear genes to produce products. In tomato, for example, the *rbc* S-3A and *rbc* S-1 genes are expressed in dark-grown leaves and change little with illumination. In contrast, *rbc* S-3B and -3C are hardly expressed in the dark but increase greatly in the light. In *Petunia* the 8 genes differ a hundred-fold in expression. Discussion of these points is given by Dean *et al.* (1989) and Manzara and Gruissem (1989).

cab *genes*

A chloroplast membrane component entirely encoded by nDNA is the light-harvesting complex of PSII (LHCII), composed of a major and 3 minor chlorophyll *a/b* binding proteins, namely LHCs, II*a*, II*b*, II*c* and II*d* of 29, *c*. 27, 29 and 21 kDa, with classical Mendelian inheritance (Bennett *et al.* 1984). These 4 are encoded by *cab* genes which have eukaryotic 5′ and 3′ flanking regions of TATA and CAAT boxes and polyadenylation sites. They form a multigene family of 3—16 genes depending on the species. In *Petunia*, the 16 genes are grouped into 5 small families in which the genes are related but the differences between families are greater. A sequence of 31—37 amino acids forms a transit peptide, that is, a sequence of amino acids at the amino-terminal end of the polypeptide chain which enables the protein to be transferred into the chloroplast envelope and thylakoid membrane. The *cab* genes produce mRNA in the nucleus which is translocated to the cytosol where it associates with ribosomal components to form free polysomes on which the precursor protein is made. On release from the polysomes the polypeptides attach (in a non-energy dependent binding) to the outer chloroplast envelope membrane: some 3—5000 molecules per chloroplast, thus allowing rapid synthesis of the protein even if transport is slow. The protein binding site is probably specific, e.g. Rubisco and LHCP precursors have different binding characteristics, pH requirements and sensitivity to proteolysis.

Genes for other thylakoid proteins

The apoprotein of the Rieske Fe—S centres, plastocyanin and ferredoxin-NADP oxidoreductase are products of *pet*C, *pet*E and *pet*F nuclear genes, respectively. The *pet*C product is made as a 26 kDa precursor, 7 kDa larger than the mature protein, on cytosolic ribosomes and then transferred into the chloroplast from the cytosol. From the deduced sequences of amino acids there are two highly conserved sequences, Cys—Thr—His—Leu—Gly—Cys and Cys—Pro—Cys—His, near the C-terminus of the protein which probably co-ordinate the iron—sulphur centre and also sequences which code for very hydrophobic amino acids which may anchor the protein in the membrane.

psa *and* psb *genes*

Photosystem I has 5 of its 8 polypeptides encoded in the *psa*D, *psa*F, *psa*E, *psa*G and *psa*H genes. They are, respectively, subunits II, III and possibly IV and VI; II is possibly the site for Fe—S centres, III is involved in the reduction of the PSI reaction centre, and the others are of unknown function. All are synthesized as larger precursors which are modified ('processed') after entering the chloroplast.

Photosystem II is characterized by having 4 of the polypeptides of the water-splitting complex, the 33, 24, 18 and 10 kDa polypeptides, coded in the nucleus by the *psb*O, *psb*P, *psb*Q and *psb*R genes. Each protein has two transit peptide sequences attached, one enabling transport across the chloroplast envelope and the other across the thylakoid enabling the water-splitting complex to be assembled in the thylakoid lumen.

atp *genes*

Coupling factor (CF_1CF_0) has the δ and II subunits nuclear encoded in the *atp*C, *atp*D and *atp*G genes, respectively. Suggestions have been made that transfer of these genes to the nucleus took place during evolution after the initial endosymbiosis. The precursor proteins are synthesized in the cytosol and transported into the chloroplast after which the transit peptide is removed. However, regulation of this process is poorly understood.

Nitrate and nitrite reductases

Nitrate assimilation in the chloroplast is intimately associated with utilization of the primary products of the light reactions (Ch. 7). Nitrite reductase (NiR) is in the chloroplast and nitrate reductase (NR) on the chloroplast envelope (p. 146). Both NR and NiR are coded for by small gene families in the nucleus and are synthesized on cytosolic ribosomes; NR remains in the cytosol but NiR is transported into the chloroplast.

Chloroplast gene expression and protein synthesis

Some chloroplast genes are transcribed monocistronically (e.g. *rbc*L and *psb*A) whilst others are polycistronic (e.g. *atp*I—*atp*A and *atp*B—*atp*E), resembling prokaryotic genes. Upstream of the genes are sequences of DNA similar to promoters of prokaryotes. The initiation sites for transcription of tobacco *rbc*L, *psb*B and *atp*B—E genes have upstream sequences (probably promoters) very similar to the −10 and −35 promoters of bacteria. However, the presence of introns particularly in tRNAs is less characteristic of prokaryotes. The chloroplast has a complete system for transcription and translation of cpDNA genes. Half of the total leaf ribosomes are in the chloroplast and half of the total leaf protein is made in there. Thus, the chloroplast is the most active compartment for protein synthesis in mesophyll cells, much more so than the mitochondria. Chloroplast genes are transcribed by RNA polymerases, of which there are two classes in the chloroplast. One class is tightly bound to DNA, the other soluble; this could be important for gene regulation. Expression of chloroplast genes may be controlled at several steps in the process and regulation is complex; the references (e.g. Bogorad and Vasil 1991a, b) should be consulted for this essentially molecular biological rather than photosynthetic problem.

Protein synthesis in chloroplasts is prokaryotic (Steinmetz and Weil 1989), e.g. ribosomes of chloroplasts are very similar to the prokaryotic 70S type in their subunit structure (50S and 30S) and sedimentation coefficients, initiation of chain elongation and the sequences and size of the rRNAs. Proteins of prokaryotic ribosomes have considerable homology with the products of chloroplast genes. The chloroplast DNA sequences have great homology with the sequences coding for ribosomal proteins in *E. coli*. These are composed of rRNA (with a $G + C$ content of 55.8 per cent) and about 60 proteins and are sensitive to prokaryotic antibiotics (e.g. chloramphenicol), but not to eukaryotic inhibitors (e.g. cycloheximide). Initiation of protein synthesis requires *N*-formylmethionine rather than methionine and factors for initiation and elongation can be interchanged with those of bacteria and remain fully functional. However, there is not total similarity between chloroplast and bacterial ribosomes; chloroplasts are affected by both prokaryotic and eukaryotic ribosomal inhibitors. This is explained by the involvement of both nuclear and chloroplastic genomes in the synthesis of plastid ribosomes. Of the 60 higher plant ribosomal proteins, only about 20 are made in the chloroplast; it is assumed but not proven that the remainder are made in the cytosol. Chloroplast and nuclear genomes co-operate intimately in the synthesis of the machinery needed to form the chloroplasts components, e.g. a chloroplast protein may be needed for synthesis of a nuclear encoded protein. Further details may be obtained from papers in Govindjee *et al*. (1989) and the reviews in Bogorad and Vasil (1991a, b).

Regulation of chloroplast protein synthesis

Protein synthesis by chloroplasts is driven by light and needs no additional energy. Low intensity light (15–30 per cent of the energy needed for full photosynthesis) is effective and blue light is more effective than red light of equivalent energy. In darkness, protein synthesis will take place if ATP is supplied, but at only half the rate of that in saturating light. The rate can be considerably increased by supplying Mg^{2+} to overcome the depletion of Mg^{2+} resulting from supplying ATP alone. There is probably a requirement for electron transport but it is indirect; the synthesis of sufficient ATP will depend on the balance with NADPH reactions. Competition for ATP may occur between CO_2 assimilation and amino acid incorporation into proteins but factors which stimulate the turnover of NADPH and increase the rate of ATP production increase protein synthesis.

Synthesis of chloroplast proteins takes place on ribosomes including polysomes bound to the thylakoid membranes by ionic charges and the nascent polypeptide chain. Binding is stimulated by light and the increase in stromal pH resulting from illumination. However, it seems that there is specificity in the function of the bound ribosomes, e.g. membrane bound polypeptides such as the D1 and PSI reaction centre proteins and subunit III of coupling factor are made on thylakoid bound ribosomes whereas Rubisco LSUs seem to be made on both the stromal and bound ribosomes. Proteins from the ribosomes may be prematurely released by environmental stresses such as low oxygen, salts and darkness.

Much of the newly synthesized protein may be rapidly degraded by proteolysis if it is defective or if other components which are essential for the structures are not available. An example of protein which is rapidly broken down and resynthesized is the D1 protein of photosystem II. (It is also called the Q_B or 32 kDa atrazine binding protein.) In photo-inhibitory conditions (p. 269), particularly in bright light, the protein is rapidly turned over. In fact, the D1 protein is the most rapidly synthesized and degraded of all chloroplast proteins. Although a major product of protein synthesis, it never accumulates when chloroplast development is blocked as do some other components of the chloroplasts. The mechanism of proteolysis of D1 is not known (Vierstra 1989).

Chloroplast genes may be activated by light (Table 10.4). Such 'photogenes' show large increases in transcripts and their protein products when illuminated; they include the D1 protein of PSII and *rbc*L for LSU of Rubisco. The latter is barely detectable in darkness and light greatly stimulates production of the mRNA. Expression of *rbc*L genes in the chloroplast is regulated primarily at the post-transcriptional level so that the stability of the chloroplast RNA and translational regulation of gene expression may be important in regulation of expression, together with promoter strength. Nuclear genes are also light activated, e.g. those for LHCII and the SSU of Rubisco. This important aspect of gene regulation in relation to the environment will be considered later. Detailed analysis of the problem is given in Thompson and White (1991).

Genes coding for photosynthetic components, such as the SSU of Rubisco, occur in tissues other than leaves of higher plants, e.g. stems and fruits, where they are expressed but to a greatly reduced extent, and in roots where they are negligible. Also, in C4 plants, genes may be present in the mesophyll and bundle sheath tissues but only expressed in one cell type; Rubisco, for example, only accumulates in the bundle sheath. Such tissue specific

Table 10.4 Light activated genes of the chloroplast

Gene	Product	Effect of light
*psb*D	protein of PSII	+
*psb*K	protein of PSII	independent, constitutive
*psb*I	protein of PSII	independent, constitutive
*rbc*L	Rubisco large subunit	+
*pet*A	cytochrome *f* apoprotein	+
*pet*B	cytochrome b_6 apoprotein	independent, constitutive
*atp*A	coupling factor α	+
*psa*A	PSI apoprotein	+
*psa*B	PSI apoprotein	?
*trn*k	tRNA genes	light independent, constitutive
*trn*G	tRNA genes	light independent, constitutive
*rps*16	ribosomal protein	light independent, constitutive

regulation of the expression is related to control DNA sequences which provide the tissue specificity.

Light receptors, phytochrome and gene activation

Regulation of gene expression by light occurs both in nuclear and chloroplast genes. Light regulates development through a system involving light perception, transduction of the energy into a signal and coupling of the signal to gene expression and regulation of products. Details of the processes may be obtained from Mohr (1986), Pratt *et al.* (1990) and Jenkins (1991). There are four main photoreceptors in plants:

1. phytochrome, which is regulated by red/far-red light.
2. the blue/UV-A absorbing photoreceptor (often called 'cryptochrome') which is also involved in phototropism and stomatal opening.
3. a UV-B receptor, important in inducing plant protective responses to damaging UV-B radiation; one of its more important features is stimulation of genes responsible for chalcone synthase production, a key point in the synthesis of flavonoids (major protective pigments against UV-B radiation).
4. protochlorophyllide reductase (p. 223).

Plastid genes dependent on light regulation are *psb*A for the D1 (32 kDa Q_B) protein of PSII, *rbc*L (LSU of Rubisco) and *pet*A, the cytochrome *f* apoprotein. Nuclear genes which are increased include the *cab* genes encoding the light-harvesting *a/b*-binding protein of the LHCII complex, *rbc*S for the Rubisco SSU (in most species examined but not barley) (Gallagher *et al.* 1984) and the ferredoxin gene *Fed*-1. However, gene expression for protochlorophyllide reductase (NADPH:protochlorophyllide oxidoreductase) and phytochrome is decreased. Species differ greatly in response to light. For example, *cab*1 gene expression is very sensitive, the mRNA being greatly increased by extremely dim light which gives only 1 per cent conversion of inactive phytochrome (P_r) to the active form (P_{fr}) and is saturated by light flux at which only 3 per cent of the P_{fr} is produced. Interestingly the *phy*A and *Pcr* genes are also very sensitive to very low fluxes but their transcripts are decreased. The very low light response is shown by barley and *Arabidopsis* but not by oat; there are differences in expression of the individual genes within a gene family and *cab*1 in wheat shows both low and normal light responses and in pea some genes show no low light response. There is, at present, no simple explanation of these complex differences. Probably they are related to the ecological behaviour of plants, as the phytochrome system may confer advantages in regulation of development. The sensitivity of plants to light flux and to the changed spectrum may be regarded (teleologically) as an advantage when seedlings grow towards the soil surface or in deep shade. Dim light enriched towards the far-red end of the spectrum predominates and signals that the plant must extend into bright light in order to photosynthesize fully. It is an advantage to prevent development of a competent photosynthesis apparatus (because of the drain on resources in the seed or other storage organ) until light is available and then a very rapid response is needed to achieve full competence.

Phytochrome is of the greatest physiological significance for plants; it is rapidly induced by low intensity light and is sensitive to low intensity and quality, principally the red/far-red ratio. It determines plant development and thus ability to respond to environment (see papers in Baker and Barber (1984)). Phytochrome is a chromoprotein with a 120−127 kDa apoprotein and a linear tetrapyrrole chromophore. It exists in dark grown tissues in a dominant form (P_r) which is physiologically inactive but can absorb red light at 665 nm absorption maximum and is then converted to P_{fr}, the physiologically active form. Only very brief (1 minute) exposure to red light of low energy is necessary: $P_r \rightarrow P_{fr}$ (Fig. 10.2). This is sufficient to trigger response of light sensitive genes. Equilibrium is achieved between P_r and P_{fr} depending on light quality; it is constant under steady state illumination but changes in the spectrum, e.g. due to growth of vegetation around an organ or plant, result in a growth response.

When P_{fr} absorbs far-red light (735 nm) it reverts back to the inactive form. Thus, the effect of red light can be reversed by immediate application of far-red light; equilibrium between the two forms is altered quickly by the red/far-red ratio, permitting regulation of photomorphogenesis and chloroplast structure, leading to such differences as in sun and shade leaves (p. 270). This rapid, low energy red/far-red reversibility provides a means of differentiating phytochrome effects from other light effects. There are, however, more complications for different forms of P_r and P_{fr} exist in tissues grown in darkness or in light; they are distinguished by molecular characteristics and have different effects on gene expression.

Phytochrome itself is nuclear encoded by the *phy* genes; *phy*A codes for the I type phytochrome and *phy*B and *phy*C coding sequences are like the type II phytochromes (based on sequence homology). Phytochrome provides an interesting example of the complex regulatory processes in plants. The *phy* genes are autoregulated, i.e. small exposure to light and formation of P_{fr} decreases expression of the gene and formation of phytochrome; in rice and oats the response is large but the effect in *Cucurbita* is much smaller. Thus, a regulatory cycle with very subtle balancing between the amount of receptor, the response of selected genes and light quality is possible, allowing rapid and large developmental and morphogenetic responses to sudden illumination. Phytochrome operates at the level of gene expression and the effects on the development of the chloroplast are many and well described (some will be considered later, see Nagy *et al.* 1988). However, details of the mechanisms by which light quality is transduced into regulation of gene activity are not well understood. Direct interaction with DNA is unlikely; P_{fr} has not been found in the nucleus. Amplification of the signal

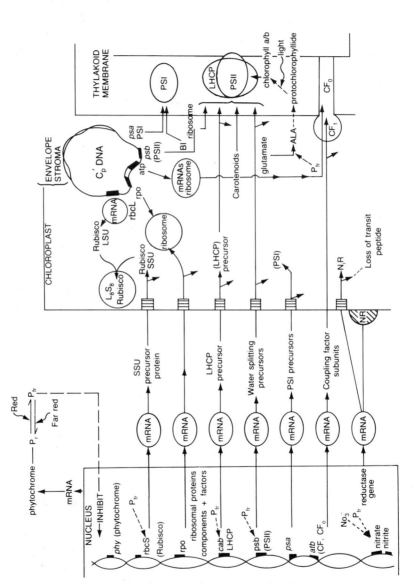

FIG. 10.2   Scheme of the interactions between nuclear and chloroplast genomes (nDNA and cpDNA) in terms of their gene products incorporated into the chloroplast stromal and thylakoid proteins. The role of phytochrome in regulation of gene expression is indicated. Also the interaction of nitrate ions in the regulation of gene expression for nitrate and nitrite reductases is shown. Details of the processes are discussed in the text (see Jenkins 1991).

is most likely, for the rapid and far-reaching effects would need more than the few P_{fr} molecules produced in very low light.

The 'triggering' reactions caused by phytochrome are strongly exhibited by tissues grown in the dark but induction depends on the gene. The transcription rates of *cab* and *rbc*S genes and that for GS2 (a plastidic glutamine synthetase) increase rapidly over 24 hours following illumination whereas products of *Fed*-1 (a gene for ferredoxin) increase rapidly in the first 2 hours and then remain constant. However, sensitivity to far-red light is retained for longer by *Fed*-1 than by *cab*, *rbc*S and GS2. There is much evidence of differences in the ability of tissues to respond to the light, e.g. production of anthocyanins by P_{fr} occurs only in subepidermal cells of mustard seedlings. Competence of the system is possibly determined by production of a 'plastid factor' which is needed for expression of nuclear genes but is stopped by chloramphenicol although it seems unlikely to be a protein itself.

There is much evidence for the interaction of the phytochrome and blue light receptors in regulation of gene expression in tissues grown in the light and transferred to darkness for a period. For example, abundance of *rbc*S and *cab* mRNAs was unaffected by the phytochrome equilibrium in pea tissues, despite the marked response of dark-grown material, but blue light did induce gene expression although it was reduced by applying far-red light. Perhaps such complex regulatory responses influenced by light quality allow full expression of tissue characteristics and enable the plant to achieve a high degree of efficiency in the normal environment. The interpretation is that blue light receptors control the red light effects, i.e. there is synergism between red and blue light (see Jenkins 1991; Grierson 1991).

Light and the organization of thylakoid membranes

Phytochrome has major effects on the expression of a number of genes coding for components of chloroplasts; this influence extends to the development and organization of thylakoids and chloroplasts. However, there is, as yet, no fully accepted time course or understanding of the mechanisms; a scheme is given in Fig. 10.2. Constitutively expressed genes allow the tissue to develop to the point where further changes depend on light. There are regulatory steps in the synthesis of chlorophyll *a* and *b*, formation of carotenoids (violaxanthin, neoxanthin and lutein) and, very importantly, in the expression of the *psb* genes of the nucleus and chloroplast and especially the *cab* nuclear genes; without the latter no PSII complex will be made. With light and P_{fr} formation *cab* and other nuclear gene products are made in the cytosol and transported into the chloroplast stroma. There further processing by proteases may take place and, if energy is available from ATP (or from the thylakoid proton energy gradient), the polypeptides are inserted into the thylakoid membranes. Assembly into the light-harvesting complex requires the correct carotenoid and, of course, chlorophyll *a* and *b*. In the absence of chlorophyll no complexes are made and the *cab* gene products are rapidly broken down. Concomitantly, P_{fr} stimulates synthesis of the water splitting and reaction centre polypeptides; these are assembled in the membrane and form the active complex. However, this simple statement hides a considerable degree of uncertainty about the undoubtedly complex mechanisms.

Similar considerations apply to the other multicomponent complexes in the membrane. Thus, photosystem I is assembled from polypeptides which are both nuclear and chloroplast coded. In darkness, chloroplasts contain small amounts of the products of the *psa*A and *psa*B chloroplast genes (the A1 and A2 protein of the core of PSI) and on illumination this hardly changes but polypeptide II, a product of the nuclear gene, is rapidly synthesized. This is followed by component III and thereafter the other proteins. Polypeptide II may act as a core for assembly of the other units.

The products of the *psa*A and *psa*B genes are synthesized on mRNA which may be at low abundance in the dark in some plants (e.g. maize) and subsequently increases in the light or are at rather large amount and do not increase markedly in the light. Transcripts of these developmentally very important components may decrease as the leaves age but gene transcription decreases even faster (see Jenkins 1991).

Synthesis of the coupling factor complex involves the transcription of nuclear and chloroplastic genes, transport of the former into the stroma and assembly into the transmembrane segment (CF_0) and the stroma unit (CF_1). Evidence for some light regulated control of ATP synthase exists but there is rather poor understanding of how the different groups of genes are co-ordinated in their expression and the final structure and stoichiometry of the complex is achieved *in vivo*.

Another type of control mechanism, which is only poorly understood, is that involving a signal from the chloroplast to the nucleus, a chloroplast factor, which 'tells' the nucleus that the chloroplast is 'ready' to receive the protein. In the case of nitrate reductase, this factor, active phytochrome (P_{fr}) and nitrate are all necessary for NR gene expression (Jenkins 1991). Without light, energy for nitrate reduction is not available in the chloroplast, so this mechanism avoids draining the plant's material resources in the construction of the system until light and nitrate ions are available; when both are, it is essential to have the light transducing and nitrate reducing components fully functional. Many of the photosynthetic system components including those encoded by *cab*, NR and to a smaller extent *rbc*S, *rbc*L, *psb*A, *psb*O and *psa*D genes exhibit diurnal rhythms in amount and activity probably related to the combined effects of light quantity and quality changes and the inherent patterns which are under complex control at many different levels of activity.

It is still unclear how the balance between the chloroplast and nuclear genomes is regulated. There are many questions: is there one or two way exchange of regulatory factors made by chloroplasts and nucleus, are common signals responsible for regulating both compartments? How is the content of Rubisco LSU and SSU genes regulated? The genes are in very different doses yet the stoichiometry of the products is fixed. There is evidence that decreased production of LSU may lead to accumulation of the SSU, triggering breakdown of the SSU and thus regulating the amount.

Photoregulation of nuclear gene expression

Nuclear gene expression in plants is regulated by a multistep series of controls in which *cis*-acting DNA sequences, control elements, and *trans*-acting gene products (probably proteins which bind to specific DNA sequences including the promoter regions) determine the initiation of transcription and are affected by light. It has been found that non-transcribed regulatory sequences confer the photoregulatory features; the transcriptional regulatory sequences are in the promoter and enhancer sequences 5' to the transcription initiation site. There are binding sites for nuclear factors which are essential for light responses.

As an example of the mechanism by which light controls gene expression, the *rbc*S gene from pea was fully expressed in petunia, with the 970 bp 5' flanking sequence determining the response to light; a −35 to −2 promoter was essential for transcription with sequences further upstream required for maximum gene expression in light. This effect of enhancer elements is seen in *rbc*S gene expression where two regions controlling expression (between −373 and −204) affect photoregulation and tissue specificity. In *Petunia* an enhancer which sequenced 3' to the start of coding also affected gene expression. It has been shown that small blocks of sequences in these regions determine the great differences in degree of expression in the petunia genes. It seems that there are positive, negative and reiterated DNA sequences in the enhancer regions so that a very subtle control of gene transcription is possible and, indeed likely, given the many environmental and plant factors which affect gene expression.

Regulation of transcription seems the most common method of controlling mRNA but post-translational processing may also occur. For example, *rbc*S mRNA decreased in seedlings of soybean but transcription decreased even faster. However, the opposite effect was seen in mature soybean plants. Translational and post-translational controls are also very important in the regulation of synthesis of the chloroplast nuclear encoded components, allowing more rapid and flexible response to environment. Light probably affects translation, protein folding and transport and protein degradation among other factors. Translational control applies to LSU together with the SSU of Rubisco in pea; their synthesis stops when light-grown plants are placed in darkness, probably because the ATP supply is interrupted. The RNAs for both the LSU and SSU remain bound to the polysomes in the chloroplast stroma and cytosol, respectively, but are inactive, although if cultured *in vitro* they are functional. In dark-grown plants no polysome-

RNA complexes are made, suggesting that initiation is prevented. Another example of light regulation of translation is given by the mRNA (transcribed from *psa* genes) for LHCP polypeptides. Transcription is blocked in darkness at the level of chain elongation and synthesis of chlorophyll *a* removes the block. Models by which internal and external factors regulate genes and the development, structure and function of the photosynthetic system are discussed by Tobin *et al.* (1984) and Thompson and White (1991).

Protein transport into chloroplasts

Proteins synthesized in the cytosol and destined for the chloroplast must be transported across the chloroplast envelope. When synthesis of the precursor protein with the transit peptide at the amino terminal end of the chain of the passenger protein is completed, the polypeptide binds to the chloroplast surface. Binding may take place at specific receptors, although this has not been substantiated nor have receptor proteins been found. Functional binding domains in the transit peptide have not been identified but there are generally three blocks of coding thought to be related to recognition, binding and transport sequences. Probably the transit peptide is recognized by membrane proteins which provide a channel through the membrane but evidence for protein translocators is poor for most systems (e.g. endoplasmic reticulum) and non-existent in chloroplasts. By analogy with the mitochondrial import of proteins it is possible that membrane contact areas (points at which the two envelope membranes appear fused) are the places where transport occurs. They may contain binding sites for precursor proteins; an antibody to the 30 carboxy terminal residues of pea Rubisco small subunit has been shown to bind to a 30−35 kDa envelope protein in the contact region. There is evidence for receptors (proteins or glycoproteins) which bind proteins and that the amino terminal sequences target proteins to particular organelles. Little is known of the conformation of the proteins during translocation but it is probable that the chains are not exposed to the hydrophobic lipid layer during passage although there may be partial unfolding.

Passage of the precursor protein across the envelope membrane, by an as yet obscure mechanism, is an energy (ATP) requiring process. There is no evidence that a transmembrane pmf is necessary for the process and it is unclear if it is ATP generated inside or outside the chloroplast which is needed; such considerations may also apply to thylakoid transport where in addition to ATP a protein component may also be required. Transport of the protein requires the presence of the transit peptide sequence on the protein, as outlined on page 208. Once the protein has passed across the envelope membranes the transit peptide is removed by a protease. One of the best studied is the transport of Rubisco SSU; by making and labelling the protein with a radioactive marker in a transcription system, import into intact, isolated chloroplasts could be demonstrated (Ellis 1984). Proteins passing one membrane system have only 1 transit peptide which is removed by proteases during or immediately after translocation, but Rubisco SSU, which only crosses the envelope has 2. The polypeptides of LHC which pass from the cytosol into the thylakoid membrane have 2 transit sequences. On entering the stroma, the first transit sequence is removed and the second in the thylakoid. Similarly, proteins of the thylakoid lumen (such as those of the water-splitting complex and plastocyanin) must pass through both the chloroplast envelope and the thylakoid membrane before getting to the site of function. They undergo two post-translational transport processes using two different transit peptides. Possibly secondary or tertiary structures may be the important feature of transit peptides. Transit peptide sequences are very diverse in length and amino acid composition, without correlation between amino acid sequences and transport characteristics. However, there must be specific recognition sequences, for foreign proteins enter chloroplasts only very slowly or not at all. Transit peptides from several nuclear encoded proteins can confer the ability to pass across the envelope even to foreign proteins; this may be a way of artificially introducing genetically modified or foreign proteins into chloroplasts or other organelles.

The mechanisms at the molecular level by which proteins recognize specific sequences and by which the large molecular components (often of a hydrophilic nature) pass the hydrophobic lipid membrane are largely unclear. Transit peptidases have been partially purified and described but their regulation remains obscure. Once in the stroma the protein is assembled, in the case of Rubisco LSU and SSU, into the holoenzyme. Proteins destined for thylakoid complexes are incorporated into the membrane; details of the mechanism are unclear — involvement of a hydrophobic section of the precursor protein and formation of membrane vesicles are possible — but other products are needed for this, e.g. chl *a* or *b* and xanthophyll carotenoids (lutein, violaxanthin or neoxanthin) are involved in the structure and thus light is required (p. 79) (Bartly *et al.* 1991).

Many interesting features of the translocation of proteins and processing remain to be answered. Do, for example, different proteins such as plastocyanin and cytochrome *f* use the same or separate domains in the membrane during import across the membrane? Perhaps plastocyanin had, at an early stage, a transit sequence for passage across the thylakoid and later acquired that for the chloroplast envelope and thus may use the original translocation path of the earlier ancestor and a later acquired one. There are also many intriguing questions as to the evolution of such a system. Was the transit peptide 'tacked onto' a pre-existing nuclear encoded enzyme enabling it to enter the symbiont before the symbionts own gene for the enzyme was eliminated in the course of evolution? Were the characteristics of the cpDNA product, e.g. inability of the polypeptide to unfold and refold if translocated, the reason why some proteins remained encoded in cpDNA and others moved to the nucleus? Perhaps the regulatory processes in the cytosol would have been disturbed by particular sequences of proteins and thus remained in the cpDNA because of the selection pressure, even if there was a positive advantage in terms of genome organization and expression for cpDNA genes to migrate into the nucleus. Better understanding of transport mechanisms into chloroplasts may provide a way of effectively directing foreign and modified proteins into the organelle to achieve a particular photosynthetic response (see Keegstra *et al.* 1989; Ellis 1991).

Protein folding and molecular chaperones

Proteins, once made on the mRNA in the form of the primary amino acid chain, must undergo folding to achieve the correct secondary and tertiary structures which are essential for function. Also, if part of a complex, they must be combined with other components, e.g. prosthetic groups and metal ions. Multiunit proteins, such as Rubisco, are then assembled in the chloroplast stroma to form the active holoenzyme (Fig. 10.3). Folding was assumed to be 'spontaneous', based on electrostatic charge and size and

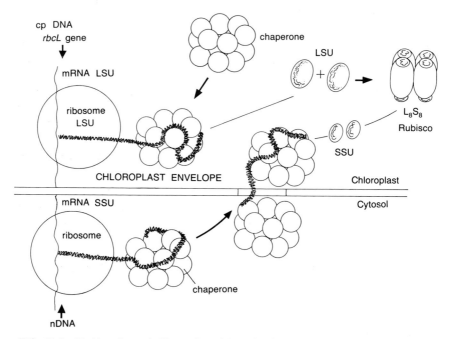

FIG. 10.3   Highly schematic illustration of the role of a molecular chaperone, a protein of many subunits, in permitting the correct folding of the primary amino acid chain after synthesis in the ribosome. The situation for Rubisco large subunit (LSU) and small subunit (SSU) illustrates the role the molecular chaperones may have in transporting proteins between cellular compartments, the correct folding of the subunits and their assembly into the mature L_8S_8 Rubisco protein.

shape of the primary chain. However, with increasing use of molecular engineering to produce foreign proteins in cells, it became clear that often newly formed proteins would not fold and assemble properly even if the primary structure were correct. An example is provided by higher plant Rubisco expressed in *Escherichia coli*. Only inactive, denatured protein was made which precipitated out in the bacteria. The reason for this is the absence of the machinery required to fold and assemble the enzyme correctly. In chloroplasts it was observed that folding of newly synthesized LSUs of Rubisco required brief, non-covalent binding with a large (720 kDa), oligomeric protein composed of 12 subunits of two types, α and β, of molecular mass 61 and 60 kDa, respectively. This abundant, soluble oligomeric protein, Rubisco LSU binding protein, has an $\alpha_6 \beta_6$ structure and is nuclear encoded; it is made as higher molecular mass precursors in the cytosol before transport into the chloroplast. It is 50 per cent homologous to a protein of *E. coli* called cpn60 or GroEL protein, which is required for folding one of the head proteins of phage virus.

The concept is that the unfolded polypeptides of the large and small subunits of Rubisco are 'chaperoned' by the binding protein (Fig. 10.3) giving the folded monomers (Ellis 1991). The LSU polypeptide in the presence of the chaperone forms a dimer; possibly these associate giving the L_8 structure of Rubisco to which the folded small subunits bind, thus forming the active L_8S_8 holoenzyme. There are many types of proteins involved in the folding of proteins in different cells. They may recognize the structure of the partly folded protein and bind to it in such a way that some deleterious but spontaneous arrangements are prevented whilst others can proceed correctly. This prevents synthesis of inactive proteins and their precipitation in cells. Possibly chaperonins are also involved in the transport of proteins from cytosol to chloroplast; one or more chaperonins may bind to the newly synthesized protein chains during or immediately after transcription and transport it to the membrane, and once in the stroma it is picked up by another chaperonin and the transit peptide removed. If the correct chaperonins are not available (as in a foreign cell) then active enzyme cannot be made or transported. Chaperonins are important in protein folding and DNA replication and appear to protect cell development from environmental stresses, for example, heat stress proteins (hsp) of different molecular mass (e.g. hsp70 and hsp60) with such function appear in temperature stressed plants.

Chloroplast development

In flowering plants, formation of the components of functional chloroplasts and their integration into structure is a complex, very well regulated process which depends upon environmental signals, particularly light. This 'photomorphogenesis' (control of development by light) requires low energy

radiation and depends on light quality. Regulation is thus through information, rather than provision of energy rich assimilates from photosynthesis.
In darkness, the photosynthetic system does not develop as in plants grown
in light, instead it undergoes skotomorphogenesis — *skoto* is the Greek for
darkness. Light is needed to start development of chloroplasts and associated
metabolic systems, e.g. chlorophyll formation (see references in Baker and
Barber (1984)).

Chloroplasts do not arise *de novo* but from proplastids which in developing
cells are of a generalized nature, *c.* 1 μm diameter with a double bounding
membrane and perforated internal membranes. Proplastids may develop
into plastids of several types, including chloroplasts, leucoplasts and
chromoplasts and this largely depends on whether the tissue is exposed to
light or darkness. Chloroplasts develop through stages in which the early
thylakoids become increasingly perforated and differentiated into the stacked
and unstacked regions of the mature chloroplast. During this phase of
growth the plastids enlarge to 5−10 μm diameter. The sequence of events
is shown most clearly in monocotyledonous plants where the basal leaf
meristem gives rise to daughter cells which grow and mature as they pass
into the upper mature part of the leaf blade (see Baker, in Baker and Barber,
1984)). Thus, there is a developmental sequence related to age (the above
sequence may take only 6 hours at 25 °C) which can be related to distinct
environmental stimuli such as light. Both cells and chloroplasts divide:
chloroplast division has been long observed by microscopy in the form of
a dumb-bell shaped chloroplast. The relative rates of the two processes
determine the number of chloroplasts per cell.

In young meristematic cells development under the influence of light
involves formation of gene products and their assembly into structural and
biochemical components of the cell. During normal development in light
the very young meristematic cells contain Rubisco and chlorophyll in small
amounts and during development these increase six- and eighty-fold, respectively, by maturity. Both Rubisco LSU and SSU increase in parallel in
tightly co-ordinated fashion, but the mechanism is obscure. There is a large
complement of 70S and 80S ribosomes per plastid in young cells; about
60 per cent of the mature state which are observed early decrease towards
maturity. The α and β subunits of coupling factor appear together very
early in development before the thylakoids are fully developed or stacked.
However, LHC is only detected during the phase of rapid granal assembly
leading to mature chloroplasts. It is interesting that the mRNA for LHC
peaks earlier than that for Rubisco. During the later formation of the
thylakoids there is massive accumulation of the lipids of the membranes
with the appearance of Δ^3-transhexadecenoic acid in phosphatidylglycerol.

In darkness the proplastids form etioplasts which lack chlorophyll and
other pigments but possess a prolamellar body, a circular lattice of
interconnecting membrane tubules derived from the inner plastid envelope
with a few single membranes extending from it. The prolamellar body is

largely composed of the protochlorophyllide oxidoreductase enzyme protein with protochlorophyllide and specific lipids of thylakoids, e.g. MGDG and DGDG; the prolamellar body may be an artefact of interrupted development when light is excluded or a normal feature, if only very transient, in usual environmental conditions. It is as if development is blocked at this stage in darkness. When the tissue is illuminated the prolamellar bodies develop into the thylakoid system with both stacked and unstacked membranes. Their formation is accompanied by chlorophyll synthesis, increase in proteins and activation of pre-existing ones and the incorporation of these into the system. Fully developed etioplasts have coupling factor components, Rubisco and other PCR cycle enzymes, but lack the functional light-harvesting complexes of the light-grown tissue; chlorophyll synthesis is required. When etioplasts are illuminated chlorophyll synthesis occurs and a functional chloroplast may result from the breakdown and reassembly of the prolamellar body, with formation of photosystems, development of the energy transducing system and lastly synthesis of enzymes of the PCR cycle.

A most important process in development of thylakoids is the formation of chlorophyll. Protochlorophyll(ide) captures light and is converted to chlorophyll(ide), the holochrome protein acting as a protochlorophyllide reductase (NADPH:protochlorophyllide oxidoreductase). Protochlorophyllide reductase is loosely bound to the membranes of etioplasts and acts by forming a photoactive ternary complex with protochlorophyllide and NADPH, which on illumination transfers H to the porphyrin giving the chlorophyllide. There is a characteristic shift in absorption from 630 to 638−650 nm resulting from binding of the protochlorophyllide to the complex and this is again shifted to 678 nm by light (resulting in one reduction step) and to 684 nm as the second light-independent reduction takes place, before chlorophyllide is released (absorption at 672 nm); these events take place with characteristic time scales. Protochlorophillide reductase and its mRNA are abundant in the prolamellar body of dark-grown chloroplasts and is actively synthesized in darkness. However, within 30 minutes of illumination the amount of protein and activity of the enzyme decreases to only 10 per cent of that in darkness, with loss of translatable mRNA. Probably regulation operates via phytochrome, which controls the amount of mRNA although there is the possibility that substrate availability is an important factor regulating enzyme activity. The absorption spectrum of chlorophyll in the developing membranes *in vivo* changes from 684 to 672 nm with the protochlorophyllide to chlorophyllide conversion, reflecting different binding of the pigments to proteins or other forms of aggregation or disaggregation, e.g. the pigment absorbing at 672 nm may be unbound chlorophyllide. The spectral changes occurring after 1 minute of bright illumination allow conversion of protochlorophyllide (absorption at 650 nm) to chlorophyllide (at 683 nm) followed by a further transition to 673 nm. This is called the Shibata shift and may correspond to the time course of phytol addition to the substrate. Normal development of primary thylakoids (which are perforated double membranes) can only proceed to maturity if the conversion of protochlorophyll(ide) to chlorophyll *a* takes place, otherwise development is halted and the plastids degenerate (Kannangara 1991).

Turnover and replacement of proteins in the chloroplast

Synthesis of proteins is not restricted to developing chloroplasts but is important in the mature system. The example of the D1 protein of PSII, which is probably the most rapidly turned-over protein in the chloroplast, has been given (p. 102, Schuster *et al.* 1989; Vermaas and Ikeuchi 1991). The mechanism causing the damage is unknown but may be related to regulation of the energy balance of reaction centres. If the quinone binding site on D1 is unoccupied then photodegradation may be more likely. D1 is probably damaged between the fourth and fifth membrane spans where a particular sequence found in other rapidly turned-over proteins occurs. The damaged polypeptide is removed and replaced in a reaction requiring ATP and involving the protein ubiquitin which 'labels' the damaged centre so that it may be identified and removed. Removal may not be by ordinary proteolysis as no breakdown products have been observed. Details of the process are not clear, for example, how the PSII complex remains stable without the D1 protein and for how long and where replacement occurs. The protein is made on ribosomes in the stroma and may enter into the reaction centre directly, the damaged PSII migrating to the stromal thylakoids from the granal lamellae or first entering the stromal membranes and then migrating in the membrane to the damaged complex. The latter seems more likely as CP43 and CP47 only occur in the granal membranes; D1 may be linked to palmitoyl groups which allow diffusion and correct insertion of the protein into the complex. Some 2 hours are required for incorporation into PSII. Environmental factors, such as bright light which causes the damage, and cold which decreases the rate of protein synthesis, increases photoinhibition.

References and Further Reading

Akoyunoglou, G. and **Senger, H.** (eds) (1986) *Regulation of Chloroplast Differentiation*, Alan R. Liss, Inc., New York.

Baker, N.R. and **Barber, J.** (eds) (1984) *Chloroplast Biogenesis, Topics in Photosynthesis*, Vol. 5, Elsevier, Amsterdam.

Bartley, G.E., Coomber, S.A., Bartholomew, D.A. and **Scolnic, P.A.** (1991) Genes and enzymes for carotenoid biosynthesis, pp. 331–45 in Bogorad, L. and Vasil, I.K. (eds), *Cell Culture and Somatic Cell Genetics of Plants*, Vol. 7B, *The Photosynthetic Apparatus: Molecular Biology and Operation*, Academic Press Inc., San Diego.

Bennett, J., Jenkins, G.I., Cuming, A.C., Williams, R.S. and **Hartley, M.R.** (1984) Photoregulation of thylakoid biogenesis: the case of the light-harvesting chlorophyll *a/b* complex, pp. 167–92 in Ellis, R.J. (ed.), *Chloroplast Biogenesis*, Cambridge University Press, Cambridge.

Bogorad, L. and **Vasil, I.K.** (1991a) *Cell Culture and Somatic Cell Genetics of Plants*, Vol. 7A, *The Molecular Biology of Plastids*, Academic Press Inc., San Diego.

Bogorad, L. and **Vasil, I.K.** (1991b) *Cell Culture and Somatic Cell Genetics of Plants*, Vol. 7B, *The Photosynthetic Apparatus: Molecular Biology and Operation*, Academic Press, Inc., San Diego.

Buetow, D.E., Chen, H., Erdös, G. and **Yi, L.S.H,** (1989) Regulation and expression of the multigene family coding light-harvesting chlorophyll *a/b* binding proteins of Photosystem II, pp. 283–319, in Govindjee *et al.* (eds), *Molecular Biology of Photosynthesis*, Kluwer Academic Publishers, Dordrecht.

Chitnis, P.R. and **Nelson, N.** (1991) Photosystem I, pp. 178–224 in Bogorad, L. and Vasil, I.K. (eds), *Cell Culture and Somatic Cell Genetics of Plants*, Vol. 7B, *The Photosynthetic Apparatus: Molecular Biology and Operation*, Academic Press Inc., San Diego.

Chitnis, P.R. and **Thornber, J.P.** (1989) The major light-harvesting complex of photosystem II: aspects of its molecular and cell biology, pp. 259–281 in Govindjee *et al.* (eds), *Molecular Biology of Photosynthesis*, Kluwer Academic Publishers, Dordrecht.

Dean, C., Pichersky, E. and **Dunsmuir, P.** (1989) Structure, evolution, and regulation of *Rbc* S genes in higher plants, *A. Rev. Plant Physiol. Plant Mol. Biol.*, **40**, 415–39.

Dyer, T.A. (1984) The chloroplast genome: Its nature and role in development, pp. 23–69 in Baker, N.R. and Barber, J. (eds), *Chloroplast Biogenesis. Topics in Photosynthesis*, Vol. 5, Elsevier, Amsterdam.

Ellis, R.J. (ed.) (1984) *Chloroplast Biogenesis*, Cambridge University Press, Cambridge.

Ellis, R.J. (1985) Synthesis, processing, and assembly of polypeptide subunits of ribulose-1,5-bisphosphate carboxylase/oxygenase, pp. 339–47 in Steinback *et al.* (eds), *Molecular Biology of the Photosynthetic Apparatus*, Cold Spring Harbor Laboratory.

Ellis, R.J. (1991) Chaperone function: Cracking the second half of the genetic code, *Plant. J.*, **1**, 9–13.

Feifelder, D. (1987) *Molecular Biology* (2nd Edn), Jones and Bartlet Publishers, Inc., Boston.

Gallagher, T.F., Smith, S.M., Jenkins, G.I. and **Ellis, R.J.** (1984) Photoregulation of transcription of the nuclear and chloroplast genes for ribulose bisphosphate carboxylase, pp. 303–19 in Ellis, R.J. (ed.), *Chloroplast Biogenesis*, Cambridge University Press, Cambridge.

Govindjee, Bohnert, H.J., Bottomley, W., Bryant, D.A., Mullet, J.E., Ogren, W.L., Pakrasi, H. and **Somerville, C.R.** (eds) (1989) *Molecular Biology of Photosynthesis*, Kluwer Academic Publishers, Dordrecht.

Gray, J.C., Hird, S.M., Wales, R., Webber, A.N. and **Willey, D.L.** (1989) Genes and polypeptides of photosystem II, pp. 423–35 in Barber, J. and Malkin, R. (eds), *Techniques and New Developments in Photosynthesis Research*, Plenum Press, New York/London.

Gray, M.W. (1991) Origin and evolution of plastid genomes and genes, pp. 303–30, in Bogorad, L. and Vasil, I.K. (eds), *Cell Culture and Somatic Cell Genetics of Plants*, Vol. 7A, *The Molecular Biology of Plastids*, Academic Press Inc., San Diego.

Grierson, D. (ed.) (1991) *Developmental Regulation of Plant Gene Expression, Plant Biotechnology*, Vol. 2, Blackie, Glasgow.

Groot, G.S.P. (1984) Chloroplast DNA of higher plants, pp. 67–81 in Ellis, R.J. (ed.), *Chloroplast Biogenesis*, Cambridge University Press, Cambridge.

Gruissem, W. (1989) Chloroplast RNA: Transcription and processing, pp. 151–91 in Marcus, A. (ed.), *The Biochemistry of Plants*, Vol. 15, *Molecular Biology*, Academic Press, San Diego.

Herrmann, R.G., Westhoff, P., Alt. J., Tittgen, J. and **Nelson, N.** (1985) Thylakoid membrane protein and their genes, pp. 233–56 in Van Vloten-Doting, L., Groot, G.S.P. and Hall, T.C. (eds), *Molecular Form and Function of the Plant Genome*, Plenum Press, New York.

Huang, A.H.C. and **Taiz, I.** (eds) (1991) *Molecular Approaches to Compartmentation and Metabolic Regulation*, Amer. Soc. Plant Physiol., Rockville.

Hudson, G.S. and **Mason, J.G.** (1989) The chloroplast gene encoding subunits of the H^+-ATP synthase, pp. 565–82, in Govindjee *et al.* (eds), *Molecular Biology of Photosynthesis*, Kluwer Academic Publishers, Dordrecht.

Jenkins, G.I. (1991) Photoregulation of plant gene expression, pp. 1–41 in Grierson, D. (ed.), *Developmental Regulation of Plant Gene Expression, Plant Biotechnology*, Vol. 2, Blackie, Glasgow.

Kannangara, C.G. (1991) Biochemistry and molecular biology of chlorophyll synthesis, pp. 302–30 in Bogorad, L. and Vasil, I.K. (eds), *Cell Culture and Somatic Cell Genetics of Plants*, Vol. 7B, *The Photosynthetic Apparatus: Molecular Biology and Operation*, Academic Press Inc., San Diego.

Keegstra, K., Olsen, L.J. and **Theg, S.M.** (1989) Chloroplastic precursors and their transport across the envelope membranes, *A. Rev. Plant Physiol. Plant Mol. Biol.*, **40**, 471–601.

Kuhlemeier, C., Green, P.J. and **Chua, N.-H.** (1987) Regulation of gene expression in higher plants, *A. Rev. Plant Physiol.*, **38**, 221–57.

Lawlor, D.W. (1991) Concepts of nutrition in relation to cellular processes and environment, pp. 1–32 in Porter, J.R. and Lawlor, D.W. (eds), *Plant Growth: Interactions with Nutrition and Environment*, Cambridge University Press, Cambridge.

Manzara, T. and **Gruissem, W.** (1989) Organization and expression of the genes encoding ribulose-1,5-bisphosphate carboxylase in higher plants, pp. 621–44, in Govindjee *et al.* (eds), *Molecular Biology of Photosynthesis*, Kluwer Academic Publishers, Dordrecht.

Maréchal-Drouard, L., Kuntz, M. and **Weil, J.H.** (1991) tRNAs and tRNA genes of plastids, pp. 169–89, in Bogorad, L. and Vasil, I.K. (eds), *Cell Culture and Somatic Cell Genetics of Plants*, Vol. 7A, *The Molecular Biology of Plastids*, Academic Press Inc., San Diego.

Mohr, H. (1986) Control by light of plastogenesis as part of a control system, pp. 623–34 in Akoyunoglou, G. and Senger, H. (eds), *Regulation of Chloroplast Differentiation*, Alan R. Liss Inc., New York.

Nagy, F., Kay, S.A. and **Chua, N.-H.** (1988) Gene regulation by phytochrome, *TIG*, **4**, 37–42.

Ohyama, K., Kohchi, T., Fukuzawa, H., Sano, T., Umesono, K. and **Ozeki, H.** (1989) Gene organisation and newly identified groups of genes of the chloroplast genomes from a liverwort *Marchantia polymorpha*, pp. 27–42 in Govindjee *et al.* (eds), *Molecular Biology of Photosynthesis*, Kluwer, Dordrecht.

Okamuro, J.K. and **Goldberg, R.B.** (1989) Regulation of plant gene expression: General principles, pp. 2–82 in Marcus, A. (ed.), *The Biochemistry of Plants*, Vol. 15, *Molecular Biology*, Academic Press, London.

Palmer, J.D. (1985) Comparative organization of chloroplast genomes, *A. Rev. Genet.*, **19**, 325–54.

Palmer, J.D. (1991) Plastid chromosomes: Structure and evolution, pp. 5–53, in Bogorad, L. and Vasil, I.K. (eds), *Cell Culture and Somatic Cell Genetics of Plants*, Vol. 7A, *The Molecular Biology of Plastids*, Academic Press, San Diego.

Pratt, L.H., Senger, H. and **Galland, P.** (1990) Phytochrome and other photoreceptors, pp. 185–230 in Harwood, J.L. and Bowyer, J.R. (eds), *Methods in Plant Biochemistry*, Vol. 4, *Lipids, Membranes and Aspects of Photobiology*, Academic Press, London.

Roy, H. and **Nierzwicki-Bauer, S.A.** (1991) Rubisco: genes, structure, assembly, and evolution, pp. 347–64 in Bogorad, L. and Vasil, I.K. (eds), *Cell Culture and Somatic Cell Genetics of Plants*, Vol. 7B, *The Photosynthetic Apparatus: Molecular Biology and Operation*, Academic Press, San Diego.

Schuster, G., Shochat, S., Adir, N. and **Ohad, I.** (1989) Inactivation of photosystem II and turnover of the D1-protein by light and heat stresses, pp. 499–510 in Barber, J. and Malkin, R. (eds), *Techniques and New Developments in Photosynthesis Research*, Plenum Press, New York/London.

Shinozaki, K., Hayashida, N. and **Sugiura, M.** (1989) *Nicotiana* chloroplast genes for components of the photosynthetic apparatus, pp. 1–25 in Govindjee *et al.* (eds), *Molecular Biology of Photosynthesis*, Kluwer, Dordrecht.

Somerville, C.R. (1986) Analysis of photosynthesis with mutants of higher plants and algae, *A. Rev. Plant Physiol.*, **37**, 467–507.

Steinback, K.E., Arntzen, C.J. and **Bogorad, L.** (1985) The physical organization and genetic determinants of the photosynthetic apparatus of chloroplasts, pp. 1–19 in Steinback *et al.* (eds), *Molecular Biology of the Photosynthetic Apparatus*, Cold Spring Harbor Laboratory.

Steinmetz, A. and **Weil, J.-H.** (1989) Protein synthesis in chloroplasts, pp. 193–227 in Marcus, A. (ed.), *The Biochemistry of Plants*, Vol. 15, *Molecular Biology*, Academic Press, London.

Sugiura, M. (1989) The chloroplast genome, pp. 133–50 in Marcus, A. (ed.), *The Biochemistry of Plants*, Vol. 15, *Molecular Biology*, Academic Press, London.

Taylor, W.C. (1989) Regulatory interactions between nuclear and plastid genomes, *A. Rev. Plant Physiol. Plant Mol. Biol.*, **40**, 211–33.

Thompson, W.F. and **White, M.J.** (1991) Physiological and molecular studies of light regulated nuclear genes in higher plants, *A. Rev. Plant Physiol. Plant Mol. Biol.*, **33**, 432–66.

Tilney-Bassett, R.A.E. (1984) The genetic evidence for nuclear control of chloroplast biogenesis in higher plants, pp. 13–50 in Ellis, R.J. (ed.), *Chloroplast Biogenesis*, Cambridge University Press, Cambridge.

Timko, M.P., Kausch, A.P., Hand, A.M., Cashmore, A.R., Herrera-Estrella, L., Van den Broeck, G. and **Van Montagu, M.** (1985) Structure and expression of nuclear genes encoding polypeptides of the photosynthetic apparatus, pp. 381–96 in Steinback *et al.* (eds), *Molecular Biology of Photosynthetic Apparatus*, Cold Spring Harbor Laboratory.

Tobin, E.M., Silverthorne, J., Stiekema, W.J. and **Wimpee, C.F.** (1984) Phytochrome regulation of the synthesis of two nuclear-coded chloroplast proteins, pp. 321–35 in Ellis, R.J. (ed.), *Chloroplast Biogenesis*, Cambridge University Press, Cambridge.

Vermaas, W.F.J. and **Ikeuchi, M.** (1991) Photosystem II, pp. 26–112 in Bogorad, L. and Vasil, I.K. (eds), *Cell Culture and Somatic Cell Genetics of Plants*, Vol. 7B, *The Photosynthetic Apparatus: Molecular Biology and Operation*, Academic Press Inc., San Diego.

Vierstra, R.D. (1989) Protein degradation, pp. 521–36, in Marcus, A. (ed.), *The Biochemistry of Plants*, Vol. 15, *Molecular Biology*, Academic Press, London.

Zelitch, I. (ed.) (1990) *Perspectives in Biochemical and Genetic Regulation of Photosynthesis*, Wiley-Liss, New York.

Zurawski, G. and **Clegg, M.T.** (1987) Evolution of higher-plant chloroplast DNA-encoded genes: Implications for structure–function and phylogenetic studies, *A. Rev. Plant Physiol.*, **38**, 391–410.

CHAPTER 11

Carbon dioxide supply for photosynthesis

Carbon dioxide, the major substrate for photosynthesis, is supplied from the medium in which an organism lives, water for aquatic bacteria, algae and some higher plants and the atmosphere for terrestrial plants. Only transport of carbon dioxide from the atmosphere to higher plant leaves is considered here although the basic principles are applicable to other situations. To understand the factors controlling the fluxes of CO_2 the relationships between mass and volume of gases, the effects of temperature and pressure and the basis of the expression of gas composition are briefly considered for it is often essential to interconvert volume, mass and pressure of a gas in determining rates of photosynthesis using gas exchange techniques.

For an ideal gas, and employing SI units, the relation between the number of moles of gas (n), volume (V, m^3) and pressure (P) in pascal (1 Pa = 1 N m^{-2}) at a temperature (T, in kelvin) is given by the ideal gas equation:

$$PV = nRT \qquad [11.1]$$

The molar gas constant (R) is 8.31 m^3 Pa mol^{-1} K^{-1}. Standard atmospheric pressure is 101 325 Pa = 1.013 bar = 760 mm mercury. A mole of gas at 273 K (0 °C) and 101 325 Pa occupies 0.0224 m^3 (22.414 dm^3 or litres in the frequently employed but non-SI unit) and contains Avogadro's number of particles (6.023 × 10^{23}).

Concentrations of a component gas in a mixture of gases may be expressed in several different ways; number of moles per cubic metre (molar concentration) or mass per unit volume (mass concentration) is employed. Also the ratio of the number of moles of the component gas as a proportion of a total number of moles, the mole fraction, is used. Mole fraction does not change when P or T are altered although the molar and mass concentrations do. Another basis of expression frequently used in photosynthetic studies and gas analysis is the volume of the component gas per volume of total gas. From eqn 11.1 it is clear that for a given gas mixture, the volume/volume and mole fractions, as well as the pressure of the component

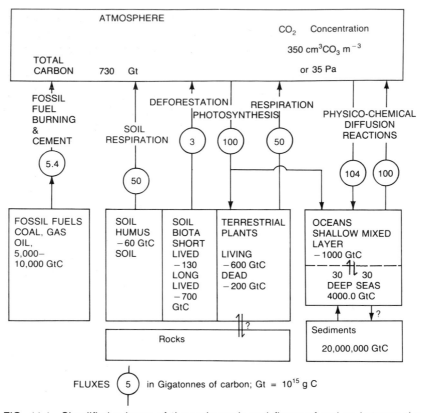

FIG. 11.1 Simplified scheme of the main pools and fluxes of carbon between the atmosphere, biosphere and geosphere. The size of each is given in giga tonne of carbon (GtC; 1 GtC = 10^{15} g C). The fluxes are also in GtC. The CO_2 concentration in the atmosphere is shown as volumes per million volumes (vpm) or as partial pressure (Pa). Data derived from the literature cited.

compared to the total pressure of the mixture, are identical and will change with, for example, T in the same way. Many units of volume concentration are used and may be confusing, e.g. $m^3 m^{-3}$, $L L^{-1}$ or $\mu L L^{-1}$; the latter is equivalent to 1 volume per million volumes and this is the frequently used (volume) parts per million, abbreviated to vpm or ppm. Volume units are also expressed as volume %, e.g. 1 vpm = 0.0001 volume %. However, SI units of volume concentration are $m^3 m^{-3}$ or preferred prefixes. Air contains about 350 cm^3 CO_2 m^{-3}, equivalent to 350 vpm or 0.035 % CO_2 and 0.21 m^3 O_2 m^{-3} or 210 000 vpm or 21 (volume) % O_2. Volumes of gas may be converted to mass and mole from eqn 11.1; air at 0 °C and 101 325 Pa contains 35 Pa CO_2 or 686 mg CO_2 m^{-3} or 15 mmol CO_2 m^{-3} and 300 g or 9.4 mol O_2 m^{-3}.

The contribution of a gas to a mixture of gases may also be described by the partial pressure. Dalton's law of partial pressures states that in a mixture of gases (which do not interact) the partial pressure of each gas is the pressure which it exerts if occupying the volume alone and the total pressure is the sum of the partial pressures. For air (ignoring rare gases)

$$P_{total} = P_{O_2} + P_{N_2} + P_{CO_2} + P_{H_2O} \qquad [11.2]$$

Partial pressure of CO_2 in air at standard T and P is currently 35 Pa (or in non SI units 350 μbar) and that of O_2 is 21.3 kPa (or 210 mbar). Partial pressure changes with T and P (eqn 11.1) and therefore must be corrected for altitude and temperature when making comparisons. As mentioned the mole fraction is equal to the partial pressure divided by the total pressure. Table 11.1 gives some conversion factors for units expressing the composition of CO_2 in air under standard conditions.

It is important to appreciate the great difference in concentration between CO_2 and O_2 in the atmosphere; O_2 is 600 times more concentrated than CO_2 which favours even inefficient oxygenation reactions compared with carboxylation reactions, for example, RuBP oxygenase compared with carboxylase. Plants must accumulate CO_2 from very dilute concentration and, at the same time, function at large O_2 concentration with large gradients of water vapour pressure between leaf and atmosphere. A leaf at 23 °C with the internal air saturated with water vapour (100 per cent relative humidity) has a vapour pressure of about 2.8 kPa compared with 1.5 kPa of air at 25 °C and 50 per cent relative humidity; this corresponds to a gradient of water content of about 12 g m^{-3}. Thus, leaves must absorb CO_2 whilst limiting loss of water vapour. Stomata function as 'control valves' which regulate these conflicting interests.

Global CO_2 balance

Carbon is held in organic compounds in living and dead biomass of plants, particularly forests (boreal, temperate and tropical), in peat and in soil

Table 11.1 Conversion factors for CO_2 and O_2 concentrations in air at 20 °C and 101 315 Pa

---

1 μl CO_2 litre^{-1} = 41.6 × 10^{-6} mol CO_2 m^{-3} = 1.83 × 10^{-3} g CO_2 m^{-3}

1 mg CO_2 m^{-3} = 0.0227 mmol m^{-3} = 0.554 μl CO_2 litre^{-1}

1 mol CO_2 m^{-3} = 44 g m^{-3} = 24.4 cm^{-3} litre^{-1}

21% O_2 = 210 cm^3 O_2 litre^{-1} = 0.21 m^3 O_2 m^{-3}

1 mol O_2 m^{-3} = 32 g m^{-3} = 24.4 cm^3 litre^{-1}

Pressure

1 μl litre^{-1} = 1.01 μbar = 0.101 Pa

340 μl CO_2 litre^{-1} = 340 μbar = 34.4 Pa

21% O_2 = 210 mbar = 21.3 kPa

---

humus of the terrestrial biosphere and also in the living and dead biomass of the oceans and the ocean floors (Fig. 11.1). There are, however, much larger quantities of carbon of plant origin in coal, oil and oil shales and natural gas. These quantities of carbon are insignificant compared to the huge sedimentary deposits of carbonates in the ocean sediments and in rocks. Also, oceans hold some 40 000 Gt of carbon as CO_2 and bicarbonate ions, which is the major store of readily remobilized carbon and an enormous buffer for the world's atmospheric CO_2. The atmosphere acts as a pool linking the carbon fluxes between these solid forms of carbon. There are very large fluxes between the oceans, continents and atmosphere; changes in their magnitude, even if quite small, affect the partial pressure of CO_2 in the atmosphere. One of the major exchange processes for carbon is in photosynthesis with a flux of 100 Gt per year of which 40 Gt is to the oceans and 60 Gt per year to the terrestrial vegetation. Standing terrestrial biomass is of the order of 600 Gt carbon. However, these large fluxes fixing carbon are almost offset by the respiration of the biomass and soil. The rate of exchange between the pools is very different; from vegetation to atmosphere the time scale is tens to hundreds of years, from soil to air probably an order of magnitude slower, between shallow seas and deep water of the order hundreds of years but from these to ocean sediments requires thousands of years.

Currently the atmosphere holds 730 Gt of carbon as gaseous CO_2 and this has increased by some 20–30 Gt since the mid-19th century; in terms of partial pressure of CO_2 this represents an increase from 27 Pa (270 vpm) in 1850 to the 1990 value of over 35 Pa (354 vpm). Measurements made since 1957 in Hawaii at Mauna Loa, an elevated site unaffected by vegetation and surrounded by oceans, show that CO_2 is increasing rapidly (Fig. 11.2) and has accelerated from about 0.7 Pa per decade early this century to about 1 Pa per decade between 1973 and 1988 although recent increases may have been even faster, 0.15 Pa per year. The Mauna Loa data shows a seasonal cycle of atmospheric CO_2 (amplitude about 0.8 Pa) which is related to the large CO_2 production and very low rates of photosynthetic uptake in the Northern hemisphere (which has a larger terrestrial landmass than the Southern) during the winter and the converse in summer. The current increase in atmospheric CO_2 is largely, if not exclusively, the result of increased human activity including burning fossil fuels and removal of forests; some 11 000 km^2 per year of tropical forest is currently destroyed. This is equivalent to 0.4–2.5 Gt carbon per year released by burning. Over the last century industrialization, based on exploitation of fossil carbon reserves, coupled with the rapid growth of human population and demand for agricultural land are rapidly increasing evolution of CO_2 and simultaneously decreasing accumulation of carbon in standing vegetation, particularly in the tropics, although carbon may be accumulating in photosynthetic products in temperate and boreal forests. Deforestation also allows soil organic matter to be oxidized, as does drainage of peat lands and marshes and their use in agriculture. All these processes contribute to increased atmospheric CO_2. Rate of consumption of fossil fuels plus production of cement has accelerated in recent decades (Fig. 11.2) and currently they release 5–6 Gt carbon per year. There is little indication of the rate decreasing in the near future, nor will it as the developing economies grow further. Indeed, increased consumption is likely and therefore further rapid increase in atmospheric CO_2. The role of photosynthesis in the removal of CO_2 from the atmosphere is clear and it is likely that the increased CO_2 will stimulate photosynthesis, thus partially absorbing the gas and increasing organic carbon. It is unclear if photosynthetic organisms will have the capacity to keep increasing their accumulation particularly if water and nutrient shortages restrict the ability of plants to grow effectively. It is also possible that increased temperatures will stimulate respiration of plants and the other biota thus offsetting the gains from enhanced photosynthesis. Currently, attempts to calculate the carbon balances are not entirely successful; indeed 50 per cent of the CO_2 evolved cannot be accounted for. Possibly it is being sequestered by the temperate and boreal forests, etc., in roots, soil humus, etc. It is unlikely

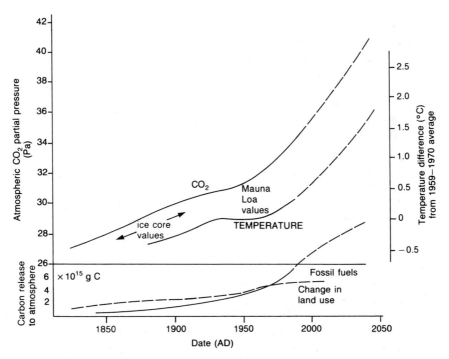

FIG. 11.2 The trends in atmospheric CO_2 partial pressure and global temperature since 1850 and projected to 2050. These changes (which are based on experimental evidence and climate modelling) are subject to uncertainty and are currently the subject of much debate. The correlation of increased CO_2 with use of fossil fuels is firmly established; the link to climate change is discussed in the text and the literature cited.

to be entering the oceans where plant growth is severely inhibited by deficiencies of nutrients.

Effects of CO_2 on world climate

Carbon dioxide absorbs infra-red radiation (p. 241); short wave radiation warms the earth and plants, which re-emit long wave infra-red radiation. This is absorbed by CO_2 and the atmosphere heats up, producing a 'greenhouse' effect, so-called because it is analogous to heating in greenhouses caused by poor transmission of infra-red compared to short-waved radiation by glass. Other gases also absorb infra-red radiation effectively, e.g. methane is 20 times more efficient than CO_2 at absorbing infra-red. Currently methane production, e.g. in anaerobic paddy fields, urban refuse disposal and by ruminants, is increasing rapidly. Increased atmospheric CO_2 and methane are expected to increase the atmospheric temperature. The extent of global warming is hotly debated, but much evidence suggests that surface temperatures are increasing. Over the last century temperatures have risen by approximately 0.7 °C and some of the hottest years of this century have occurred in the 1980s. These observations may also be linked with increased droughts in many areas, e.g. the Sahel. Factors controlling climate are complex; some feedback processes may reduce potential climate change. For example, a warmer atmosphere increases evaporation, resulting in more clouds which have a very large albedo, thus reflecting radiation back into space and so decreasing the input of energy. The role of vegetation in these complex processes cannot be modelled accurately with current methods although it is clearly important to place vegetation and photosynthetic processes in their correct, central role between the geosphere and atmosphere. Mathematical modelling of carbon fluxes in the earth from earliest times (Berner and Lasaga 1989) suggest that 100 million years ago the atmosphere contained 20–30 times more CO_2 than today. However, by 60 million years ago there was probably only twice the present concentration. Atmospheric CO_2 has fluctuated considerably, for example, there was an increase of CO_2 some 40 million years ago associated with its release by volcanic activity caused by sea-floor spreading and continental drift. In the much more recent past, the last ice age, for example, CO_2 content of the atmosphere was probably very small. Thus, plants have faced substantial, if slow, changes in CO_2 and climate during their evolutionary history. Changes in climate may affect natural vegetation and agriculture and may seriously disrupt the stability of ecosystems; it is therefore important to assess the changes.

Solubility of CO_2 and O_2 in water

Photosynthesis consumes forms of CO_2 dissolved in water, as enzymatic reactions proceed only in the aqueous state. Ribulose bisphosphate carboxylase uses dissolved CO_2, and PEP carboxylase bicarbonate ions.

Solubility of CO_2 and O_2 depends on solvent temperature, partial pressure of the gas, and the chemical nature of gas and solvent. For dilute solutions, Henry's law states that at constant temperature the volume or mass of gas dissolved in a given volume of liquid is directly proportional to the pressure. Table 11.2 gives the solubility of CO_2 and O_2 in water. For air 6.51 cm^3 O_2 dissolve in 1000 cm^3 or 0.291 mol m^{-3} (291 μM). For CO_2 at a partial pressure of 34 Pa and with solubility of 0.888 m^3 CO_2 m^{-3} water, 0.29 cm^3 dissolve in 1000 cm^3 or 12 mmol CO_2 m^{-3} (12μM) (see Table 11.3).

Gases are less soluble, generally, at warm temperatures than cold (Table 11.2). Solubility of gases is inversely proportional to the concentration of solutes; in complex biological fluids proteins and salts decrease the solubility of all gases similarly.

Carbon dioxide in solution

In pure water over 99 per cent of CO_2 is in solution, but some is hydrated to carbonic acid, H_2CO_3, which dissociates to give the bicarbonate ion,

Table 11.2 Solubility of CO_2 and O_2 in pure water at 101 325 Pa pressure of the gas as a function of temperature

$T(°C)$	mol CO_2 m^{-3}	m^3 CO_2 m^{-3}	mol O_2 m^{-3}	m^3 O_2 m^{-3}
0	76.4	1.71	2.17	0.049
10	51.4	1.19	1.68	0.038
15	43.1	1.02	1.50	0.034
20	36.5	0.87	1.36	0.031
25	31.1	0.76	1.23	0.028
30	26.7	0.67	1.12	0.026
40	21.2	0.53	1.01	0.023

Table 11.3 Solubility of CO_2 and O_2 in water when in equilibrium with air containing 32 Pa (320 μbar) of CO_2 and 21 kPa (21 per cent) O_2 at atmospheric pressure as a function of temperature

$T(°C)$	μM CO_2	cm^3 CO_2 $(m^3$ water$)^{-1}$	μM O_2	m^3 O_2 $(m^3$ water$)^{-1}$	Molar ratio O_2/CO_2
0	23	515	458	0.0102	19.9
10	16	371	356	0.0079	22.2
20	12	290	291	0.0066	24.3
25	10	244	263	0.0059	26.3
30	9	223	245	0.0055	27.2
40	7	179	216	0.0048	30.9

After Šesták, Čatský and Jarvis (1971)

HCO_3^-, and H^+, decreasing pH:

$$CO_2 \; \rightleftharpoons \; CO_2 + H_2O \; \rightleftharpoons \; H_2CO_3 \; \rightleftharpoons \; H^+ + HCO_3^-$$
$$\text{(gas)} \qquad \text{(dissolved)} \qquad \text{(solution)} \qquad \text{(solution)}$$

\hfill [11.3]

Formation of carbonate ions, CO_3^{2-}, from $HCO_3^- \rightleftharpoons CO_3^{2-} + H^+$ is not considered further. If H_2CO_3 or HCO_3^- are removed by chemical reactions then the apparent solubility of CO_2 changes. As H_2CO_3 in solution is only 1/400 of the other components it may be neglected. The pH of solutions of CO_2 in water depends on the molar concentration of CO_2, the gas phase CO_2 partial pressure, the dissociation constant (pK) and bicarbonate ion concentration. This is expressed in the Henderson–Hasselbalch equation:

$$pH = pK + \log [HCO_3^-] - \log [CO_2] \qquad\qquad [11.4]$$

The pH determines the balance between CO_2 dissolved and the bicarbonate ion concentration; H^+ ions drive the reaction in favour of CO_2 (by the law of mass action) so that in acid solutions there is little bicarbonate (acids are used to remove bicarbonate from solution and produce CO_2).

Changes in CO_2 and bicarbonate concentration are important to photosynthetic organisms, algae, for example, grow in acid or alkaline waters differing greatly in temperature. With increasing acidity and temperature the concentrations of bicarbonate ions and dissolved CO_2 decrease, and with them the supply of substrate for photosynthesis. Also, within photosynthesizing tissues the cellular fluids such as the chloroplast stroma have a variable pH which, together with temperature, influences the CO_2 concentration. The solubilities of CO_2 and O_2 and their ratio in equilibrium are shown in Table 11.3. Solubility of O_2 is not affected by pH so the CO_2/O_2 ratio rises greatly with increased alkalinity. As temperature increases, solubility of O_2 decreases less than that of CO_2 so the ratio of O_2/CO_2 increases substantially.

CO_2 and bicarbonate equilibria and photosynthesis

The chloroplast stroma is alkaline during illumination and bicarbonate ions predominate, but Rubisco uses CO_2 as substrate, not HCO_3^-. This is shown by supplying the enzyme with $^{14}CO_2$ or $H^{14}CO_3^-$ at high pH and low temperature (10 °C), where HCO_3^- formation from CO_2 is very slow. With $^{14}CO_2$ the reaction can proceed and ^{14}C is incorporated into 3PGA but $H^{14}CO_3^-$ is not used. By adding the enzyme carbonic anhydrase which increases the equilibration rate greatly, $H^{14}CO_3^-$ is used as well, as it is converted to $^{14}CO_2$. During photosynthesis, in order to avoid starving Rubisco of substrate CO_2, the rate of supply of CO_2 must match the rate of reaction. The rate of conversion of bicarbonate to CO_2 in alkaline conditions is slow, so that both the CO_2 concentration in solution and the rate of supply to Rubisco could limit assimilation at high pH. However, the rate of conversion is increased a hundred-fold in tissues by carbonic anhydrase; as CO_2 is depleted it is rapidly produced from bicarbonate. Conversely if CO_2 is available in solution but HCO_3^- is needed (by PEP carboxylase) then the rate of HCO_3^- formation is also increased by carbonic anhydrase.

Carbonic anhydrase, a protein of 180 kDa molecular mass and containing zinc, is found in many photosynthetic tissues, often in very large amounts, particularly in the chloroplast and at cell membranes. It also occurs in non-photosynthetic tissues and in animals. The enzyme, which has an extremely large turnover number (10^6 s^{-1}) facilitates the diffusion of CO_2 by speeding up the formation of HCO_3^- in the cytosol and maintaining a large gradient of CO_2 concentration. Also, carbonic anhydrase allows HCO_3^- to act as a buffer (albeit small) to provide CO_2 if the supply temporarily fails. In the chloroplast, carbonic anhydrase prevents depletion of CO_2 and probably maintains a high partial pressure of CO_2 at the active site of Rubisco, only slightly below that of the intercellular spaces, and minimizes the effects of low atmospheric CO_2. Plants or algae grown in high concentrations of CO_2 contain less carbonic anhydrase than those grown in low concentrations. Also, chloroplasts of C3 plants contain more carbonic anhydrase than C4 plants, which have a CO_2 concentrating system, although the enzyme is active in the C4 mesophyll. Algae, particularly in alkaline natural waters, have carbonic anhydrase at the cell surface which increases the rate of supply of CO_2 to the cells.

Carbon dioxide concentrating mechanisms

Metabolic systems which increase the effective CO_2 concentration for Rubisco are familiar from discussion of the C4 and CAM mechanisms. The advantages are apparent even though CO_2 diffuses to Rubisco active sites in leaves mainly in air and only over short distances in the liquid phase of the cell. In aquatic algae and higher plants CO_2 and bicarbonate ions must diffuse through a water film around the surfaces of cells into tissues. As diffusion of CO_2 in air is 10^4 times greater than in water, the problem of CO_2 supply in aquatic environments is much greater than in the aerial environment. In acid water the total carbon content is small and the bicarbonate$-CO_2$ equilibrium lies far towards CO_2.

Aquatic organisms, e.g. cyanobacteria, green algae and aquatic higher plants (e.g. *Elodea*), are limited by the carbon supply for photosynthesis. In some organisms active transport systems which increase the CO_2 concentration in cells have developed; their effect is similar to that of C4 photosynthesis (p. 184). The concentrating mechanism is an active metabolic pump which transports bicarbonate ions and leads to internal concentrations 40$-$1000-fold greater than in the medium. The pump in *Anacystis* is a 42 kDa polypeptide on the cytoplasmic membranes, using energy from photosynthesis and ATP. Probably an exchange of bicarbonate and CO_2 with H^+ and Na^+ occurs. Carbonic anhydrase may be combined with the pump to maintain the required rate of supply of substrate for photosynthesis under adverse conditions of pH, etc.

Cyanobacteria often grow in environments where the CO_2 concentration is 5−20 times smaller than the K_m (CO_2) of Rubisco. A bicarbonate pump may operate at the plasmalemma, possibly with Na^+ ions involved in exchange. Rubisco within the cell is enclosed in a protein shield (in which carbonic anhydrase may also be involved) with the small subunits of Rubisco possibly forming part of the shield. This carboxysome may function to increase the CO_2 in the shield; bicarbonate is pumped into the shield and converted to CO_2 (carbonic anhydrase speeding up the processes), thus increasing the CO_2 in the carboxysome so that Rubisco can function efficiently. Such mechanisms are very important adaptations to extreme environments and may be induced by the availability of CO_2.

Carbon dioxide movement to photosynthetic cells

Carbon dioxide in the chloroplast stroma is removed by the RuBP carboxylase reaction and a gradient of CO_2 develops across the chloroplast envelope, cytosol, cell membranes and walls to the intercellular spaces and, via the stomata, to the ambient air (Fig. 11.3). The gradient is the driving

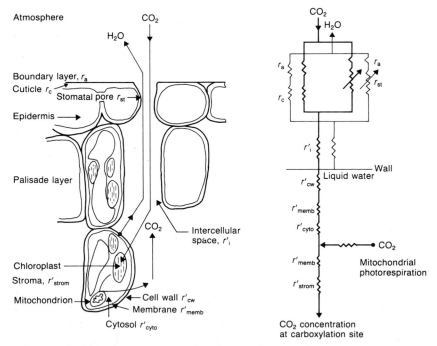

FIG. 11.3 Pathway for water vapour loss from, and carbon dioxide entry into a leaf in the light, shown as an electrical analogue with resistances denoting sites of restricted diffusion i.e. small conductances (conductance is the reciprocal of resistance) related to leaf structure.

force for CO_2 diffusion, but the rate at which this occurs to the reaction site depends on the conductances (reciprocal of resistance) to CO_2 diffusion in the gas and liquid phases in the leaf and atmosphere and on external CO_2 concentration. In the atmosphere CO_2 diffuses toward and O_2 away from the leaf during illumination and the fluxes are reversed in darkness; water vapour diffuses away from the wet internal surfaces in the leaf to the dry atmosphere under most conditions. In the turbulent air, gas concentrations are uniform, due to rapid mixing by mass flow. In the intercellular gas spaces and also in the liquid spaces in the cell wall and cytosol, the movement of gases is by diffusion, not by mass mixing. In the 'boundary layer' surrounding the cell or leaf there is incomplete mixing by mass flow and gases move partially by diffusion. Diffusion is the movement of molecules of a substance from higher to lower concentration and diffusion coefficients depend on molecular species, the medium, and temperature and pressure. Graham's law states that the diffusion coefficient for a gas in air is approximately equal to the reciprocal of the square root of the molecular mass. Thus, H_2O (mass 18) and CO_2 (mass 44) have, in air, diffusion coefficients of 0.257 and 0.16 cm^2 s^{-1}, respectively, so that the ratio of D_{H_2O}/D_{CO_2} is approximately 1.6 so it is possible to convert conductances for water vapour to those for CO_2 by this factor. In the boundary layer over a leaf with the transition from diffusion to turbulence, simple diffusion does not apply and the rate of diffusion of water vapour is approximately 1.37 times that of CO_2. Diffusion coefficients increase for all gases by approximately 1.3 between 0 and 50 °C so the ratios are almost constant over the physiological temperature range. The magnitude of the diffusion coefficients and their ratio are important in calculating fluxes and for determining the diffusion of one molecular species from measurements of another. Diffusion depends upon the temperature and pressure of the medium and is very rapid over short distances; it is faster in gases than liquids. Diffusion of CO_2 in air (0.16 cm^2 s^{-1}) is four orders of magnitude greater than in water (0.16 $\times$ 10^{-4} cm^2 s^{-1} at 15 °C). Diffusion of gases and ions in the liquid phases of cell walls and cytosol is particularly important for it slows the flux of materials and may limit the rate of photosynthesis.

Measurement of leaf water vapour and CO_2 exchange

Fluxes of CO_2, O_2 and H_2O between a leaf and surrounding atmosphere may be measured in several ways both separately and, more usefully in combination, to provide understanding of their interaction and dependence on the plant characteristics. Physiological measurements are generally made by enclosing a leaf of known projected surface area (L, m^2) (Fig. 11.4) in a transparent chamber through which air (or another gas) flows (mol s^{-1}). The air is stirred to maximize boundary layer conductance. Water vapour pressure is measured by a suitable sensor, e.g. dew point hygrometer or capacitive resistance sensors, in the air entering and leaving the chamber

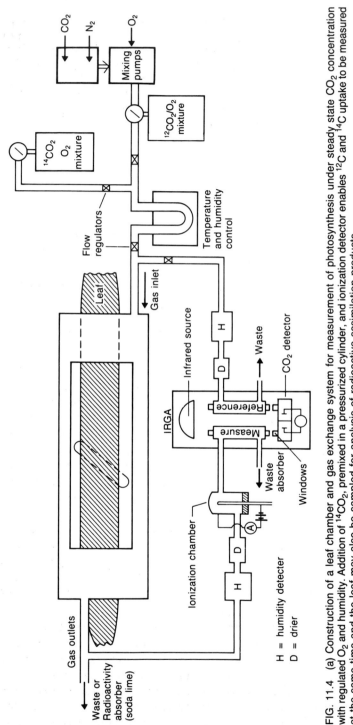

FIG. 11.4 (a) Construction of a leaf chamber and gas exchange system for measurement of photosynthesis under steady state CO_2 concentration with regulated O_2 and humidity. Addition of $^{14}CO_2$, premixed in a pressurized cylinder, and ionization detector enables ^{12}C and ^{14}C uptake to be measured at the same time and the leaf may also be sampled for analysis of radioactive assimilation products.

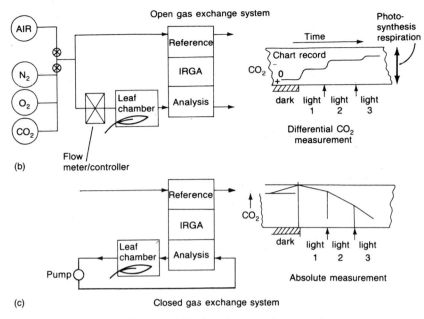

FIG 11.4 (continued) (b) Open and (c) closed gas exchange systems

and the water vapour flux (mol m^{-2} s^{-1}) is calculated from the differences in water vapour pressure, flow rate and leaf area. The mean water vapour in the chamber air is calculated from the inlet and outlet humidity and that in the intercellular spaces is determined by measuring the leaf temperature and assuming that air in the leaf is at saturation at that temperature. The boundary layer is estimated by using wet replica leaves or from the leaf energy balance. This basic approach is also used in porometers which rapidly measure water vapour flux from the leaf surface, leaf temperature and humidity of the air to compute conductance. Carbon dioxide exchange may be measured at the same time as water vapour using infra-red analysers (IRGA) to determine the concentration of CO_2 in the air entering and leaving the leaf chamber; from this, the flow rate and leaf area, the assimilation is determined. Infra-red gas analysis depends on the absorption of infra-red radiation by CO_2 over the range $2.5-25$ μm. The main peaks of CO_2 absorption are at 2.7, 4.2 (principal band) and 17 μm. An IRGA is constructed of 2 tubes, with windows at each end made of fused silica or calcium fluoride, which are transparent to infra-red radiation. Radiation from a source is passed through the end windows along the tube to a detector; CO_2 molecules in the tubes absorb radiation and so radiation reaching the detector is inversely proportional to the number of CO_2 molecules in the light path. Differences in radiation absorption caused by differences in CO_2 concentration in the tubes is detected by the instrument and may be calibrated to provide a measure of CO_2 concentration of air entering and leaving a leaf chamber. Water vapour also absorbs infra-red radiation (in 3 major bands at 2.5–2.8 μm, which overlaps with one CO_2 band, 5–8 and 18–28 μm). As water vapour is at much greater concentration than CO_2 in the normal atmosphere, it must be removed or special optical filters used to block the water absorption bands. IRGAs are sensitive to differences in partial pressure of 0.01 Pa CO_2 over a range of 0–100 Pa; thus leaf CO_2 exchange is routinely measured in flowing gas streams.

Oxygen pressure may be measured by paramagnetic detectors or zirconium chemical cells but these methods are generally too insensitive to detect the small fluxes in O_2 from leaves against the very large background of O_2. In small volume, closed chamber polarographic oxygen electrodes are sufficiently sensitive to determine the exchange but have not been generally applied to open gas exchange systems.

The type of system shown diagrammatically in Fig. 11.4b is an open system, for the gas flows through the chamber, is of constant composition and no recirculation of the gas occurs. The rate of CO_2 assimilation and the composition of the atmosphere are constant, providing light and other features of the plant's environment are not altered. The difference between the CO_2 in the gas entering and leaving the chamber is used to calculate the exchange rate. By altering the CO_2 in the gas stream entering the chamber, CO_2 response curves can be constructed and also by changing irradiance, light responses can be determined. Closed gas exchange systems, in contrast, (Fig. 11.4c), recirculate the air over the leaf so the composition of the atmosphere changes with time and the leaf experiences different humidity and CO_2 and its rate of photosynthesis decreases with CO_2 depletion from which the photosynthetic rate is calculated. This type of system may also be used to determine, for example, the CO_2 compensation point by allowing the leaf to come to an equilibrium with respect to CO_2 in the atmosphere; it is difficult to seal closed systems against the inflow of CO_2 from outside and allowances must be made for this. Yet another approach for measuring photosynthesis is to use a closed system but to inject gas containing CO_2 into the chamber to maintain a gas phase of required, constant composition; the rate of injection of CO_2 provides a measure of photosynthesis. This method rests on the use of sensitive, accurate flow monitors and controllers which respond to the IRGA. Maintenance of constant humidity in closed gas exchange systems is more difficult than in open systems and is achieved by circulating part of the chamber air through a desiccant.

Measurement of photorespiratory CO_2 efflux may be made by allowing a leaf in an open gas exchange system to reach steady state net photosynthesis with $^{12}CO_2$ and then switch to a gas of identical composition except that it contains $^{14}CO_2$ of known specific activity. The depletion of $^{14}CO_2$ from the gas in the first 15–60 s after passage over the leaf (measured with a flow-through ionization chamber or scintillation detector of small volume) allows the gross flux of CO_2 into the leaf to be calculated from the ^{14}C

depletion and the specific activity. During this short period ^{14}C enters the PCR cycle and the glycolate pathway but very little ^{14}C is evolved in photorespiration, hence allowing gross photosynthesis to be determined.

Determination of the number of photons required for the fixation of a given amount of CO_2 requires accurate estimation of the absorption of photons by the leaf, independent of reflection or transmission of radiation. True quantum yield (see p. 49) is determined in an integrating or Ulbrich sphere. This device has a highly reflective inner surface (coated with Eastman white reflecting paint, for example) which allows the absorbed light to be measured without the effect of reflectance or transmittance. A chamber enabling the CO_2 exchange of the leaf to be measured as described earlier is placed in the integrating sphere and photosynthesis measured over a range of irradiances within the sphere. Light enters the chamber from light guides and is reflected to give diffuse radiation. The irradiance at the sphere surface is determined by quantum sensors and the difference between the irradiance measured with and without the leaf in the chamber is used to determine the net photon absorption. This is then related to CO_2 exchange close to the light compensation point. From this the true quantum yield is obtained. Details of techniques may be obtained from Sestak *et al.* (1971), Pearcy *et al.* (1989) and Hashimoto *et al.* (1990).

Pathway of water vapour flux and stomatal conductance

Evaporation from the water saturated cell surfaces determines the vapour pressure in the intercellular spaces. As this cannot be measured directly it is taken as the saturated vapour pressure of pure water at the bulk leaf temperature. Mesophytic leaves are essentially isothermal throughout. Only under extreme wilting is the vapour pressure below saturation. For a leaf in a gas exchange chamber (Fig. 11.4) with air of controlled, unsaturated humidity entering and passing over the surface (as in a steady state open gas analysis system) the rate of transpiration (E, mol H_2O m^{-2} s^{-1}) for a leaf of plan area L (m^2) is given by the difference in the water vapour content of air entering and leaving the chamber times the flow rate. With a molar flow into the chamber of v_e (mol s^{-1}) and a mole fraction of water m_e (mol H_2O mol^{-1}) and v_o and m_o the flow and mole fraction of the air leaving the chamber:

$$E = \frac{v_o m_o - v_e m_e}{L} \qquad [11.5]$$

A correction is required for the addition of water vapour to the air. The corresponding correction for exchange of CO_2 by a photosynthesizing leaf is negligible.

Conductance of the pathway for water vapour for evaporating water surfaces to the bulk atmosphere is complex as Fig. 11.3 shows. Assuming that it is made up of the stomatal conductance (g_s), cuticular conductance

(g_c) and a boundary layer conductance (g_a) the total conductance is

$$g_l = g_s + g_c + g_a$$

The g_l may be calculated as the constant of proportionality between the vapour concentration gradient from the inside of the leaf to the bulk air and the transpiration rate. This is a form of Fick's law where the flux is proportional to the gradient and inversely proportional to the resistance (the reciprocal of conductance) of the path. Thus:

$$g_l = \frac{E \, (1 - m)}{(m_i - m_a)} \qquad\qquad [11.6]$$

where

$$m = \frac{(m_i + m_a)}{2}$$

and m_i and m_a are the mole fractions of water vapour in air inside the leaf and in the ambient air, respectively. The units of conductance are $mol \ m^{-2} \ s^{-1}$. To measure the boundary layer conductance a replica leaf surface (e.g. wet paper) with no stomata or cuticle is substituted. However, it is difficult to determine the true conductances on a complex structure like the leaf where stomata are distributed often non-uniformly on both sides; if the boundary layer conductance is large then these errors are relatively small.

Boundary layer thickness and therefore conductance at the leaf surface varies with wind speed, surface dimensions and characteristics such as hairiness. At wind speeds of $0.5-10 \ m \ s^{-1}$, g_a is $0.2-4 \ mol \ m^{-2} \ s^{-1}$. Effects of g_a on transpiration and CO_2 assimilation depend on its magnitude relative to other conductances in the system. In general g_a is large compared to g_s (2 cf. $0.4 \ mol \ m^{-2} \ s^{-1}$ with open stomata) and so has little control over water or CO_2 flux. To minimize the effects on the boundary layer (and keep it constant) during measurements of the fluxes and calculation of stomatal conductance, leaf chambers are vigorously stirred. The waxy cuticle is a very effective barrier to water loss with small conductance (g_c, $0.005 \ mol \ m^{-2} \ s^{-1}$ or smaller) but it also prevents the entry of CO_2. Plants appear to be in an evolutionary impasse, with large surface area required for light and CO_2 capture but losing water as a consequence. Desiccation is a major limitation to growth of terrestrial plants which have not developed a material allowing diffusion of CO_2 but not water vapour.

Stomatal pores in the cuticle offer a high conductance pathway (g_s) for CO_2 flux into the leaf, but allow H_2O to escape. Cuticular and stomatal pathways operate in parallel and both are in series with the boundary layer (Fig. 11.3). Stomata occur on leaf surfaces in variable numbers ($0-3 \times 10^8 \ m^{-2}$) on both surfaces (amphistomatous) or only on upper (ab-) or lower (adaxial) surfaces. The ratio of total stomatal pore area to leaf surface is about 1 per cent so that water vapour diffusion out of, and CO_2 diffusion into, the leaf occur only through a very small part of the leaf area; the maximum g_s of mesophytic C3 crop plants is $1.2-0.4$ mol m^{-2} s^{-1} for H_2O vapour, but many trees and C4 plants have smaller maximum stomatal conductance than C3 plants. The stomatal pores are bounded by guard cells which regulate the width and area of the aperture via changes in turgor pressure, controlling CO_2 entry and, possibly of more importance for land plants, water vapour loss. Conductance of closed stomata is of the order of 0.01 mol m^{-2} s^{-1} and probably approaches cuticular conductance. As the maximum g_s is much greater than cuticular conductance (g_c) the efflux of water and influx of CO_2 occurs via the pores. Conditions which encourage stomatal opening — adequate water supply, bright light and high humidity — stimulate rapid photosynthesis; some aspects of stomatal physiology are discussed on page 246.

Carbon dioxide exchange and conductances

The pathway for CO_2 entry and exit from the inside of the leaf is shown in Fig. 11.3. In the light CO_2 is removed from the intercellular spaces by the metabolic processes and diffuses into the leaf from the air. In darkness, respiration produces CO_2 which diffuses from the cells into the intercellular spaces and thence into the atmosphere. The net CO_2 assimilation rate (P_n) of leaf area L (m^2) is determined in an open gas exchange system (Fig. 11.4) from the difference in CO_2 content of the air entering and leaving the chamber and the flow rates:

$$P_n = \frac{v_o c_o - v_e c_e}{L} \qquad [11.7]$$

where v_e and v_o are the flow of air (mol s^{-1}) and c_e and c_o are the mol CO_2/mol air entering and leaving the chamber, respectively. Then P_n has units of mol CO_2 m^{-2} s^{-1}. The same considerations about the differences in flow into and out of the chamber apply as in discussion of water vapour exchange.

The conductance of the leaf surface and the rate of photosynthesis determine the concentration of CO_2 in the intercellular spaces c_i. It is important to know this value because the CO_2 concentration at the cell surface is the effective supply of CO_2 which the assimilating cells experience within the leaf (see Ch. 4). Thus, it is possible to relate metabolic processes to CO_2 more directly than if related to atmospheric CO_2. Calculation of c_i is from:

$$c_i = c_a - \frac{P_n}{(g_s' + g_c' + g_a')} \qquad [11.8]$$

where $g_s' + g_c' + g_a'$ are the stomatal, cuticular and boundary layer conductances to CO_2. These conductances to CO_2 are calculated from the corresponding conductances to water vapour from the ratio of the diffusion coefficients (see section 239) by the factor 1.6. However, in the boundary layer due to the transition from turbulent to diffusional transfer, $g_a' = g_a'/1.37$. The CO_2 concentration of the intercellular, as opposed to substomatal, cavities depends on the geometry of the mesophyll air spaces and the conductance of the pathway; this is generally large ($4-0.8$ mol m^{-2} s^{-1}) and probably constant except in severely wilted leaves. Leaves with large photosynthetic rate have relatively large internal spaces allowing rapid diffusion of gases and a large cell surface to leaf surface ratio ($20-40$ in many mesophytes) which minimizes the gradients and provides large surface area for CO_2 absorption.

Extending the Fick's law concept to the flux of CO_2 from the cell surface to the site of fixation by Rubisco, we may consider the movement as across a series of 'resistances' in the liquid phases of the cell wall, cell membrane, cytosol, chloroplast membranes and stroma to the enzyme active site. These, in combination, may be expressed as a mesophyll conductance which is currently difficult to estimate because the CO_2 concentration at the enzyme site is unknown. Thus, the gradient is not known. It was once assumed that the CO_2 concentration in the chloroplast was the same as the compensation concentration but this cannot be so for P_n is then zero (see p. 260). Another view was that the conductances of the liquid phase (which are essentially determined by metabolic processes) were large so that the chloroplast CO_2 was very close to the air CO_2, although clearly some gradient must exist otherwise CO_2 would not diffuse to Rubisco. Current evidence is that the CO_2 content of the chloroplast is some 30 per cent below that of the intercellular space under normal atmospheric conditions. The relation between P_n and c_i is discussed at length in Chapter 12. Here we may mention that the carboxylation efficiency (μmol CO_2 assimilation m^{-2} s^{-1} Pa^{-1}) may be calculated from the slope of the relation between P_n and c_i when P_n is zero. If the liquid phase conductances to CO_2 in the cell were infinite then this would be a measure of metabolic efficiency, e.g. enzyme reactions and diffusion of metabolites in the cell. Because c_i may not reflect chloroplast CO_2 concentration then the relation includes the supply of CO_2. Carboxylation efficiency is related to metabolism, with C3 plants having smaller efficiency than C4 plants and it depends on the amount of Rubisco in the tissue, e.g. nitrogen deficiency decreases both Rubisco amount and activity and also the efficiency in C3 and C4 plants.

Stomatal structure and function

Stomata are microscopic pores ($c.$ $15-30$ μm long in *Vicia faba*) in the outer epidermal surfaces of plants, particularly the leaves of terrestrial plants. Carbon dioxide diffuses into the interior of the tissue and water vapour diffuses out through the stomata. Stomata are chemically and osmotically driven turgor operated valves which respond rapidly to environmental

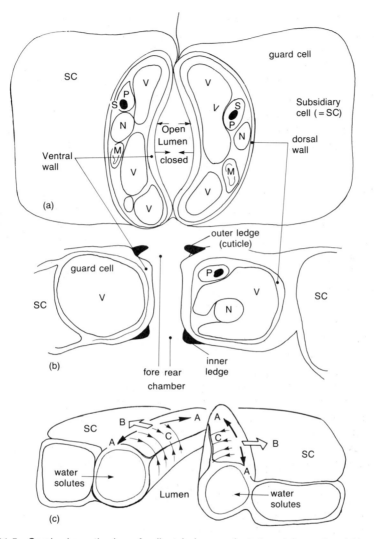

FIG. 11.5 Semi-schematic view of a dicotyledonous plant stomatal complex: (a) in plan view indicating cell contents; (b) in section. V, vacuole; N, nucleus; P, plastid with S, starch; M, mitochondrion; (c) forces in the stomatal apparatus generated when water and solutes enter the guard cell. Microtubules in the ventral walls, orientated as indicated (C) coupled with the thickenings prevent expansion into the lumen but allow it (B) into the subsidiary cells (SC) because the walls are thin and not braced longitudinally. Expansion is possible in direction A.

conditions. Specialized guard cells (Fig. 11.5a) which contain chloroplasts, mitochondria and large continuous vacuoles, bound the pore and are capable of movement, increasing the pore area when conditions are conducive to CO_2 fixation, e.g. in light, and when the water supply is adequate. They thus regulate CO_2 and H_2O fluxes and, via the latter, protect the plant from desiccation and are very important in plant survival strategies. Stomata were probably a prerequisite for plant invasion of dry land for they are found in fossils of *Rhynia*, a Psilophyte, present day examples of which have similar stomatal structure.

The form and structures of a stomatal complex are illustrated in Fig. 11.5a. The kidney-shaped guard cells of bean are typical of many dicotyledonous plants, whereas dumb-bell shapes are characteristic of grasses and other monocotyledons. It is important to understand the structure of the guard and subsidiary cell of the complex in order to analyse how movement of the guard cells and pore aperture are regulated. Guard cells are firmly anchored at each end and linked to the subsidiary cells by rather thin walls (Fig. 11.5b). In contrast the outer and inner ventral walls are thick with cuticular ridges; when the pore is closed these provide an effective seal. The cell walls are constructed of a cellulose matrix with micro fibrils running around the cell diameter of the outer and ventral walls (Fig. 11.5c). An analogy is that of radial steel-belted motor car tyres. When the guard cells absorb water via the thin walls connecting them to the subsidiary cells which is the pathway for rapid exchange of solutes and water, they swell and the forces generated result in the dorsal wall expanding into the subsidiary wall. However, because the ventral wall is thickened it does not expand longitudinally (pore and guard cell length remain almost constant between opening and closing) although the outer and inner more dorsal regions can expand between the tubules. Thus, the guard cell bows and the pore opens, the aperture increasing from zero to some $10-15$ μm. During opening guard cell volume may increase by 40 per cent from the volume of 4 pL per guard cell when closed. The first phase of opening (Spannungsphase) involves inflation of the guard cells and a large increase in turgor. Osmotic potentials increase by $4-6$ MPa with opening (compared to a range of $1-2$ MPa in epidermal and mesophyll cells generally) and water potentials decrease similarly, thus pulling water from the epidermis and increasing turgor greatly. Turgor generates the forces which distort the guard cell walls and open the pore.

What is responsible for the changes in turgor? Experimental determination of the amounts of ions and solutes in guard cells has provided the basis of the model in Fig. 11.6. Clearly metabolism regulates the processes, which involve energy and ion and organic solute fluxes resulting in marked changes in concentration between the closed and open states. These determine the passive water movement into the guard cell. During opening in bean, the K^+ ion concentration rises some $6-10$-fold (by 300 mM, i.e. K^+ increases from 0.3 pmol per guard cell in closed to 2.5 pmol in open stomata). This occurs within $0.5-2$ hours so large K^+ fluxes — of the order of 30 pmol cm^{-2} s^{-1} — occur. For every 1 μm increase in aperture the K^+ increases by 25 mM. Chloride ions also increase six-fold. Stomatal closure is faster than opening (15 minutes) and the reverse fluxes are correspondingly rapid.

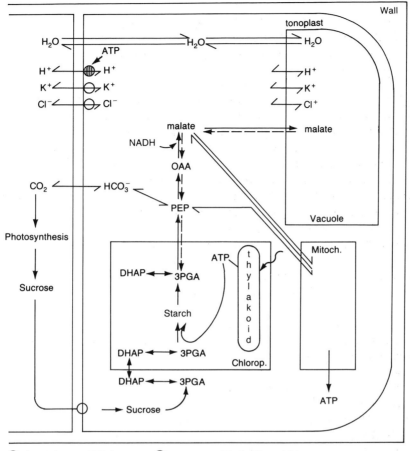

FIG. 11.6 Simplified scheme of ion fluxes and generation of organic metabolites in a stomatal guard cell which leads to accumulation of osmotica in the vacuole, increased turgor and pore opening.

During opening the membrane potentials in guard cells become more negative and there is a flux of protons (H^+) from the guard cells so that their pH rises from 5.1 to 5.6.

Associated with these changes are those involving organic compounds. Guard cells contain starch, indeed it was one of the earliest observed components shown to change with stomatal activity; closed stomata contain abundant starch and open stomata contain much less. From this arose the theory of starch hydrolysis as providing the osmoticum required for turgor generation and therefore opening. It is now known that malate also increases greatly — ten-fold — with opening, from 50 to 500 mM. Changes in this compound are related to the observed large amounts and activities of PEP carboxylase. However, there is rather little Rubisco or other PCR cycle enzymes and reduced amounts of chlorophyll, suggesting that guard cell chloroplasts are not very active in CO_2 fixation. In the light they probably do not synthesize starch but the products go to make malate. In darkness guard cells contain relatively little osmoticum and have low turgor, the walls are not stretched and the pore is closed. The relatively large content of starch in the plastids probably results from synthesis using malate which is abundant in the light and is converted to starch in darkness.

On illumination of stomata (when water is adequate) starch is remobilized using ATP from the thylakoid reactions or from mitochondria (blue light may stimulate their metabolism) to form PEP via glucose and fructose-6-phosphates and 3PGA and DHAP. PEP reacts with HCO_3^- (catalysed by PEP carboxylase) giving OAA which is reduced by malate dehydrogenase, using NADH, to malate. This enters the vacuole and has osmotic effects. At the same time K^+ and Cl^- are pumped into the cytosol via a plasmalemma ATPase and then into the vacuole via the tonoplast. Protons are exported from the cell. These processes generate additional osmotic potential and the turgor required for opening during the 'motorische phase'.

Regulation of stomatal operation is complex; clearly it must be flexible to be effective at maintaining plant homeostasis in the face of multiple environmental stimuli. Light has a major effect. Blue light, at low intensity, stimulates the fluxes of ions and metabolism; it probably operates via flavins and may set the system so that it can respond to other signals including additional response to white light. Carbon dioxide also influences stomatal behaviour. Large concentrations decrease stomatal aperture whereas low concentrations increase aperture but it seems that the CO_2 effect is rather indirect, although clearly involved in the regulation of HCO_3^- concentration and thereby the PEP carboxylase reaction and generation of malate. Probably CO_2 operates indirectly on the mechanism so that internal CO_2 of the leaf is maintained approximately, but not absolutely, constant under many conditions. Optimization of net photosynthesis and water loss is shown by stomata but quantitative aspects of the regulatory mechanisms are still to be analysed. Both the light and CO_2 responses are overridden by the water balance of the tissue. Water stress, with decreased leaf turgor, prevents opening probably because the necessary guard cell turgor cannot be generated. There is also rapid hormonal regulation of stomatal movement. Abscisic acid and farnesol, both sesquiterpenoids, are present in normal tissues and may have a role in the regulation of stomatal behaviour but in water-stressed plants their synthesis is greatly increased either via the C15 pathway from mevalonic acid or by the C40 route via carotenoid breakdown. The +ABA, increasing above a threshold, blocks the K^+ transporter on the plasmalemma and prevents opening. It thus acts as a long-term 'off switch' with high survival value for the plant when water loss exceeds supply.

Nutrition also affects stomata, e.g. inadequate phosphate decreases stomatal opening and reduces sensitivity to other environmental stimuli, but the mechanisms are not understood. Atmospheric humidity has also been shown to regulate stomatal action; dry air frequently causes pore closure, possibly because loss of water from the subsidiary cells upsets the turgor balance of the guard cells. There are many intriguing questions still remaining about the mechanisms of stomatal regulation and control by environment including what determines the behaviour of CAM plant stomata which open in darkness and close in the light.

Stomatal control of water and CO_2 exchange

Regulation of leaf gas exchange involves a balance between assimilation and water loss. For a given ambient CO_2, humidity and energy balance, light determines the assimilation rate and c_i decreases until a steady state is achieved between stomatal conductance, water loss and CO_2 uptake. If the leaf loses water and turgor decreases then stomatal conductance decreases and with it c_i and also P_n if it is not on the saturation part of the response curve. If c_i is saturating for P_n then stomatal closure may occur with reduction in water loss but no effect on P_n. Under conditions favourable for many plants, e.g. abundant water, light and moist air, stomata may open so that c_i rises and P_n increases. There is a complex, dynamic control system which maintains CO_2 fixation whilst preventing excessive water loss. Probably these factors are genetically determined by evolutionary selection of the mechanisms. For example, plants of very droughted environments may have inherently small assimilation rates (limited metabolic capacity) and operate with small stomatal conductance, whereas plants of well watered, nutrient rich environments may have large capacity and productivity but lose water and therefore be susceptible to drought conditions. Stomatal closure for prolonged periods is unusual in mesophytes of good environments, possibly because it induces photoinhibition and other forms of metabolic damage and increases photorespiration. Thus, stomata may regulate, indirectly, energy balances of cells providing protection for the biochemical mechanisms as well as regulating water and carbon fluxes.

Photosynthetic discrimination against carbon isotopes

Atmospheric CO_2 is composed of the stable isotopes of carbon, 98.9 per cent ^{12}C and 1.1 per cent ^{13}C. The thermodynamic and kinetic properties of compounds containing the different isotopes differ due to mass, and they participate to different extents in physical and chemical reactions, that is, there is discrimination, and the proportion of isotopes in products of the reactions changes. The $^{13}C/^{12}C$ ratio gives important information about the reaction and can be used as a tracer for the flow of carbon.

The ratio is determined by mass spectrometry of CO_2 (organic matter is first oxidized) and compared with a standard of known ratio, usually a belemnite (fossil shell) earth from the Peedee formation in South Carolina, USA and called the PDB standard. The ratio of unknown to standard isotope distribution is $\delta^{13}C$:

$$\delta^{13}C(\%_{00}) = \frac{(^{13}C/^{12}C) \text{ unknown substance} - (^{13}C/^{12}C) \text{ standard}}{(^{13}C/^{12}C) \text{ standard}} \times 1000$$

[11.10]

Negative $\delta^{13}C$ shows ^{13}C enrichment; atmospheric CO_2 is $-7\%_{00}$. Plant material $\delta^{13}C$ ranges from -10 to $-40\%_{00}$; C3 plants have -22 to $-34\%_{00}$ (mean $-27\%_{00}$), C4 plants -10 to $-18\%_{00}$ (mean $-13\%_{00}$) and CAM plants are in the range of C3 and C4 plants. These differences arise as Rubisco discriminates more strongly ($-28\%_{00}$) against $^{13}CO_2$ than PEP carboxylase ($-9\%_{00}$) due to the chemical reaction; Rubisco uses CO_2 and PEPc HCO_3^- ion. In C4 plants, CO_2 is concentrated in the bundle sheath and $^{13}CO_2$ is fixed more effectively than from the atmosphere. CAM plants use PEPc and also RuBP carboxylase directly, hence the wide range of $\delta^{13}C$ values. Measurement of $\delta^{13}C$ provides confirmation of the metabolic origin of organic carbon; it suggests that very old samples of organic carbon were derived from photosynthesis some $3.5-4$ billion years ago, thus setting this process very early in earth's biological history. Also $\delta^{13}C$ discrimination shows that ferns and bryophytes from the Carboniferous period had C3 photosynthesis, and distinguishes between sucrose from sugar cane (C4) and sugar beet (C3) as well as between artificial and natural vanilla. Isotope discrimination is affected by the physiological state of plants, for example, small stomatal conductance increases the utilization of CO_2 from the intercellular spaces and decreases discrimination against ^{13}C. Small stomatal conductance also reduces water loss so that discrimination is related to water use efficiency. By determining $\delta^{13}C$ composition (which is relatively rapid) for total plant dry matter accumulation over long periods it is possible to select plants with high water use efficiency.

References and Further Reading

Badger, M.R. (1987) The CO_2-concentrating mechanism in aquatic phototrophs, pp. 220–75 in Hatch, M.D. and Boardman, N.K. (eds), *The Biochemistry of Plants*, Vol. 10, *Photosynthesis*, Academic Press, London.

Ball, J.T. (1987) Calculations related to gas exchange, pp. 445–76 in Zeiger, E., Farquhar, G.D. and Cowan, I. (eds), *Stomatal Function*, Stanford University Press, Stanford.

Berner, R.A. and **Lasaga, A.C.** (1989) Modelling the geochemical carbon cycle, *Sci. Amer.*, **26**, 54–61.

Bolin, B. (ed.) (1981) *Carbon Cycle Modelling*, Scentific Committee on Problems of the Environment (Scope 16), John Wiley and Sons, Chichester.

Coombs, J., Hall, D.O., Long, S.P. and **Scurlock, J.M.O.** (eds) (1985) *Techniques in Bioproductivity and Photosynthesis* (2nd edn), Pergamon Press, Oxford.

Cure, J.D. and **Acock, B.** (1986) Crop responses to carbon dioxide doubling: A literature survey, *Agric. For. Meteor.*, **38**, 127–45.

Farquhar, G.D., Ehleringer, J.R. and **Hubrick, K.T.** (1989) Carbon isotope discrimination and photosynthesis, *A. Rev. Plant Physiol. Plant Mol. Biol.*, **40**, 503–37.

Field, C.B., Ball, T. and **Berry, J.A.** (1989) Photosynthesis: principles and field techniques, pp. 209–53 in Pearcy, R.W., Ehleringer, J., Mooney, H.A. and Rundel, P.W. (eds), *Plant Physiological Ecology, Field Methods and Instrumentation*, Chapman and Hall, London.

Gregory, R.P.F. (1989) *Biochemistry of Photosynthesis* (3rd edn), Wiley, Chichester.

Gribben, J. (ed.) (1986) *The Breathing Planet (A New Scientist Guide)*, Blackwell Ltd, Oxford.

Hashimoto, Y., Nonami, H., Kramer, P.J. and **Strain, B.R.** (1990) *Measurement Techniques in Plant Science*, Academic Press Inc., San Diego.

Houghton, R.A. and **Woodwell, G.M.** (1989) Global climate change, *Sci. Amer.*, **260**, 18–26.

Kellogg, W.W. and **Schware, R.** (1981) *Climate Change and Society*, Westview Press, Boulder, Colorado.

Lange, O.L., Nobel, P.S., Osmond, C.B. and **Ziegler, H.** (eds) (1983) *Physiological Plant Ecology IV. Ecosystem Processes: Mineral Cycling, Productivity and Man's Influence*, Springer-Verlag, Berlin.

Lemon, E.R. (1982) CO_2 and plants. The response of plants to rising levels of atmospheric carbon dioxide, *American Association for the Advancement of Science, Symposium 84*, Westview Press, Baltimore.

Liss, P.S. and **Crane, A.J.** (1983) *Man-made Carbon Dioxide and Climate Change. A Review of Scientific Problems*, Geo Books, Norwich.

Long, S.P. (1985) Leaf gas exchange, pp. 453–99 in Barber, J. and Baker, N.R. (eds), *Topics in Photosynthesis*, Vol. 6, Elsevier, Amsterdam.

Mannion, A.M. (1991) *Global Environmental Change*, Longman Scientific and Technical, Harlow.

Mansfield, T.A. (1985) Porosity at a price: The control of stomatal conductance in relation to photosynthesis, pp. 419–52 in Barber, J. and Baker, N.R. (eds), *Topics in Photosynthesis*, Vol. 6, Elsevier, Amsterdam.

Mansfield, T.A., Hetherington, A.M. and **Atkinson, C.J.** (1990) Some current aspects of stomatal physiology, *A. Rev. Plant Physiol. Plant Mol. Biol.*, **41**, 55–75.

Marshall, B. and **Woodward, F.I.** (eds) (1985) *Instrumentation for Environmental Physiology*, 238 pp., Cambridge University Press, Cambridge.

Outlaw, W.J.Jr and **Harris, M.J.** (1991) Water stress, stomata, and abscisic acid, pp. 447–61 in Abrol, Y.P. *et al.* (eds), *Impact of Global Climate Changes on Photosynthesis and Plant Productivity*, Oxford and IBH Publishing Co. PVT., New Delhi.

Pearcy, R.W., Ehleringer, J., Mooney, H.A. and **Rundel, P.W.** (eds) (1989) *Plant Physiological Ecology, Field Methods and Instrumentation*, Chapman and Hall, London.

Schlesinger, W.H. (1991) *Biogeochemistry. An Analysis of Global Change*, Academic Press, San Diego.

Šesták, Z., Čatský, J. and **Jarvis, P.G.** (eds) (1971) *Plant Photosynthetic Production, Manual of Methods*, Dr W. Junk N.V., Publishers, The Hague.

Slater, R.J. (ed.) (1990) *Radioisotopes in Biology. A Practical Approach*, IRL Press, Oxford.

Troughton, J.H. (1979) $\delta^{13}C$ as an indicator of carboxylation reactions, pp. 140−9 in Gibbs, M. and Latzko, E. (eds), *Encyclopedia of Plant Physiology* (N.S.), Vol. 6, *Photosynthesis II*, Springer-Verlag, Berlin.

Umbreit, W.W. (1957) Carbon dioxide and bicarbonate, pp. 18−27 in Umbreit, W.W., Burris, R.H. and Stauffer, J.F. (eds), *Manometric Techniques* 3rd edn, Burgess Publishing Co., Minneapolis.

Zeiger, E., Farquhar, G.D. and **Cowan, I.R.** (eds) (1987) *Stomatal Function*, Stanford University Press, Stanford.

CHAPTER *12*

Photosynthesis by leaves

The rate of CO_2 assimilation per unit area of leaf is an important characteristic of higher plants and integrates all the biochemical and biophysical processes. The potential rate of CO_2 assimilation depends on the development of an effective metabolic system; the actual rate is a complex function of the photosynthetic system's interaction with environmental conditions. This chapter considers the nature of the responses of gross (P_g) and the net (P_n) photosynthesis of C3 and C4 plants to the principal environmental factors, light, CO_2, temperature, nutrition and water, and relates them to the metabolic and physiological structures and functions of the photosynthetic system. The approach draws on understanding of the basic mechanisms (Woodrow *et al.* 1990) described in earlier chapters. The information will be used to consider models, both qualitative and semiquantitative, of the photosynthetic system which have been developed to link the many facets of this complicated process (Farquhar and von Caemmerer 1982). Despite the complexity and great number of species and environmental combinations which have been described, certain features of the photosynthetic system are common to all plants and are a key to understanding their function and efficiency (Lawlor 1991) in given environments. This is a central problem in ecology and may be partially solved if the factors determining photosynthetic productivity are understood. Currently there is much effort given to understanding how photosynthetic rates are related to composition of the leaf tissue (Stitt 1991) and to environment, e.g. light (shade plants compared to sun plants, p. 270) and nutrient supply (e.g. nitrogen supply affects rates of photosynthesis; Evans 1989). Early studies of the relation between assimilation and characteristics of the plant attempted to correlate P_n with composition and structure, e.g. pigment or protein content, cell surface area, stomatal frequency (see Heath 1969). However, failure of this approach has led to the concept that the rate of photosynthesis by the intact leaf is the result of integrated cellular biophysics, e.g. diffusion of cell metabolites or substrates, and biochemistry, e.g. enzyme activities (Woodrow *et al.* 1990; Bowes 1991; Stitt 1991). This approach is, however, still *in embryo*. Current concepts are that the photosynthetic rate depends on the properties of the combined energy-capturing and transducing systems (light reactions, electron transport, etc.) in the thylakoids, acting in series with the processes taking place in the chloroplast stroma (enzymology of the PCR cycle) together with those consuming assimilates as well as on the factors regulating CO_2 supply to the chloroplast. These processes are considered to be 'balanced' to achieve optimization of the rates of each step so that they are not greatly in excess or deficient. This complex problem of regulation in a multistep process, is considered in greater detail later in this chapter. Models of the system consider only parts of it and are still research tools rather than applicable to plant production directly. Mathematical approaches have been used to link photosynthesis to environmental factors and plant composition, e.g. curves describing how P_n depends on radiation and CO_2. Also statistical correlations, e.g. regressions of P_n on N content of leaves are valuable for calculation of photosynthetic productivity of vegetation (see Thornley and Johnson 1990, reference in Chapter 13) and for practical purposes where generalized photosynthetic responses are required, but are not considered further here. Mechanistic models may provide, in the long term, better understanding and predictive capacity than generalized 'black-box' or statistical approaches.

Net CO_2 exchange of whole leaves depends on the balance between CO_2 uptake in gross photosynthesis (P_g) and the production of CO_2 by photorespiration (R_1) and respiration by other processes, predominantly the tricarboxylic acid cycle. This was generally referred to as 'dark' respiration (R_d) although as it proceeds in the light as well as in darkness it is increasingly called 'day' respiration (see p. 169, Ch. 1 for discussion and Graham 1980). Net photosynthesis (P_n) is thus

$$P_n = P_g - (R_1 + R_d) \qquad [12.1]$$

The rate of P_g is determined by irradiance (photon flux and spectral characteristics), CO_2 and O_2 concentration (or partial pressure) and by temperature and depends upon the species of plant as well as on the conditions of growth, e.g. nutrient and water supply.

Response of photosynthesis to environment

Irradiance

Photosynthetic rate per unit of leaf area of C3 plants is a function of incident photosynthetically active radiation (PAR) (Fig. 12.1). The typical curve is often described by a rectangular hyperbola, rising steeply from darkness with increasing light until a plateau is reached, i.e. further increase in photon flux does not increase P. In darkness P_g is zero and CO_2 is generated by respiration (mainly TCA cycle) and escapes from the tissue to the

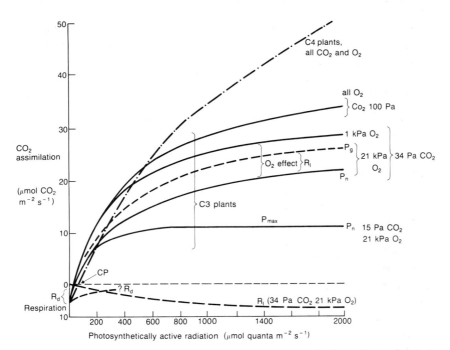

FIG. 12.1   Response curve of gross (P_g) and net (P_n) photosynthesis in C3 and C4 plant leaves with irradiance at different CO_2 supply; in bright light (>1500 μmol quanta m^{-2} s^{-1}) assimilation may decrease due to photoinhibition. Dark respiration (R_d) occurs in both types but photorespiration (R_l) is much greater in C3 plants particularly in high O_2. Light compensation (cp) and maximum rate of photosynthesis (P_{max}) are shown. The initial linear slope gives the photochemical efficiency. The smaller P_g in C3 species in air (estimated from $^{14}CO_2$ uptake) than in low O_2 is caused by consumption of RuBP by the oxygenase reaction in high O_2. Schematic after information from references.

atmosphere. The rate of R_d in darkness may not be the same as in the light but convincing experimental evidence and quantitative measurements of fluxes are lacking (see Ch. 8). The rate of R_d increases with warmer temperatures and may be greater when leaves contain large amounts of carbohydrate. This may involve respiration by the 'alternative pathway' in mitochondria which 'burns off' carbohydrates and thus helps to regulate cell metabolism by allowing the PCR cycle to operate and perhaps reduce the energy load on the photosynthetic system. The mechanisms and consequences of different forms and rates of respiration for effective photosynthesis under a range of conditions require further analysis.

As irradiance increases there is a decrease in the evolution of respiratory CO_2 from the leaf. There are two possible reasons for this: gross photosynthesis increases and R_d may be inhibited although photorespiratory CO_2 evolution also increases during this phase. Note that the ratio of R_l/R_d is not constant but depends upon the magnitude of the photorespiratory flux, which is a function of the behaviour of Rubisco, O_2 partial pressure and temperature whereas control of R_d is probably linked mainly to temperature and only indirectly to assimilate supply and thus to assimilation. With increasing photon flux a point is reached where $P_g = (R_l + R_d)$ and so P_n is zero. This is called the light compensation point, although as it varies with many factors, it should not be regarded as a fixed value. Shade plants, for example, generally have smaller compensation points than sun plants (20 compared to 80 μmol quanta m^{-2} s^{-1}). Other environmental factors also influence the light compensation point, such as nutrition and water stress.

With increased irradiance P_n increases linearly, as CO_2 fixation and electron transport processes are proportional to the photons captured. In dim light and with smaller P_n, leaves are not limited by CO_2 or stomatal conductance, g_s, as increasing the availability of CO_2 does not increase P_n. The slope of the curve at $P_n = 0$ is a measure of the photochemical efficiency, Φ, or quantum yield (mol CO_2 mol photon^{-1}). This is a measure of the ability of the photosynthetic system to utilize photons for CO_2 fixation at the greatest efficiency when the flux is zero. The true quantum yield is calculated (p. 243) with only photons absorbed by the leaf. If light incident upon the leaf is used as a measure of photon flux then only an apparent quantum yield is obtained; this is not a measure of the true efficiency of the system because light reflected and transmitted by the tissue is included in the energy supposedly required for CO_2 assimilation. The methods of measuring photosynthetic rates and quantum yield are described in Chaper 11. C3 leaves in air have Φ of 0.05, that is 1 mol CO_2 is fixed per 20 photons absorbed. Eliminating R_l with low O_2 partial pressure increases Φ to 0.07 or about 1 CO_2 per 14 photons because the effects of the oxygenase reaction on CO_2 fixation are eliminated. However, as P_n increases and particularly at small stomatal conductance, c_i, i.e. CO_2, rather than light becomes limiting and eventually further increase in irradiance has no effect on P_n (see p. 245 for the dependence of internal CO_2 partial pressure on atmospheric CO_2 and on stomatal and boundary layer conductances). With g_s for CO_2 of 0.4 mol m^{-2} s^{-1} and 30 Pa CO_2 and 21 kPa O_2 in the atmosphere, the internal CO_2 in a C3 leaf becomes limiting at P_n of about 30 μmol m^{-2} s^{-1} when irradiance is about 1000 μmol quanta m^{-2} s^{-1}. If the PAR is increased to 1500 μmol m^{-2} s^{-1} or greater there may be progressive decrease in P_n. This may be due to processes such as decreased stomatal conductance resulting from induction of water stress — water loss under such conditions is large and difficult to avoid — but also to photoinhibition (p. 98). Photoinhibition results from excessive energy loads on the light-harvesting apparatus and large rates of electron transport when the capacity to use the electrons is limited, leading to damage to PSII and particularly the D1 protein of the reaction centre (see p. 80). It may result from the inability of the non-photochemical quenching processes in the thylakoids (involving the zeaxanthin cycle, see p. 44) to dissipate the excess energy. The capacity to utilize the energy captured for CO_2 fixation depends on the adaptation of the plant to particular conditions; some are particularly susceptible to photoinhibition in very bright light as a consequence of genetic limitations, e.g. shade plants.

Light saturated rates of photosynthesis (P_{max}) reflect the ability of the photosynthetic apparatus to utilize light to fix CO_2. It is of considerable importance to understand the characteristics of the plant which determine this maximum assimilation for ultimately this is the limiting factor in the efficiency of plant production when other environmental conditions are satisfactory. Electron transport *per se* probably does not limit CO_2 assimilation. The maximum rate of electron transport (J) between the water oxidizing process and the main acceptor for electrons, $NADP^+$ depends upon the rate of formation of ATP (as the plastoquinone cycle in the thylakoid links J to ATP synthesis) and on the availability of $NADP^+$, which is a function of the pool sizes of the pyridine nucleotides and their turnover, in itself dependent upon the CO_2 fixation. In the discussion of chlorophyll *a* fluorescence (p. 99) it was shown that the energy in the PSII reaction centre could be utilized by the photochemical quenching processes, e.g. CO_2 and NO_3^- reduction, and by non-photochemical quenching, e.g. by dissipation via carotenoids. With increasing PAR the amount of energy captured rises but the capacity of photochemical quenching does not rise in proportion because of limitations in other parts of the system. Therefore, non-photochemical quenching rises and electron flow increases suggesting that electron flow is not a limiting factor in CO_2 assimilation in very bright light. However, some caution is advisable for clearly there are species differences and also growth conditions may affect the plant's responses. To use the example of C3 shade plants again, these have rather limited capacity for electron transport but have a large antenna for photon capture; hence the system may be limited by electron transport. In practice such plants often have very small capacity for CO_2 fixation also and thus regulation may be distributed between electron transport and carboxylation and they are genetically obligate shade species.

Light responses of C3 plant photosynthesis under normal atmospheric conditions are greatly dependent upon Rubisco oxygenase activity. Decreasing O_2 to 2 kPa and increasing CO_2 to 70–100 Pa increases photosynthesis, Φ and P_{max}, substantially, because the reduction of oxygenation makes more RuBP available for CO_2 assimilation at a given ATP and NADPH supply. A large increase in P_n may result, for example, because almost 25 per cent of the carbon fixed by gross photosynthesis is lost as photorespiration at 25 °C and the effective increase in P_n is then 40–50 per cent.

Carbon dioxide

When a typical C3 leaf in PAR of 1200 μmol m^{-2} s^{-1} is exposed to partial pressures of CO_2 from 0 to about 100 Pa, a response curve (Fig. 12.2) results. Rather than expressing the relation of photosynthesis to the atmospheric CO_2 pressure, the effect of stomatal conductance is removed

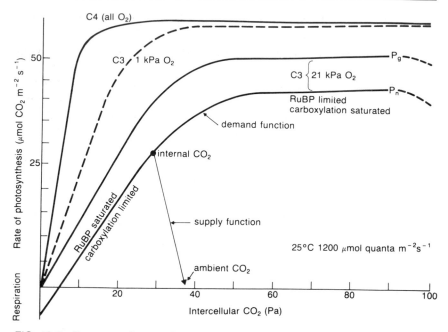

FIG. 12.2 Response of gross (P_g) and net (P_n) photosynthesis to CO_2 concentration and oxygen for C3 leaves, which respond to O_2, and C4, which do not. The 'demand function' is the requirement for CO_2 and depends on the characteristics of RuBP carboxylase/oxygenase; RuBP non-limited and limited parts of the response curve are indicated.

by calculating the internal CO_2 as described in Chapter 11. With no CO_2 entering the leaf chamber, respiration dominates and CO_2 is evolved by the mesophyll cells in the light and escapes into the intercellular spaces, increasing c_i, and thence into atmosphere. The rate of CO_2 evolution from the leaf is smaller than the leaf cells as some of the CO_2 evolved is reassimilated. It has been estimated that 50 per cent or more of evolved CO_2 is reassimilated but it is difficult to quantify the rate of dark or photorespiration under the normal atmospheric conditions, for if the O_2 partial pressure is decreased to limit photorespiration then this affects the behaviour of the entire system, e.g. electron flow to the O_2 in the Mehler reaction. Isotope exchange provides a method of estimating respiratory fluxes (Ch. 11).

With increasing CO_2, P_g rises until $P_g = R_1 + R_d$ when P_n is zero. This is the CO_2 compensation point which — as with the light compensation point — is not a fixed value. In C3 leaves at 1 kPa CO_2 this value is about 0–1 Pa CO_2 and in 21 kPa O_2 about 4–5 Pa O_2. The slope of the curve relating photosynthesis to CO_2 at the compensation point is called the carboxylation efficiency and is a measure of the ability of the system to assimilate CO_2 at maximum efficiency and independent of light limitation. Carboxylation efficiency is about $2-2.5$ μmol m^{-2} s^{-1} Pa^{-1} in 2 kPa O$_2$ and $1-2$ μmol m^{-2} s^{-1} Pa^{-1} in 21 kPa O$_2$.

The curves relating photosynthesis to irradiance and to c_i are called demand functions (Fig. 12.2); they represent the requirement for CO$_2$ resulting from photosynthetic metabolism. The supply function is the rate at which CO$_2$ is available to the system, determined by stomatal conductance combined with atmospheric CO$_2$ concentration. As discussed in Chapter 11 the internal CO$_2$ partial pressure inside leaves is substantially smaller (c. 0.7 times) than the pressure in air. As CO$_2$ at the leaf surface increases and c_i rises, the supply function satisfies the demand function and photosynthesis rises. However, as limitation by carboxylation or electron transport increases and the demand function reaches a plateau, so the increase of P in response to c_i becomes less marked. In an atmosphere of 2 kPa O$_2$, O$_2$ saturation, P_{max}, is achieved at about $20-30$ Pa CO$_2$ with rate about 40 μmol m^{-2} s^{-1}. However, there is a marked effect of O$_2$ on these values; 21 kPa CO$_2$ increases the compensation point as already discussed and decreases P_{max} from 40 to 30 μmol CO$_2$ m^{-2} s^{-1} and increases the CO$_2$ partial pressure at which stimulation is achieved, from c. 20 to $40-60$ Pa. This is due to photorespiration and the characteristics of Rubisco. It is important to appreciate that the photosynthetic apparatus of C3 plants is not saturated with CO$_2$ in the current atmosphere. However, plants are expected to respond to the increased CO$_2$ in the future atmosphere and may have already done so over the last century. With CO$_2$ increase from c. 28 to 35 Pa over that period gross photosynthesis of vegetation may have increased by $10-15$ per cent; vegetation may therefore have acted as a sink for some of the excess carbon generated by industrialization, etc. (see Ch. 11).

Temperature

Temperature responses of plants are very dependent upon the species and growth conditions. Despite this, a general response of photosynthesis to temperature is evident although temperatures at which particular responses are observed shift. Generally photosynthesis is slow at cooler temperatures, increasing to a maximum (which may be fairly broad or rather narrow) as temperatures rise and then, at yet higher temperatures, falling steeply to a point at which the tissue is killed (Fig. 12.3). Despite much research no clear picture of the mechanism of response to mild or extreme temperatures has developed. The primary effect is via changed rates of enzyme reactions, metabolite transport and diffusion; at low temperatures rates of these processes are slower than at high. An example: the rate of supply of phosphate ions to chloroplasts at low temperature probably limits photosynthesis and prevents a response to O$_2$ in C3 plants. There is evidence that the Q_{10} of processes can change with the growth conditions,

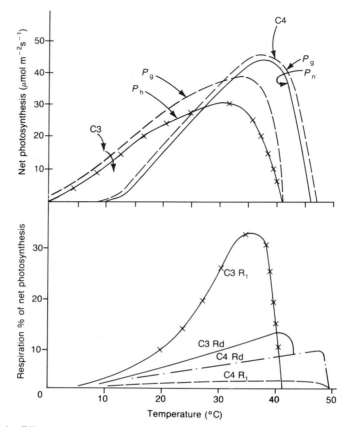

FIG. 12.3 Effects of temperature on (a) the net CO_2 assimilation rate and the effects of O_2 partial pressure for C3 and C4 plants and (b) the effects of temperature on 'dark' and photorespiration in C3 and C4 plants as a proportion of net photosynthesis. Highly schematic and generalized from information in the quoted literature.

e.g. that for electron transport in thylakoids from cold adapted winter rye (*Secale cereale*) plants was 1.53 compared to 1.85 for unadapted and the photosystems differed in temperature response also (for discussion see Oquist and Martin (1986)). In plants able to tolerate cold, growth at low temperatures increases the contents of chlorophylls, carotenoids, plastoquinones and cytochromes, but generally the proportions remain very similar (e.g. the ratio of light-harvesting complexes to photosystems). Lipids of thylakoids may also alter with the temperature at which they develop. Some evidence for a decreased MGDG/DGDG ratio and increased unsaturated fatty acids in cold-grown leaves of different species suggests that this may be a general response, although thylakoids seem to conserve their structure and composition rather more than other plant membranes. Enzymes of photosynthesis may be modified in leaves grown at different temperatures. Rubisco of rye, for example, shows changes in kinetic constants related to alterations in the exposed SH groups on proteins; at low temperature these effects result in greater efficiency *in vitro*. Cold seems to alter the capacity for photosynthesis, increasing the content of 'machinery' per unit leaf area and perhaps the ratio of some key components, e.g. plastoquinone or other electron transport components and improving some enzyme characteristics. Components of protective metabolic sequences might increase if there were a need to dissipate energy from the photosynthetic apparatus in cool conditions to avoid photoinhibition but this appears not to have been examined. Respiration of plants, both R_d and R_l, is small in cool conditions and often large accumulations of carbohydrate result from continued photosynthesis in the cold partly as a consequence of the low respiration and partly because of reduction in organ growth.

The larger content of photosynthetic components in plants grown in cool conditions will allow rapid response to improved temperature environment. Photosynthesis of such plants often has a relatively broad response to temperature and an optimum beyond the normal growth range. The type of response outlined suggests that regulation of the gene expression and development of the photosynthetic system is such that over a very wide range its characteristics are very tightly controlled to give a stable, efficient system well protected against many aspects of the environment.

High temperatures generally lead to smaller production of photosynthetic components per unit area and decreased proportions of electron transport components compared to enzymes but, as at the other temperature extreme, ratios of components do not change drastically and the rate of photosynthesis remains relatively constant over a wide range. The carbohydrate economy of plants grown in warm conditions is very different to that in the cold. Respiration is greatly stimulated at high temperatures and also growth of organs, so that the whole plant may be more carbohydrate limited than in the cold. Although the R_d increases it is a relatively small fraction of photosynthesis; photorespiration becomes an increasingly large part of assimilation at higher temperatures. The importance may be seen in the response of quantum yield of C3 plants to temperature. In small O_2 partial pressures they have quantum yields of about 0.08 mol CO_2 mol photon^{-1} over the range 10–40 °C. However, in 21 kPa O_2 the value decreases progressively to half that at 40 °C. Photorespiration is a small part of p_n in cool conditions, but with increasing temperature, as the photosynthetic rate rises, R_l increases more than P_g (particularly above 30 °C) more than offsetting gross photosynthesis, resulting in a small increase in P_n and a marked drop in Φ. These responses are a consequence of the Rubisco reactions; oxygenation increases relative to carboxylation with increased temperature in large measure because the solubility of CO_2 is smaller, relative to O_2, at higher temperatures so favouring oxygenation over carboxylation. Also R_d of non-photosynthetic organs increases as temperature rises, thus placing a substantial burden on the photosynthetic production of assimilates. As a result plants not acclimated to high temperature may eventually die of assimilate starvation.

Tropical and subtropical plants and C4s have much higher temperature optima than many temperate plants which are predominantly C3s and are more tolerant of heat than cold. Many factors contribute to their differences in response to extremes of temperature, including growth of leaves and of organs which consume photosynthate. The photosynthetic system of plants which cannot adjust to cold may be inhibited because specific processes become damaged. Membrane instability may be a cause of chilling damage; the mechanisms by which this is overcome in some species but not others are not yet understood. Chilling results in phase changes in the lipids of membranes and disruption of membrane stability and functions, e.g. PSII. This may be at the level of accumulation of high energy intermediates in the electron transport chain and photoinhibition which is characteristic of chilling injury. Slower metabolism at low temperatures decreases the use of light reaction products and thus increases the possibility of photochemical damage (particularly in bright light) which occurs as the energy load on the photosystems increases and the energy dissipating mechanisms, e.g. carotenoid 'safety-valve' or superoxide dismutase (SOD) and peroxidase detoxification mechanisms, do not have the capacity or are too slow to remove the energy or products. Perhaps the plant is genetically unable to produce the required intermediates or they are not made in adequate amounts because of the growth conditions. Other plants may suffer lesions during development of organs under low temperatures which reduce expression of their photosynthetic capacity, e.g. chlorophyll synthesis is prevented in maize by temperatures below $8-10\ ^{\circ}\mathrm{C}$.

The other temperature extreme is heat. At high temperatures (the range and response will depend on species and growth conditions), instability of membranes and denaturation of enzymes change photosynthesis and increase respiratory demands on the assimilate supply. Prevention of proper development of organs with high temperatures may also contribute to the inability of the plant to form effective photosynthetic structures. Gene expression, protein synthesis and assembly of proteins into the correct forms may be affected and heat shock responses, with accumulation of chaperonin proteins (see Ch. 10) may occur. These changes reflect the inability of tissues to make and assemble protein complexes. Possibly chaperonins have a role as protein protective agents or in synthesizing protective metabolites by which plants avoid heat damage. These factors are outside the realm of the present discussions of photosynthesis but are vitally important from an ecological perspective. Adaptations to the growth conditions may involve overcoming several limitations in growth, development and metabolism in order to form mechanisms by which plants adjust to particular ecological conditions.

C3 and C4 plant responses to environment

There are considerable differences between C3 and C4 plants in the way they respond to environment, related to the differences in photosynthetic mechanisms and their biochemical and physiological characteristics. Consider the response to light. C4 leaves have substantial R_d in darkness but with increasing irradiance and thus increased gross photosynthesis, CO_2 evolution decreases. Quantum yield (mol CO_2/mol photons) of C4 plants may be smaller than that of C3 plants in dim light because of the greater requirement for ATP in C4 than in C3 metabolism. Thus, C4 plants are less efficient in dim light than C3 plants even under photorespiratory conditions. However, when irradiance reaches $200-400$ μmol m^{-2} s^{-1} the advantage of the C4 mode of photosynthesis exceeds the disadvantages and photosynthesis continues to increase up to very high irradiance. C4 photosynthesis is, of course, insensitive to CO_2 over a wide range and similarly with O_2 so Φ remains constant (0.05) over a range of CO_2 conditions from about $10-20$ Pa to $80-100$ Pa (Fig. 12.4). Thus, although the photochemical efficiency of C3 plants exceeds that of C4 in low O_2, as oxygen increases so more CO_2 is required to overcome the oxygenase induced photorespiration whereas C4 plants are insensitive above CO_2 of $5-10$ Pa (Fig. 12.4b).

C4 plants are relatively inefficient at low temperatures as the majority are of warm, bright, dry habitats. Few C4 species are adapted to cooler,

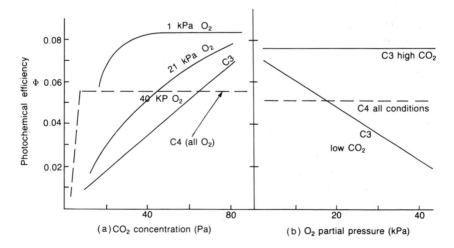

FIG. 12.4 Photochemical efficiency of C3 and C4 leaves (a) as a function of CO_2 and O_2 partial pressure. (b) Photochemical efficiency of C3 and C4 leaves as a function of oxygen partial pressure and CO_2. Highly schematic; modified from Osmond *et al.* (1980).

low irradiance environments. One is *Spartina townsendii* a temperate salt marsh grass; the features which allow it to occupy its ecological niche are not understood. The low temperature intolerance of C4 plants, shown by the decrease in photochemical efficiency (Fig. 12.5) involves the processes already discussed, for example, damage to membranes and inhibition of chlorophyll synthesis. In hot, dry environments the major advantage of the C4 syndrome is clearly shown. Whereas C3 photosynthesis decreases beyond about 30−35 °C C4s have an optimum generally substantially above 30 and may not be damaged until 45−50 °C. Also their photochemical efficiency remains constant over a wide range.

Another advantage of the C4 syndrome is striking insensitivity of C4 plants to low partial pressures of CO_2 and to high O_2 over a very large range; at current ambient CO_2 C4s have a substantial advantage over C3s. Also, the decreased CO_2/O_2 ratio in leaves, resulting from progressive closure of the stomata in response to water stress which is so frequent in hot, dry conditions, has little effect on C4 photosynthesis. Whereas C3 plants must keep large stomatal conductance (0.4−0.4 μ mol m^{-2} s^{-1}) to prevent CO_2 depletion and stimulation of Rubisco oxygenase activity relative to carboxylase, C4s have smaller stomatal conductances which can decrease considerably without substantial effects on carboxylation. This also avoids the detrimental effects of decreased CO_2/O_2 solubility ratio with increasing temperature. The advantage of C4 metabolism, its independence of O_2 and CO_2 in the atmosphere and to warmer temperatures, is due to the CO_2 'pump' driven by PEP carboxylase which results in large partial pressure

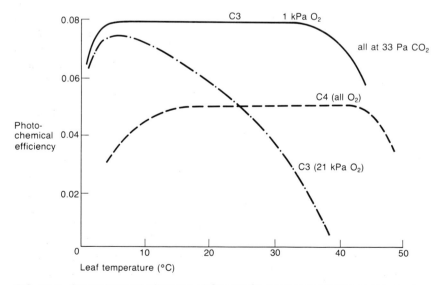

FIG. 12.5 Photochemical efficiency of C3 and C4 species in relation to temperature at air partial pressure of O_2 and CO_2 or at low CO_2 and normal oxygen; generalized responses based on information from Osmond *et al.* (1980).

of CO_2 in the bundle sheath where Rubisco functions under saturating conditions thus avoiding photorespiration (Ch. 9). Hence, C4s form a larger proportion of vegetation under hot, dry conditions with water deficits. This is shown by the increase in the proportion of C4 species in the grasslands of the North American great plains in hot compared to cooler seasons. However, the ability of C3 plants to adapt to very high temperatures should not be overlooked, e.g. the C3 *Larrea divaricata* is able to grow in the extreme heat of Death Valley, California. The C4 mechanism explains many of the differences between C3 and C4 plants, although the long-term ecological adaptations involve other metabolic and physiological mechanisms.

Photorespiration and environmental factors

Photorespiratory CO_2 release by C3 plants is a major aspect of their carbon and energy budgets, and depends upon the temperature and CO_2/O_2 ratio. Photorespiration is estimated, most accurately, as the difference between gross and net assimilation, the former measured as $^{14}CO_2$ uptake and the latter as $^{12}CO_2$ assimilation under steady state conditions where the rate of photosynthesis is constant and the internal carbon fluxes also. The methods are described in Chapter 11. However, respiratory CO_2 is probably partially reassimilated in the tissue before it escapes to the atmosphere; this 'short circuit' causes an underestimation of the rate of R_1. Other methods used to measure efflux of carbon in R_1 include measuring the rate of CO_2 release from a leaf immediately after darkening following a period of photosynthesis. There is rapid CO_2 evolution, in the 'post-illumination burst', which lasts for up to 5 minutes before decreasing to the rate associated with dark respiration. It arises from the continued flux of carbon through the glycolate pathway after the light reactions and P_g have stopped and decreases with cessation of the photosynthetically produced glycolate flux. R_1 is also measured by feeding intermediates of the glycolate pathway and measuring CO_2 or $^{14}CO_2$ efflux. Inhibitors of the pathway have also been used to stop R_1; the accumulation of intermediates or changes in CO_2 production are then measured to estimate R_1. However, these methods suffer to some extent by altering the steady state conditions, and often decrease photosynthesis and therefore R_1. Photorespiration is about 25–30 per cent of P_n in air and the proportion increases with larger O_2 and smaller CO_2 partial pressures, high temperatures and water stress. The increase is due to the increased ratio of Rubisco oxygenase to carboxylase, α (p. 133), so with internal CO_2 concentration in the leaf of about 30 Pa and 21 kPa O_2, α is about 0.4 and $R_1 = \alpha/2 \times P_g$. At high assimilation rate (40 μmol m^{-2} s^{-1}) R_1 is 8 μmol m^{-2} s^{-1} and P_n is 32 μmol m^{-2} s^{-1}, that is, R_1 is 25 per cent of P_n.

Conditions causing smaller stomatal conductance, such as water stress, may decrease P_g and decrease the CO_2/O_2 ratio in the tissue and increase

α, so the proportion of R_1 to P_n rises. C4 plants have similar rates of P_g and P_n and are affected little by conditions. Increasing temperature increases α, partly because the solubility of CO_2 is less than that of O_2. R_d also increases, so dark respiration offsets photosynthesis.

CO_2 and light compensation

The CO_2 compensation point has special significance as an indicator of photorespiratory activity in plants. This has been discussed in the case of C3 and C4 metabolism in Chapter 9. Carbon dioxide compensation point of C3 leaves is small in cool temperatures with low O_2, but increases as temperature and oxygen rise, particularly when O_2 exceeds the point where the oxygenation/carboxylation ratio, α, exceeds 2 and temperatures are greater than 35 °C. In contrast, C4 plants have small compensation concentrations independent of O_2 and temperature up to higher values. Light compensation of C3 leaves is achieved in very dim light in low O_2, high CO_2 and cool temperatures but increases as they increase. C4 plants have higher light compensation in those conditions but it remains small up to much higher O_2 and temperature conditions than C3 plants. Below light compensation, respiration exceeds assimilation and the CO_2 partial pressure increases for both C3 and C4 leaves to very large values in closed systems. When assimilation is inhibited, for example, by water stress, the compensation values also rise, even at bright light, in C3 and C4 species. Compensation values are relatively easy to measure and reflect the photorespiratory characteristics of the plant; they have therefore been used to select for variations in R_1. Although the majority of plants examined fall into the high (C3) or low (C4) groups, some intermediate C3/C4 types have been found which are of considerable value for analysis of the evolution of the C4 syndrome (see p. 188).

Photosynthetic response to environmental stresses

A plant growing successfully over a long period, i.e. surviving, in a particular environment may be regarded as broadly adapted to the physical and chemical characteristics of that environment. They are genetically capable of forming the structures required for photosynthesis and effective growth. However, the plant's genetic ability to adapt its limits, although in the evolutionary time scale the genotype may be modified to achieve a new balance with the environment. In the short term, environmental conditions outside the plant's capacity to maintain its functions at the normal — and presumably relatively large — efficiency, constitute 'stresses' which result in 'strain' to the plant's physiological and metabolic system. All environmental factors can alter to the extent that they become stresses although the degree of change in the environment which the plants must experience to be adversely affected will depend on the plant's characteristics.

If a plant has mechanisms permitting it to adapt to new conditions without loss of productivity (total biomass or reproductive structures, for example) or decreased efficiency (amount of plant component produced per unit of resource used) then the plant is effectively not subject to a stress. The physical/chemical features of the environment may be regarded as neutral. It is the plant's ability to respond which effectively determines the stress—strain relationship.

The main stresses encountered by plants are associated with extremes of temperature, light (both large and small energy flux or changes in spectral characteristics such as increased UV-B radiation), water supply and chemical composition of the environment. In the latter are included nutrients required for growth (e.g. nitrogen, phosphorus, potassium), those which damage growth (e.g. sodium ions in saline conditions) or environmental pollutants (e.g. SO_2, heavy metals).

Temperature stress

Temperature has been discussed in relation to the photosynthetic rates of C3 and C4 plants (p. 188), including brief mention of its role in determining tissue composition and the relative rates of carboxylation and oxygenation of Rubisco. Extreme cold and heat damage the photosynthetic system at many different levels of organization and, because of the interacting processes in metabolism, cause and effect are difficult to disentangle. Detailed discussions of the effects of temperature stresses on metabolism may be found in the references (Alscher and Cumming 1990).

Light stress

Besides its role as an absolute requirement for photosynthesis, radiation affects other processes, for example, plant development (Ch. 10). Also the light-generated toxic oxygen radicals cause damage to the photosynthetic system (photoinhibition) by destruction of reaction centre proteins in thylakoids and by attacking lipids forming lipid peroxides which are themselves very destructive of membrane components. This type of response to adverse environments is common and prevents normal electron transport and CO_2 fixation. The mechanisms of adaptation to radiation stress may involve formation of protective compounds, for example, carotenoids of the chloroplast zeaxanthin cycle (p. 44). In the case of UV-B radiation, which is damaging to many plants and may increase as a consequence of global climate change, plants form large quantities of flavonoid compounds which absorb and dissipate the energy and are thus protective. Adaptation to such adverse radiation is very species specific and may depend on the genetic ability to produce flavonoids and other protectants (Teramura and Sullivan 1991).

Dim light is a stress as well as bright light but for rather different reasons, because the available energy is insufficient for normal growth of non-adapted plants which suffer from deficient assimilation. Conversely, shade plants which are adapted to dim light may be severely photoinhibited on exposure to bright light for even relatively short periods. Dim light may also cause photomorphogenetic responses, aspects of which are considered in Chapter 10.

Sun and shade plants Metabolism and characteristics of C3 plants, their light-harvesting systems, chemistry of CO_2 fixation and rate of photo-synthesis have been presented as if without variation. Even within the distinct types of C4 and CAM plants, uniformity of response has perhaps been implied. However, this is not the case. C3 and also C4 plants show great diversity of response to the environment, particularly to the major determining factor in plant growth — light. Distinction may be made between plants with high rates of photosynthesis and growth in very intense light, so-called 'sun plants', which are inefficient, in dim light with poor photosynthesis and survival, and 'shade plants' which photosynthesize and survive in dim light but are unable to function efficiently in bright light (low maximum rates of photosynthesis and photochemical damage). Those plants clearly in either category are obligate sun or shade plants, genetically adapted and incapable of adjustment to the other extreme condition. However, many species show flexibility in response to light intensity; they are facultative sun or shade species lacking the ability to adapt to the extremes.

Sun plants, which include many crops and plants of tropical regions, achieve maximum rates of photosynthesis greater than 30 μmol CO_2 m^{-2} s^{-1} and respiration rates in darkness of 2 μmol CO_2 m^{-2} s^{-1}. Shade plants may have photosynthesis rates less than 10 μmol CO_2 m^{-2} s^{-1} at light intensity perhaps one-tenth of sun species and may be damaged by light intensities above half that of full sunlight.

Extreme adaptation to low light is shown by plants from the floor of tropical forests, where the photosynthetically-active radiation (PAR) may be less than 3 per cent of the radiation at the top of the forest canopy and greatly enriched in wavelengths of green, far red and infra-red, which are not absorbed by the foliage above. Sunflecks, small spots of intense light, which pass between the leaves in the canopy, contain over half the PAR in rainforests. Many taxonomically different plants are able to grow in poorly illuminated habitats; species which have been studied in detail such as *Alocasia* and *Cordyline* are from such environments. Even in temperate conditions, dim light within the canopy of a single tree is also related to the development of shade leaves.

Anatomically, shade plants have characteristics which increase light absorption. Sun plants have thick palisade and spongy mesophyll tissue so that there is a 2—3-fold increase in the number of cell layers. Mesophyll cells are large and thick-walled, and there is less intercellular air space and a greater (up to five-fold) surface area of cells to total leaf than shade plants. Leaf thickness appears to be an important variable in many plants and allows flexibility in the use of light and CO_2, etc., without changing the other physiological properties of the leaf.

Shade and sun plants and leaves of the same plant from different illuminations differ in chloroplast membranes and light-harvesting and electron transport mechanisms. Chloroplasts are usually more numerous in mesophyll cells of shade plants, and arranged near the upper leaf surface whilst cells of the lower mesophyll have few chloroplasts. However, as the number of cell layers is smaller, shade plants often have less chlorophyll per unit leaf area. Thylakoid membranes are stacked into many grana with many lamellae and less stromal lamella. *Alocasia* grown in dim light has four times as many granal partitions per stack as spinach grown in bright light, thus increasing the area of light-capturing membrane. Grana are often irregularly orientated which may increase capture of diffuse or variably orientated light.

Shade plants may contain four to five times more chlorophyll a and b per unit volume of chloroplast and have a higher b/a ratio than sun plants because the light-harvesting complex (LHC) increases. Shade enhances the capacity for light capture and energy transfer to the reaction centres. However, the capacity of the electron transport chain in shade plants is not increased, as there is relatively much less (one-fifth) cytochrome f, plastoquinone, ferredoxin and carotenoid per unit of chlorophyll than in sun plants. Shade plants have, therefore, more light-collecting apparatus, but a smaller complement of electron carriers than sun plants. In dim light the rate of electron transport is limited by the number of photons falling on the leaf; it would be no advantage for shade plants to produce a large capacity electron transport chain. With a pool of plastoquinone receiving electrons from several PSII reaction centres, this rate limiting step is minimized in the shade plants, except when the plant is exposed to light of a brightness outside its normal range.

Sun plants have less developed thylakoid systems, fewer granal stacks and partitions and less LHC so they are less efficient at absorbing light energy at low photon flux than shade plants and so have lower quantum yield. Electron transport, or photophosphorylation, may be limiting at very large photon flux for many species. Hence, the larger capacity of electron transport acceptors, particular plastoquinone. Electron transport in sun plants may be 15–30 times faster than in shade species (uncoupled rate). However, the quantum yields of electron transport to DCIP are very similar in very dim light, so there is no difference in photochemical efficiency, that is, the reaction centres are able to transfer electrons at the same rate. Yet the uncoupled electron transport rates are saturated at 600–800 μmol quanta $m^{-2} s^{-1}$ for sun plants and 50–100 μmol quanta $m^{-2} s^{-1}$ for shade species. Clearly the main differences between the two groups of plants is the capacity of the light-harvesting system and of electron transport. The light absorbing system in shade plants makes them very effective at gathering the light available and passing it to the reaction centres, especially in dim light, but they are limited in bright light by the rate of electron transfer; sun plants in contrast are very efficient at transporting electrons but not at gathering weak light.

Of course it is not electron transport *per se* which determines CO_2 assimilation but NADPH and ATP synthesis. Shade plant photosynthesis may be limited by NADPH synthesis; possibly extensive granal stacking is essential to obtain sufficient rates of electron flow to reduce $NADP^+$. Cyclic electron flow could drive ATP synthesis to match $NADPH^+$ production and the extensive grana may restrict the coupling factor to a small area of thylakoid to increase the ΔpH gradient near CF_1. Sun plants probably have adequate $NADP^+$ reduction, but may be limited by ATP supply; this applies also to all C4 plants. Cyclic electron transport then may produce the required ATP without $NADP^+$ reduction. The photosynthetic systems of strongly illuminated shade plants may be irreversibly damaged by very intense light, whereas sun plants are apparently insensitive. Slow movement of electrons through plastoquinone at high rates of electron flow in bright illumination causes 'backing up' of electrons and reaction centres cannot use excitation energy so that the high energy states of chlorophyll accumulate and damage increases. Photosystem II is more sensitive to photoinhibition than PSI. The structure of shade plant thylakoids makes them more easily damaged. Possibly the structure of the photosystems, or carotenoid complement, which reduces the energy load and provides a 'safety valve' is inadequate in shade plants. Sun plants have relatively smaller light-harvesting systems so are inefficient in weak light, but the greater number of cell layers and the greater capacity of the electron transport chain for electron flow and excess energy dissipating systems contribute to their greater efficiency and capacity to assimilate CO_2 in intense illumination.

Assimilation is expected to be limited by CO_2 supply when photosynthesis is rapid; sun plants have more stomata per unit area of leaf and larger stomatal conductances than shade species. Internally the cell surface area is greater, increasing the conductance of the CO_2 supply pathway. Sun plants also have more enzyme capacity in a greater stromal volume. There is more Rubisco per unit of chlorophyll; as this enzyme is a major protein of leaves, the larger protein content of sun leaves compared with shade plants is expected. In very brightly illuminated sun and shade plants, with saturated rates of electron transport, the capacity of the enzyme system might be insufficient to exploit all the NADPH and ATP synthesis. However, this remains to be tested; there is a parallel between rate of photosynthesis and the amount of Rubisco under such circumstances.

Dark respiration in shade plants is much less than in sun plants, $c.\ 0.15$ compared with $2\ \mu mol\ m^{-2}\ s^{-1}$. Also, photorespiration is probably less because although the RuBP oxygenase to carboxylase ratio of the protein is similar, the CO_2/O_2 ratio in the tissue is more favourable and dimly lit environments are cooler. Light compensation, the photon flux at which respiratory CO_2 production and photosynthetic CO_2 assimilation are equal, and net photosynthesis is zero, is very much smaller in shade than sun plants. Of course the ratio of respiration to CO_2 assimilation at saturating light intensities is smaller in sun plants compared with shade plants and so the growth rate of sun plants is much greater than shade species. These features may be genetically determined for maximum efficiency in use of resources, such as nitrogen for protein production. However, the degree of change is usually restricted, and most species cannot adapt fully to the extremes, presumably due to genetic regulation. The complex regulation of nuclear and chloroplast DNA encoded genes for components of the photosynthetic system is described in Chapter 10.

Water stress

The rate of water loss and onset of stress depends on many features of the soil−plant−atmosphere continuum (see Alscher and Cumming 1990). If water loss from plants exceeds water uptake then water content, turgor pressure and water and osmotic potentials of cells in leaves and other tissues decrease. One of the earliest responses is a decrease in stomatal conductance resulting from decreased turgor, which reduces the rate of transpiration, slowing dehydration of the tissue and also decreasing growth of organs so that less leaf area is formed under drought. Decreased leaf area together with smaller stomatal conductance are the major ways of decreasing loss of water. As a consequence, however, photosynthesis also decreases due to smaller leaf area per plant and decreased conductance for CO_2 diffusion into the leaves. The reasons for decreased photosynthesis under stress are more complicated than simply reduction of stomatal conductance. There is much evidence for metabolic inhibition as well. In the first phase of water loss from a leaf, with reduced stomatal conductance, the internal CO_2 partial pressure decreases because assimilation is not substantially inhibited so the major site of regulation is the stomata. With longer exposure to, and greater degree of, stress (particularly low turgor and decreased osmotic potential which are internal to the plasmamembrane and therefore likely to be the components of changed water balance perceived by the cellular contents) mesophyll cells progressively lose the ability to photosynthesize. Water stress decreases assimilation over a wide range of CO_2 partial pressures, i.e. the demand function decreases with stress, so despite stomatal closure c_i remains large or even increases. Analysis of the relative stomatal and mesophyll limitations under these conditions indicates that despite stomatal closure, control of CO_2 assimilation resides in the mesophyll. There is currently concern that this effect may result from incorrect estimation of c_i in stressed leaves because stomatal closure occurs irregularly across the leaf surface ('patchy stomata') invalidating the calculation. Evidence supporting this includes reversal of stress effects by very large CO_2 partial pressures but in some cases stomatal patchiness is absent.

The decrease in photosynthesis has been ascribed to the effects of water stress on several photosynthetic processes. Rubisco, for example, is affected by the decrease in CO_2 partial pressure and the proportion of photorespiration relative to gross photosynthesis increases under stress although with severe stress respiration is probably TCA cycle respiration rather than photorespiration. Water stress may decrease Rubisco activity but it is probably not the primary site of action. Lest it be thought that the effect of stress is only through R_1, C4 plants suffer decreased photosynthesis when water stressed despite their smaller stomatal conductances, lower c_i and virtual absence of photorespiration.

One process which seems particularly sensitive to decreasing turgor, or more likely increasing osmotic concentration in the chloroplast, is photophosphorylation. The ATP content of stressed leaves is generally decreased and this is probably the cause of the decreased content (slow synthesis?) of RuBP in the chloroplast which inhibits photosynthesis. Under these conditions electron transport and reduction of $NADP^+$ continue at a substantial rate and a large ratio of NADPH to ATP results. Fluorescence increases substantially at low turgor where CO_2 assimilation is strongly inhibited and there is decreased photochemical and increased non-photochemical energy quenching in thylakoids with progressive increase in stress. If this situation continues for long, dissipation of high energy states may fail to prevent progressive damage to PSII (photoinhibition) and related destruction of the thylakoids. At very low turgor, leaves in the stressed, highly reduced state accumulate large amounts of characteristic secondary metabolites, for example, amino acids (particularly proline), betaine glycine and the plant hormone abscisic acid (ABA). These may protect the photosynthetic mechanism under stress conditions, e.g. abscisic acid induces stomatal closure. However, such changes may not have a functional role; possibly they represent the accumulation of end products of disturbed metabolism, for example, synthesis of proline requires NADPH which may stimulate its accumulation and ABA may be a product of the breakdown of carotenoids (Young and Britton 1990). Clearly the photosynthetic system cannot tolerate such metabolic imbalances for long and leaves become damaged and senesce rapidly under stress conditions. The accumulation of neutral osmotica such as proline may minimize exposure of the photosynthetic system to decreased osmotic potential and protect cells and membranes against increased concentration of ions.

Nutrient deficiency

Plants require many different chemical elements from their environment in order to form leaves and photosynthesize effectively (Lawlor 1991). With such complex systems, responses depend on the particular nutrient.

However, several general points should be noted. The effects of deficiency will depend on the supply relative to the demand by the plant, for if we consider the development of the photosynthetic system as controlled by the genetic processes then the rate at which a nutrient must be supplied is determined; deficient supply will decrease the production of photosynthetic components. The time at which the nutrient is required also imposes demands upon the supply from the storage capacity of the plant and the environment. The effect of a particular nutrient will depend on the step in the photosynthetic system at which it operates, e.g. nitrogen (needed for proteins and many other components, e.g. chlorophyll) will affect the composition widely whereas iron (in Fe−S centres) may have specific effects on electron transport (Terry and Rao 1991).

Nitrogen is a major nutrient and photosynthesis requires considerable investment of this element; light-harvesting complexes form 50 per cent of the protein in thylakoid membranes and Rubisco 50 per cent of total soluble protein in leaves, so these two components alone require much nitrogen. In addition, nitrogen is needed for production of leaf area which determines, together with the rate of photosynthesis per unit area, total plant production (Ch. 13). Nitrogen deficiency decreases the content of proteins, e.g. Rubisco, and chlorophyll per unit area of leaf resulting in decreased photosynthesis. However, the proportion of components of the photosynthetic system remains relatively constant even with severe deficiency so maintaining quantum yield even if the carboxylation efficiency decreases. Many processes in the plant place demands on nitrogen supply and deficiency often induces remobilization of nitrogenous compounds from older leaves for use in developing organs, accelerating senescence. Rubisco has been regarded as having a storage function because it is used in this way, but it is not unique in being a source of N. Sacrificing older leaves may benefit the whole plant even if photosynthetic carbon assimilation capacity is lost.

Phosphate supply affects many aspects of photosynthetic function and other plant processes, so it is difficult to ascribe specific metabolic lesions. Deficiency during growth decreases the number and sizes of leaves and their content of proteins and pigments but, as with nitrogen, the proportion of photosynthetic components remains rather constant over a wide range of supply of the nutrient. One of the main effects of phosphate deficiency is to decrease the ATP content of tissues and thus to reduce the regeneration of RuBP. Under such conditions electron transport may not decrease as much as CO_2 assimilation so the pyridine nucleotides are highly reduced as shown by increased fluorescence. However, the regulation of metabolism with phosphate deficiency also involves changes to enzyme activity (e.g. low P decreases Rubisco activity). Several processes may limit photosynthesis, depending on the conditions.

These brief examples show how the photosynthetic system responds to different forms of environmental stress depending on the nature of the stress, the point in the system at which it operates and the severity. All affect some aspect of photosynthesis; future analysis will continue with quantitation of the responses to particular stresses so that a proper assessment of controls is obtained. This may allow the stress effects to be minimized.

Regulation of CO_2 assimilation

Despite the depth of understanding of the photosynthetic mechanisms, it is still not established what limits and regulates the overall rate of photosynthesis under a range of conditions. If the process were a linear sequence then the rate of the slowest step (be it the supply of materials or energy or the rate of an individual chemical reaction) would determine the overall rate; this is Blackman's law of limiting factors introduced in 1906. Increasing the rate of the slowest process increases the overall rate until another factor becomes limiting as shown (Fig. 12.2) by the effects of CO_2 on P_n in bright light. In reality, photosynthesis is a network of metabolic events, with complex regulatory mechanisms, controlled by external factors (light) and internal factors (substrates and effectors).

Complex metabolic systems adjust to conditions, optimizing the supply of materials and energy to the demands for the products, both within and between the different parts ('subsystems') of the whole system. Light-harvesting and electron transport and the enzymatic reactions of the PCR cycle are subject to complex controls, both feedback and feed-forward, that is, the product of a reaction modifies the processes leading to further reactions, or stimulates later ones, so increasing the overall rate. The modification may be both stimulation or inhibition. Such processes are non-linear functions of particular inputs to the system and show complex responses, which appear to provide long term metabolic stability over a wide range of conditions in relation to growth, reproduction, etc. In complex systems several factors may interact, each being close to the 'control point' at which the system is 'designed' to work or, in the case of plants, selected in evolution to function optimally. Interaction between several processes provides subtle control of the rate of the overall process. As conditions within the system or external to it alter so one or other of the control processes regulates until the system returns to a long term equilibrium, with perhaps somewhat different factors controlling the overall rate, yet each still close to the optimal value.

In the very short term (from fractions of seconds to hours) assimilation is controlled by the supply of light, CO_2, etc., as well as by enzyme amount and kinetics. The overall rate is governed in the long term by the amount of components of the system, for example, of electron transport carriers in thylakoid membranes and the amount of membrane or Rubisco. These are determined by conditions during the growth and development of the leaf which change its structure and composition and capacity for photosynthesis. As an example, inadequate nitrate supply impairs synthesis of enzymes (e.g. Rubisco) and therefore CO_2 assimilation. It also inhibits formation of structural proteins and pigments, and therefore decreases the formation of thylakoids, slowing electron transport and light-harvesting. Deficient phosphorus supply inhibits assimilation by decreasing inter-mediary metabolism and the formation of assimilates. Iron deficiency inhibits processes in which iron is an electron carrier, for example, in the Rieske Fe−S centres in thylakoids.

Photosynthesis is not generally a linear function of nutrient supply; deficient plants may grow much less but maintain those leaves which are formed with sufficient and balanced nutrient to allow a viable system to develop and therefore permit assimilation to take place at quite high capacity. However, such leaves may senesce rapidly as materials are remobilized to younger organs. This shortens the assimilation life of the leaf and, with severe deficiencies, impairs assimilation. Long term effects of nutrition and other conditions on development of the photosynthetic apparatus and its function are not well understood, despite the agronomic importance.

Control of the rate is a property of several or many steps ('co-limitation') of the whole system so that an optimum performance is achieved, where most efficient use of energy, substrates, etc. occurs (see Woodrow *et al.* 1990; Stitt 1991). These processes may operate in such a way that control is distributed widely. One set of processes limits in a given environment but as conditions change control shifts to other processes. A leaf optimized to some average light, for example, may be considered to have balanced electron transport and carboxylation processes; on transfer to weak light this balance is impaired by energy limitation and so electron transport related processes become the limiting step, e.g. RuBP supply restricts CO_2 fixation and Rubisco is not fully used. Conversely very bright light would result in large rates of electron transport and RuBP synthesis and Rubisco would then become limiting.

Light response curves (Fig. 12.1) may be interpreted in this way as the balance between RuBP synthesis and Rubisco activity. In darkness the RuBP pool is small and activity of Rubisco is very small, perhaps reduced by nocturnal inhibitors. Increasing irradiance stimulates electron transport, RuBP content rises and Rubisco activation also and CO_2 assimilation increases. However, during this phase RuBP concentration may rise to over four times the concentration of Rubisco enzyme sites so RuBP regeneration may not be the determining factor. Further increase in irradiance results in a decrease in c_i and a situation is reached where Rubisco activity, electron transport and CO_2 supply are in dynamic equilibrium. At saturating light, electron transport may result in a large rate of RuBP synthesis and the Rubisco activity may limit photosynthesis. Thus, control has passed from RuBP regeneration to carboxylation by Rubisco. Leaves operating under normal conditions tend to have c_i lower than c_a, Rubisco is fully activated but not working at full capacity because of CO_2 limitation, and electron transport is possibly close to the condition where photochemical quenching and non-photochemical quenching are poised so that energy dissipation and potential damage to the system may be minimal. Such concepts are supported by data and quantitative models of the photosynthetic mechanism and its functions (Farquhar and von Caemmerer 1982).

A consequence of this type of regulation is that the leaf can respond to increased CO_2 and has considerable flexibility in utilization of energy, yet is not rigidly linked to any one particular growth condition. Hence, it is also possible that changing the proportion of components may not result in a direct response of photosynthesis. An example is that leaves grown with very large nitrogen supply often contain a large amount of Rubisco which does not result in significantly increased photosynthesis. However, given suitable light and CO_2, this additional enzyme still increases carbon assimilation compared to leaves with deficient Rubisco with potential advantage to the plant. Direct demonstration of the non-linearity between components of the system and its function is shown (Stitt 1991) by using antisense DNA to block expression of the rbcS gene for Rubisco small subunit. A decrease of up to 50 per cent of Rubisco achieved by this method does not reduce photosynthesis in light of 340 μmol m^{-2} s^{-1} at 30 or 100 Pa CO_2 but at 10 Pa CO_2 photosynthesis is regulated by Rubisco (Stitt 1991). Analysis of such systems requires application of control theory which will account for the changing responses to modifications of the system. For example, control of the flux, F, through a given pathway by a particular enzyme, E, may be expressed as a control coefficient, C, which expresses the fractional change (dF) in F, caused by a fractional change (dE) in E. In the example, modified Rubisco content at low CO_2 gave a value of C = 1, at 33 Pa CO_2 $C \rightarrow 0.1$ and at 100 Pa CO_2 $C \rightarrow 0$. Under bright conditions greater control might be exerted by Rubisco at higher CO_2 partial pressures. Thus, photosynthesis operates for plants grown in particular conditions close to the point where several factors operate, i.e. at the transition between RuBP regeneration and Rubisco limitation. Long-term adjustments to conditions may then involve regulation of the capacity of particular steps in the process rather than changes to the activity of existing amounts of components. The genetic constitution provides the framework within which the system functions, subject to environmental constraints. Understanding these regulatory processes and links to the expression of genes is an exciting area of photosynthesis (Kacser and Porteous 1987).

Metabolic models of photosynthesis

Light and CO_2 response curves may be analysed using models that combine electron transport and synthesis of NADPH and ATP with models of enzyme characteristics to calculate RuBP synthesis. This is related to the characteristics of Rubisco and the equations governing CO_2 supply to

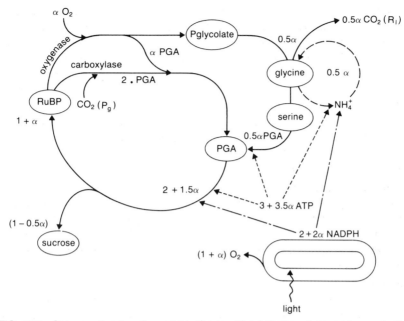

FIG. 12.6 Scheme of carbon fluxes (a) in the combined photosynthetic carbon reduction and photorespiratory carbon cycles in a C3 leaf. P_g is gross photosynthesis, R_1 the rate of photorespiration, α the RuBP oxygenase/carboxylase ratio which depends on the O_2/CO_2 ratio at the Rubisco active site. This Figure should be compared with Fig. 8.4 which shows the glycolate pathway. The stochiometry of the carbon fluxes is indicated. (See Farquhar and von Caemmerer 1982.)

show what determines the dynamics of assimilation with changing environmental conditions. Biochemical processes are abstracted into mathematical equations, describing the biochemistry as a function of the conditions, to simulate the photo- and carboxylation efficiencies of leaves: the conclusion is that photosynthesis of C3 leaves is controlled by carboxylation at low CO_2 in air but by the rate of RuBP synthesis at high CO_2 partial pressures. The equations describe assimilation in relation to light (photon flux), I, CO_2 and O_2 partial pressure, C and O respectively, and temperature, which is not further discussed here. As photosynthesis is driven by light the modelling starts with the photochemical processes. Rate of electron transport, J, from water to $NADP^+$ increases linearly with increasing I until the capacity of the system to transport electrons is saturated at J_{max} (as observed experimentally). A relation between J_{max} determined on isolated thylakoids and the *in vivo* rate enables an estimate of J to be made for different I.

As J_{max} is principally a function of e^- and H^+ transport through the plastoquinone pool, so shade-grown plants, with less PQ and reduced electron transport capacity compared to those grown in bright light, have smaller J_{max}. Thus photosynthesis of low-light plants may be limited by electron transport in bright light. NADPH is produced by reduction of $NADP^+$ by 2 e^- and depends on the concentration of the acceptor and the $NADP^+$-ferredoxin reductase enzyme. With adequate acceptor and enzyme the rate depends on J, but as the concentrations decrease so the rate is determined by the smallest component. Measurements suggest that enzyme activity is unlikely to limit NADPH production.

ATP production is more complex, because three modes of electron flow coupled to photophosphorylation provide flexibility. In linear electron flow, J and ATP synthesis are related, as $3 H^+$ are consumed per ATP and $2 H^+$ are transported into the thylakoid per electron. In cyclic photophosphorylation one electron gives $1 H^+$ and synthesis of 1 ATP is related to one-third of cyclic electron flow, J_c, as $3 e^-$ gives $3 H^+$ and 1 ATP. ATP synthesis depends on the concentrations of ADP and CF_1 enzyme active sites; as with the rate of NADPH synthesis the rate of ATP production depends on the minimum value of J, ADP and CF_1. Evidence suggests that there is adequate enzyme capacity with the exception of plants grown in low light exposed to bright light. A large concentration of ATP + ADP is needed to maintain large rates of RuBP synthesis by ribulose-5-phosphate kinase. Thus RuBP regeneration depends on the efficiency of the PCR cycle, the amount of PGA available and the supply of NADPH and ATP which are a function of light.

The rate of photosynthesis is also dependent upon the behaviour of Rubisco. The rate of RuBP carboxylation, A, when the enzyme is saturated with RuBP, depends on CO_2 and O_2 and the enzyme characteristics only, given by Farquhar and von Caemmerer (1982) as:

$$A = V_{c_{max}} \times \frac{C - \Gamma^\star}{C + K_c (1 + O/K_o)} - R_d \qquad [12.2]$$

where $\Gamma^\star$ is the CO_2-light compensation value and is equal to $(0.5 V_{o_{max}} K_c O)/(V_{c_{max}} K_o)$. It is related to α, the oxygenase to carboxylase ratio by $2 \Gamma^\star/C$. C and O are partial pressures of, and K_c and K_o are Michaelis–Menten constants for CO_2 and O_2 respectively; R_d is 'dark' respiration in the light and $V_{c_{max}}$ and $V_{o_{max}}$ are velocities of RuBP carboxylase and oxygenase, respectively. When CO_2 pressure is low the enzyme is RuBP saturated; removing O_2 stops competition with CO_2 and increases assimilation. However, in high CO_2 *in vivo*, particularly with small photon flux, the supply of RuBP may be insufficient to saturate all the enzyme sites and so A depends on the rate of electron transport and therefore light, as it clearly does, not only on the partial pressure of CO_2 and O_2 as eqn (12.2 above) suggests. By combining the equation giving the RuBP-saturated carboxylation rate with that describing the unsaturated rate, the CO_2 partial pressure at which the transition from RuBP saturation to RuBP limitation occurs may be determined. There are several factors regulating it, e.g. light, capacity of the electron transport pathway, photorespiration. The transition occurs at higher CO_2 in bright light than dim, with low compared to high O_2.

Leaves in air have an internal CO_2 partial pressure dependent on photosynthetic rate and stomatal conductance. The transition from RuBP limitation to RuBP saturation occurs close to the normal CO_2 value in the leaf, 25−27 Pa CO_2. This is below ambient with the CO_2 in the chloroplast probably at 15−16 Pa. Possibly there is a balance, determined genetically and therefore subject to evolutionary selection pressure, between the use of light and CO_2, the loss of water (determined by stomatal conductance) and the costs of making the leaf and its biochemical components. The photosynthetic system seems to be co-limited by many factors all operating close to an optimum for the system, if not for the individual components.

Leaf composition and assimilation

To determine what steps in the photosynthetic process limit the overall rate, the maximum rates of partial process may be calculated from their characteristics, as far as they are known. The approach is only semi-quantitative and uses values determined *in vitro* which may not be applicable *in vivo*. However, it is a useful illustration of how the system functions. The maximum rates of processes are calculated from the amount of pigments or enzyme per m^2 of leaf (this basis is adopted because it determines the interception of light in most plants), from the number of enzyme active sites, and the number of moles of substrate converted in unit time. This is a function of the binding characteristics of the enzyme sites, that is, rates of binding substrate, of reaction and of release of products (the turnover time (measured in seconds; the reciprocal is the turnover number, s^{-1})). For 1 m^2 of leaf:

$$\frac{\text{mol substrate}}{\text{converted } s^{-1}} = \frac{\text{mol}}{\text{enzyme}} \times \frac{\text{active enzyme}}{\text{sites (mol enzyme)}^{-1}} \times \frac{\text{mol substrate}}{\text{(mol active sites)}^{-1} s^{-1}}$$

[12.3]

Complex parts of the system, as well as single enzymes, may be treated in this way by substituting the measured amounts of the components and the rate at which reactions occur.

Light absorption and transduction

Chlorophyll captures photons and energy is transferred at rates much greater than 10 events s^{-1}. A leaf containing 560 μmol chlorophyll m^{-2} (see Table 4.1 (Ch. 4)) could trap 5600 μmol photons $m^{-2} s^{-1}$ or twice the number of PAR photons incident upon a leaf at noon in the brightest environments. With a photon flux of 1500 μmol quanta $m^{-2} s^{-1}$ and with efficiency of 3 photons per e^- (allowing for inefficiency in capture of some wavelengths) the potential rate of e^- transport would be 500 μmol $m^{-2} s^{-1}$. With 4 e^- per CO_2 (this efficiency is about 12 photons per CO_2 fixed) the rate of CO_2 fixation would be 100 μmol CO_2 $m^{-2} s^{-1}$ or twice the rate observed for C3 plants under ideal conditions (50 μmol CO_2 $m^{-2} s^{-1}$). This light capture is not the limiting factor in assimilation at high photon flux. Even leaves 50 per cent deficient in chlorophyll because of Mg^{2+} or nitrogen shortage can capture sufficient photons in normal light for photosynthesis, but more severe chlorosis limits assimilation. Increasing the chlorophyll content of leaves, e.g. by large application of nitrogenous fertilizers, may not increase light capture. Shade plants have less chlorophyll per unit area of leaf than sun plants but more per reaction centre, so allowing greater efficiency of light capture but restricting the capacity of the photosynthetic apparatus. If the number of reaction centres of PSII and PSI is each about 1 per 300 chlorophylls, then with 560 μmol chlorophyll m^{-2} leaf this is approximately 2 μmol reaction centres m^{-2}. If turnover time is between 100 and 1000 s^{-1}, 200–2000 μmol m^{-2} s^{-1} of electrons could enter the transport chain; the lowest rate is probably an underestimate as turnover time is much faster. With 4 e^{-} needed per CO_2 the potential rate is between 50 and 500 μmol CO_2 m^{-2} s^{-1}, up to 10 times greater than the maximum rates of photosynthesis so the concentration of reaction centres probably does not limit photosynthesis. Reaction centre concentration is greater in bright-light adapted plants, allowing faster photochemistry; photosynthesis of shade plants may be limited in bright light by the concentration of reaction centres.

Electron transport is rate limited by the transfer of electrons via plastoquinone (p. 81). Leaves contain about 10 mol of PQ per mol of PSII and PSII reaction centres, that is 20 μmol m^{-2} leaf. An electron passes from PQ to PSI in about 20 ms, a turnover number of 50 s^{-1}, giving a potential rate of 20 μmol PQ m^{-2} $\times$ 50 s^{-1} or 1000 μmol e^{-} transferred m^{-2} s^{-1}. With 4 e^{-} required per CO_2, assimilation is potentially of the order of 250 μmol CO_2 m^{-2} s^{-1}, five times greater than the fastest observed rates in good conditions and unlikely to be limiting. Plants grown in bright light have more PQ and other electron transport components per unit area than those from shade; this is probably an adaptive mechanism to increase the rates of electron transport. In bright-light grown *Atriplex triangularis* the capacity of photosystem-driven electron transport was 3–4 times that of dim-light grown plants, but the amount of PQ and photosystem increase by only 1.3 times. It is not understood why leaves do not form excess PQ and electron transport components under all conditions. Perhaps the energetic and material costs are too great in dim light, whereas in bright light the extra energy capture and photosynthesis are greater than the investment in photosynthetic machinery.

Rate of NADPH synthesis

The rate of e^{-} transport controls NADP^{+} reduction if NADP^{+} supply is adequate. The concentration of NADPH is about 0.1 mM in leaves and if all is in the chloroplast stroma (volume 1.4 $\times$ 10^{-5} m^{3} m^{-2} leaf) the amount per unit area of leaf is 1.4 μmol m^{-2}. With a maximum rate of assimilation of 50 μmol CO_2 m^{-2} s^{-1} and with 2 NADPH per CO_2 assimilated the amount of NADPH required is 100 μmol m^{-2} s^{-1} and the turnover time will be about 16 ms. Reduction of $NADP^+$ requires 2 e^- and the potential rate of electron transport is 20 mmol $m^{-2} s^{-1}$ so the potential is 10 mmol of $NADP^+$ reduced $m^{-2} s^{-1}$. With 1500 μmol quanta $m^{-2} s^{-1}$ or 500 μmol e^- $m^{-2} s^{-1}$ this rate of NADPH synthesis is $2\frac{1}{2}$ times greater than that required for observed CO_2 fixation.

However the rate of $NADP^+$ reduction is possibly much slower as it is an enzymatic reaction. The concentration of NADP reductase in the stroma is 80 μmol or about 1 μmol m^{-2}; with a turnover time of 1 ms (1000 s^{-1}, rather fast for many complex enzymes) and 2 e^- per NADPH, the amount of $NADP^+$ reduced would be about 500 μmol $m^{-2} s^{-1}$ or about 250 μmol CO_2 $m^{-2} s^{-1}$, greatly in excess of the observed values. A less realistic turnover rate of 10 ms (i.e. turnover number 100 s^{-1}) would give a rate of 25 μmol CO_2 $m^{-2} s^{-1}$, considerably less than the highest rates of photosynthesis measured. However, with rapid turnover and a high concentration of the enzyme, NADPH synthesis is unlikely to be a major limitation in assimilation.

Photophosphorylation

The potential rate of ATP synthesis may be calculated from the enzyme complement and proton transport. Estimates of CF_1 are 0.42 mg CF_1 mg chl^{-1} or about 1 mol CF_1 per 840 mol chl in bright light; there may be less in dim light. With 0.56 mmol of chlorophyll m^{-2} of leaf there is 0.67 μmol CF_1 m^{-2} of leaf. With a calculated turnover time of 4 ms (turnover number 250 s^{-1}) synthesis of ATP will be $0.67 \times 250 = 168$ μmol ATP $m^{-2} s^{-1}$. Three ATP are required for assimilation of 1 CO_2 so the rate of CO_2 assimilation would be $168/3 = 56$ μmol $m^{-2} s^{-1}$, close to the maximum rates of CO_2 assimilation in C3 leaves. This is an upper limit to the rate, for fewer CF_1 per chlorophyll would decrease the rate of assimilation. Also ATP is required for other processes. Allowing one extra ATP the rate of synthesis is decreased to 42 μmol $m^{-2} s^{-1}$, inadequate for rapid photosynthesis. Some estimates give more CF_1 per unit of chlorophyll, increasing the potential rate of ATP synthesis. However estimates of the *in vitro* rate of photophosphorylation (cyclic or non-cyclic) are small (250 μmol ATP $(mg chl)^{-1} h^{-1}$ equivalent to 35 μmol ATP $m^{-2} s^{-1}$) and cannot match the required rates of 150 μmol ATP $m^{-2} s^{-1}$ for high rates of photosynthesis. Neither electron flow nor H^+ transfer limit ATP synthesis. With 2 mmol e^- $m^{-2} s^{-1}$ and 2 H^+ per e^-, the potential H^+ flux is 4 mmol H^+ $m^{-2} s^{-1}$ and with 3 H^+ per ATP gives 1.3 mmol ATP $m^{-2} s^{-1}$. A requirement of 3 ATP per CO_2 would give about 400 μmol CO_2 $m^{-2} s^{-1}$, greatly in excess of the observed rate. It is the enzyme reaction rate which limits synthesis of ATP. Shade plants have a smaller CF_1/chlorophyll ratio than sun plants and their photosynthesis saturates at smaller rates; with 1 CF_1/1800 chlorophylls the maximum rate of ATP synthesis would be more than halved and also the maximum rate of photosynthesis, as observed. Therefore, given the amount and turnover time of CF_1 in leaves, the rate of ATP synthesis is a greater potential limitation on the overall rate of assimilation than light capture, electron transport or NADPH synthesis.

Rate of RuBP synthesis

Calculating the rate of activity of PCR cycle enzymes is complex as turnover times and amounts are not well defined, but several control points have been identified. Fructose bisphosphatase and sedoheptulose bisphosphatase are two main control points regulated by light. Rates of these enzyme reactions are sufficient for fast photosynthesis. Activity of phosphoribulokinase and other PCR cycle enzymes probably does not limit assimilation as rates are in excess of the measured rates of photosynthesis.

Carboxylation and oxygenation by Rubisco

Rubisco has a low affinity for CO_2 (12–20 μM for C3 plants), is competitively inhibited by O_2 and the amount of enzyme is important; concentration is 0.5 mM (giving the very considerable concentration of 275 g protein litre^{-1}) or 7 μmol Rubisco per m^{-2} leaf. Each mole of Rubisco has eight potentially active sites giving 56 μmol active sites m^{-2} leaf or 4 mM concentration. The enzyme has a turnover time of 0.5 s (turnover number 2 s^{-1}) so the potential rate of carboxylation is 112 μmol CO_2 m^{-2} s^{-1}, greater than the maximum rates of CO_2 assimilation under the best conditions, but potentially limiting. However, not all sites are active and this decreases the potential activity under normal atmospheric conditions. More importantly the CO_2 concentration in leaves is about 10 μM so the maximum rate of carboxylation will be smaller than the potential, similar to measured rate of P_n (net rate of photosynthesis). Decreasing O_2 decreases R_1 and the competition for RuBP and increases assimilation by *c*. 40 per cent so the limitation to photosynthesis is in the enzyme characteristics. The high rate of carboxylation is achieved at the expense of very large concentrations of protein and consequently demands much energy and nitrogen.

These simple calculations suggest that rates of ATP synthesis and Rubisco activity are potentially limiting under different conditions. At large photon flux and with high CO_2 and without photorespiration ATP may be limiting; some C4 chloroplast modifications (e.g. agranal chloroplasts) may increase ATP synthesis. Even with photorespiration ATP may limit, as it is consumed for synthesis of RuBP which is not used for CO_2 fixation.

Analysis of carbon dioxide and light response curves

It is useful to reconsider the response curves in the light of the previous discussions. Without CO_2, C3 plants in 21 kPa O_2 and bright light photorespire carbon from storage. Electron transport to $NADP^+$, and ATP and RuBP synthesis are limited by the rate of turnover of the $NADP^+$ and ATP pools; the electron transport chain and NADP are probably highly reduced and energy charge high. Long periods under these conditions may damage leaves if excessive energy is not dissipated, particularly if R_1 (photorespiration) is prevented. At the CO_2 compensation concentration α is 2 and the flux of CO_2 from the glycolate pathway equals P_g. With further increase in CO_2 concentration P_g increases and also rates of NADPH, ATP and RuBP synthesis, but these are still in excess of demand and the carboxylation reaction is RuBP saturated but limited by CO_2. As CO_2 concentration increases further, the enzyme becomes more CO_2 but less RuBP saturated, and P_n increases slowly as RuBP synthesis limits assimilation under CO_2 saturated conditions. Carbon dioxide supply is limiting when the carboxylation rate ('demand function') exceeds the rate of CO_2 supply ('supply function') and internal CO_2 concentration decreases. Even with large stomatal conductance internal CO_2 concentration is smaller than ambient CO_2 in air, and P_n is limited by CO_2 supply in bright light. Possibly this enables the photosynthetic system to function at optimum efficiency without large fluctuations in rate.

Net photosynthesis is affected by temperature through RuBP synthesis and the carboxylation reaction. At low temperatures (below 5 °C in many C3 and 10 °C in C4 plants) RuBP synthesis is limiting. It increases with warmer temperatures. However, P_g of C3 plants in air is offset by increasing R_1. Comparison of P_n versus internal CO_2 response curves for leaves under different conditions or between species enables the causes of decreasing P_n to be analysed. Decrease in P_{max} indicates deficient synthesis of RuBP, and change in the initial slope of the relationship shows altered carboxylation efficiency, because of inadequate amount of RuBP carboxylase, for example. Measurement of conductance in relation to external CO_2 supply shows the limitation imposed by diffusion processes. The CO_2 response curves for C4 leaves may be analysed in the same way but understanding of the relationships is not as advanced as for C3 leaves.

Compartmentation in photosynthesis

Photosynthetic metabolites in the cell occur in particular compartments or 'pools' which may be in solution in chloroplasts or other organelles or bound to enzymes. Influx and efflux of metabolites determines the amount of material in the pools. Information about the size of pools and fluxes is essential for the analysis of assimilation in relation to environment, genotype, etc. Radiotracers (e.g. ^{14}C) may be used to analyse the size of pools, and the direction and magnitude of the fluxes, if the total radioactivity and specific radioactivity (SA = (^{14}C in pool/total C in pool)) of materials in the pools is known. SA is determined experimentally by feeding $^{14}CO_2$ for a given time, rapidly killing the tissue, extracting and separating the metabolites and measuring the amount of ^{14}C and ^{12}C and their ratio. Analysis of photosynthetic carbon fluxes may be made with leaves photosynthesizing with $^{12}CO_2$ at the steady rate (i.e. constant P_n with time) and then rapidly supplying $^{14}CO_2$ of known SA without changing P_n. In the first seconds of labelling the ^{14}C in 3PGA increases and, after a period which depends on the rates of flux into and out of the pools and on the pool size, saturates at, or close to, the SA of the feeding gas (i.e. 3PGA has an SA relative to the feeding gas of 1). With time, metabolites and storage compounds derived from 3PGA acquire label. Changes in the SA of compounds with time show the metabolic sequences. Rates of flux from one compound to another may be calculated from:

$$\text{flux per unit time} = \frac{1}{\text{SA of precursor}} \times \frac{1}{\text{time}} \times {}^{14}C \text{ in product} \quad [12.4]$$

It is possible, for example, to show that the flux of carbon into glycine of the glycolate pathway in leaves is much greater in air than in 1 kPa O_2, that glycine exists in two pools (at saturation the SA of glycine is less than that of the feeding gas) and that serine in leaves is not derived exclusively from the glycolate pathway. Mathematical analysis of compartmental systems is sophisticated and provides a powerful tool for understanding photosynthetic metabolism and factors controlling it.

Leaf fluorescence and photosynthesis

The relation between variable chlorophyll *a* fluorescence emission from the PSII complex and the biochemical and other mechanisms which are responsible for the changes in the fluorescence signal is discussed in Chapter 5. Analysis of the transient changes with time in fluorescence (Kautsky curves), measured experimentally on dark-adapted intact leaves, has proved a very valuable way of understanding the relation between light harvesting, electron transport, thylakoid energetics and the processes of CO_2 fixation and oxygen evolution (Lichtenthaler, 1988; Walker and Osmond, 1989). A fluorescence induction curve is shown in Fig. 5.9 and interpreted there; the transients are related to the rate constants of the electron transfer steps and to the rates of formation of O_2 and pH in the thylakoids and to enzyme activation and the onset of CO_2 assimilation. With only Q_A^- present, fluorescence is maximum, F_m; this may be determined by blocking electron transport at Q_A with herbicides such as DCMU (p. 99) or by giving a very powerful light pulse to saturate electron transfer. The latter is an advantage as it is reversible and can be repeated on a physiologically normal system. The difference $F_m - F_0$ (the basal fluorescence of the system) is the variable fluorescence, F_v, and F_v/F_m expresses the efficiency of exciton capture by 'open' reaction centres of PSII and is very constant (0.83) for intact, unstressed leaves of a wide range of plants, equivalent to $9-10$ quanta per O_2 absorbed and to 0.85 electrons transported through PSII if the two photosystems receive equal proportions of excitons. Environmental factors, such as water stress affect the availability of CO_2 or metabolic processes and decrease F_v/F_m and application of herbicides also. Inverse relationships are observed between photochemical reactions, e.g. CO_2 assimilation, and fluorescence. However, because of the complex nature of electron flow, its coupling to photophosphorylation and the different uses made of the electrons (only a proportion are consumed in CO_2 reduction), fluorescence is not directly related to any one process. The behaviour of fluorescence depends on changes at different sites in the thylakoid, e.g. the inhibitor phlorizin increases fluorescence by blocking H^+ flow through CF_0-CF_1 and slows electron transport and therefore CO_2 assimilation, yet if the trans-thylakoid ΔpH is destroyed by uncouplers, e.g. gramicidin, then electron transport may increase (provided acceptors are available) so decreasing fluorescence together with CO_2 assimilation, which cannot proceed because of ATP deficiency.

Although the Kautsky fluorescence induction curve gives much valuable information on thylakoid energetics, the method is limited to the induction period because the leaf has to be adapted to darkness and the analysis of energetics and routes of energy dissipation thus applies to conditions quite different from the illuminated condition with active, steady-state photosynthesis. Also it is difficult to study processes as a function of the

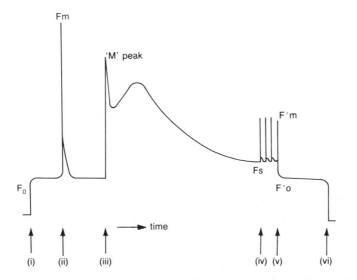

FIG. 12.7 Fluorescence transient following illumination of a leaf which has been in darkness. A typical modulated fluorescence induction curve, showing (i) modulated light on after darkness, (ii) a pulse of saturating light, (iii) actinic light on, (iv) four pulses of saturating light at 15 s intervals at steady state fluorescence (F_s), (v) actinic light off and far-red light on, (vi) all illumination off. The fluorescence parameters shown are discussed in the text together with their use in describing thylakoid energetics and quenching coefficients.

actinic light energy, thus restricting studies to constant irradiance. Also wavelengths of the actinic light which interfere with the fluorescence detection, particularly with strong background illumination, must be removed.

Modulated fluorescence techniques allow measurement under normal actinic light intensities with steady-state photosynthesis. The method uses a weak pulsed (100 Hz) measuring beam (685 nm) from light-emitting diodes. The fluorescence induced is measured in phase with suitable electronic circuitry and amplifiers. This permits the fluorescence signals to be measured independently of the actinic light source and thus show the effects on the system of the normal conditions. In Fig. 12.7 the course of modulated fluorescence during induction and the measurements of fluorescence parameters under steady-state assimilation conditions is shown. The F_0 and F_m values are the fluorescence yields for the dark-adapted leaf and F_s steady-state fluorescence in actinic light and F_0' and F_m' are for the light-adapted condition. By illuminating the leaf with a brief (2 s) burst of white light (20 mmol quanta m^{-2} s^{-1}) to saturate the light harvesting system and close all PSII reaction centres (Q_A fully reduced) the value of F_m under fully activated photosynthetic conditions (F_m') may be obtained without the use of chemical inhibitors of Q. A value of F_0 (F_0') may be obtained for the same conditions by stopping the actinic light and giving a burst of weak far-red light, which allows electrons to flow from Q to PSI in the fully oxidized state.

A number of important fluorescence parameters may be calculated, e.g. F_v/F_m, which is related to the efficiency of excitation energy capture (photon yield) of open PSII reaction centres (φ_e). The routes by which the signal is 'quenched' may be determined as described in Chapter 5; the two main components (although they are related because both are determined at the same time by the rate constants of all the processes using the excitation energy; Havaux *et al.* 1991) are photochemical quenching, q_P; and non-photochemical quenching, q_N. The former is related to CO_2 assimilation, nitrate reduction, etc. The latter has several components, including the energy-dependent quenching, q_E, which is related to the proton concentration in the thylakoid lumen. This may lead to increased thermal de-excitation and dissipation of excitons by carotenoid epoxidation reactions. Some 70 per cent of variable fluorescence is quenched by q_E when the PSII reaction centres are closed in the fully light adapted state. Other processes quenching fluorescence are slower mechanisms such as the state transition, q_T, which involves energy transfer ('spill-over') from PSII to PSI associated with phosphorylation of the light harvesting complexes, and photoinhibitory quenching, q_I, related to damage at the reaction centres.

To determine q_P and q_N during photosynthesis in normal light the difference between the actual modulated fluorescence at steady state (F_s) and the maximum variable fluorescence from a saturating light pulse (F_m')

applied to the leaf at F_s is a measure of the proportion of the traps which are open. For q_p:

$$q_P = \frac{F'_m - F_s}{F'_m - F'_0} = \varphi_{PSII}/\varphi_{PSII(max)} \qquad [12.5]$$

so q_P approaches 1 as the PSII centres open and this is the upper limit for the amount of photochemical work that the system could do.

References and Further Reading

Alscher, R.G. and **Cumming, J.R.** (eds) (1990) *Stress Responses in Plants: Adaptation and Acclimation Mechanisms*, Wiley-Liss, New York.

Atkins, G.L. (1969) *Multicompartment Models in Biological Systems*, Methuen and Co., London, Halsted Press, Division of John Wiley and Sons, New York.

Baker, N.R. (1991) The relationship between photosystem 2 activity and CO_2 assimilation in leaves, pp. 379–89 in Abrol, Y.P., Wattal, P.N., Govindjee, Ort, D.R., Gnanam, A. and Teramura, A.H. (eds), *Impact of Global Climatic Changes on Photosynthesis and Plant Productivity*, Oxford and IBH Publishing Co. PVT. Ltd, New Delhi.

Baker, N.R. and **Long, S.P.** (eds) (1986) *Photosynthesis in Contrasting Environments*, *Topics in Photosynthesis*, Vol. 7, Elsevier, Amsterdam.

Berry, J. and **Björkman, J.** (1980) Photosynthetic response and adaptation to temperature in higher plants, *A. Rev. Plant Physiol.*, **31**, 491–543.

Bowes, G. (1991) Growth at elevated CO_2: Photosynthetic responses mediated through Rubisco, *Plant, Cell Environ.*, **14**, 795–806.

Burke, J.J. (1990) High temperature stress and adaptation in crops, pp. 295–310 in Alscher, R.G. and Cumming, J.R. (eds), *Stress Responses in Plants: Adaptation and Acclimation Mechanisms*, Wiley-Liss, New York.

Coombs, J., **Hall, D.O.**, **Long, S.P.** and **Scurlock, J.M.O.** (eds) (1985) *Techniques in Bioproductivity and Photosynthesis* (2nd edn), Pergamon Press, Oxford.

Evans, J.R. (1989) Photosynthesis and nitrogen relationships in leaves of C_3 plants, *Oecologia*, **78**, 9–19.

Farquhar, G.D. and **Sharkey, T.D.** (1982) Stomatal conductance and photosynthesis, *A. Rev. Plant Physiol.*, **33**, 317–45.

Farquhar, G.D. and **von Caemmerer, S.** (1982) Modelling of photosynthetic response to environmental conditions, pp. 549–87 in Lange, O., Nobel, P.S., Osmond, C.B. and Ziegler, H. (eds), *Encyclopedia of Plant Physiology* (N.S.), Vol. 12B, *Physiological Plant Ecology II*, Springer-Verlag, Berlin.

Geiger, D.R. (1986) Processes affecting carbon allocation and partitioning among sinks, pp. 375–88 in Chronshaw, J., Lucas, W.J. and Giaquinta, R.T. (eds), *Phloem Transport*, Alan R. Liss, New York.

Graham, D. (1980) Effects of light on 'dark' respiration, pp. 526–80 in Davies, D.D. (ed.), *The Biochemistry of Plants*, Vol. 2, *Metabolism and Respiration*, Academic Press, New York.

Havaux, M., **Strasser, R.J.** and **Greppin, H.** (1991) A theoretical and experimental analysis of the q_p and q_N coefficients of chlorophyll fluorescence quenching and their relation to photochemical and non-photochemical events, *Photosynthesis Res.*, **27**, 41–55.

Heath, O.V.S. (1969) *The Physiological Aspects of Photosynthesis*, Stanford University Press, Stanford.

Heath, R.L. and **Preiss, J.** (eds) (1985) *Regulation of Carbon Partitioning in Photosynthetic Tissue*, American Society of Plant Physiologists, Baltimore.

Kacser, H. and **Porteous, J.** (1987) Control of metabolism: What do we have to measure? *TIBS*, **12**, 5–14.

Lambers, H., Neeteson, J.J. and **Stulen, I.** (eds) (1986) *Fundamental, Ecological and Agricultural Aspects of Nitrogen Metabolism in Higher Plants*, Martinus Nijhoff Publishers, Dordrecht.

Lawlor, D.W. (1991) Concepts of nutrition in relation to cellular processes and environment, pp. 1–28 in Porter, J.R. and Lawlor, D.W. (eds), *Plant Growth: Interactions with Nutrition and Environment*, Society for Experimental Biology Seminar Series 43, Cambridge University Press.

Lichtenthaler, H.K. (ed.) (1988) *Applications of Chlorophyll Fluorescence in Photosynthesis Research, Stress Physiology, Hydrobiology and Remote Sensing*, Kluwer Academic Publishers, Dordrecht.

Lorimer, G.H. and **Andrews, T.J.** (1981) The C_2 chemo- and photorespiratory carbon oxidation cycle, pp. 330–74 in Hatch, M.D. and Boardman, N.K. (eds), *The Biochemistry of Plants*, Vol. 8, *Photosynthesis*, Academic Press, New York.

Oquist, G. and **Martin, B.** (1986) Cold climates, pp. 237–93 in Baker, N.R. and Long, S.P. (eds), *Photosynthesis in Contrasting Environments*, Elsevier, Amsterdam.

Osmond, C.B., Björkman, O. and **Anderson, D.J.** (1980) *Physiological Processes in Plant Ecology. Towards a Synthesis with Atriplex*, Springer-Verlag, Berlin.

Pearcy, R.W. (1990) Sunflecks and photosynthesis in plant canopies, *A. Rev. Plant Physiol. Plant Mol. Biol.*, **41**, 421–53.

Prasil, O., Adir, N. and **Ohad, I.** (1992) Dynamics of photosystem II: mechanism of photoinhibition and recovery processes, pp. 295–348 in Barber, J. (ed.), *Topics in Photosynthesis Vol II, The Photosystems: Structure, Function and Molecular Biology*, Elsevier, Amsterdam.

Schuster, G., Shochat, S., Adir, N. and **Ohad, I.** (1989) Inactivation of photosystem II and turnover of the D1-protein by light and heat stresses, pp. 499–510 in Barber, J. and Malkin, R. (eds), *Techniques and New Developments in Photosynthesis Research*, Plenum Press, New York/London.

Šesták, Z., Čatský, J. and **Jarvis, P.G.** (eds) (1971) *Plant Photosynthetic Production, Manual of Methods*, Dr W. Junk, The Hague.

Sharkey, T.D. (1989) Evaluating the role of Rubisco regulation in photosynthesis of C3 plants, pp. 435–48 in Walker, D.A. and Osmond, C.B., (eds), *New Vistas in Measurement of Photosynthesis*, The Royal Society, London.

Stitt, M. (1991) Rising CO_2 levels and their potential significance for carbon flow in photosynthetic cells: Commissioned review, *Plant, Cell Environ.*, **14**, 741–62.

Teramura, A.H. and **Sullivan, J.H.** (1991) Field studies of UV-B radiation effects on plants: Case histories of soybean and loblolly pine, pp. 147–61 in Abrol, Y.P., Wattal, P.N., Govindjee, Ort, D.R., Gnanam, A. and Teramura, A.H. (eds), *Impact of Global Climatic Changes on Photosynthesis and Plant Productivity*, Oxford and IBH Publishing Co. PVT. Ltd, New Delhi.

Terry, N. and **Rao, I.M.** (1991) Nutrients and photosynthesis: Iron and phosphorus as case studies, pp. 55–79 in Porter, J.R. and Lawlor, D.W. (eds), *Plant Growth:*

Interactions with Nutrition and Environment, Society for Experimental Biology, Seminar Series 43, Cambridge University Press.

von Caemmerer, S. and **Edmondson, D.L.** (1986) Relationship between steady-state gas exchange, *in vivo* ribulose bisphosphate carboxylase activity and some carbon reduction cycle intermediates in *Raphanus sativus, Aust. J. Plant Physiol.*, **13**, 669−88.

von Caemmerer, S. and **Evans, J.R.** (1991) Determination of the average partial pressure of CO_2 in chloroplasts from leaves of several C3 plants, *Aust. J. Plant Physiol.*, **18**, 287−305.

Walker, D.A. and **Osmond, C.B.** (eds) (1989) *New Vistas in Measurement of Photosynthesis*, The Royal Society, London.

Woodrow, I.E. and **Berry, J.A.** (1988) Enzymatic regulation of photosynthetic CO_2 fixation in C3 plants, *A. Rev. Plant Physiol. Plant Mol. Biol.*, **42**, 351−71.

Woodrow, I.E., Ball, J.T. and **Berry, J.A.** (1990) Control of a photosynthetic carbon dioxide fixation by the boundary layer, stomata and ribulose-1,5-bisphosphate carboxylase, *Plant, Cell Environ.*, **13**, 339−47.

Young, A. and **Britton, G.** (1990) Carotenoids and stress, pp. 87−112, in Alscher, R.G. and Cumming, J.R. (eds), *Stress Responses in Plants: Adaptation and Acclimation Mechanisms*, Wiley-Liss, New York.

Zeiger, E. (1983) The biology of stomatal guard cells, *A. Rev. Plant Physiol.*, **34**, 441−75.

CHAPTER *13*

Photosynthesis, plant production and environment

Photosynthetic response to the environment, with its complex diurnal and seasonal changes in weather — temperature, radiation, rainfall, etc., plus variation in other environmental conditions, e.g. nutrient supply — determine the gross assimilation of carbon and nitrogen by plants. Great differences exist between environments but there is considerable flexibility in photosynthetic characteristics within and between species, for they evolved under changing conditions and have the genetic potential to adjust, at least partially, to conditions. The ecological behaviour of plants is determined by photosynthetic productivity coupled to use of the photosynthetic products in growth and development. Future increase in atmospheric CO_2 and climate change will require that species respond if they are to survive. However, the extent to which plants can adapt, and the ecological consequences, will depend on many factors, e.g. the particular combinations of CO_2, UV-B radiation resulting from ozone depletion, temperature, and are very hard to predict. By examining the way in which photosynthetic and other processes of plant communities respond to their environment under the present very diverse conditions, the causes of differences in production in different current and future climates may be assessed.

Mass of standing vegetation in different habitats varies widely, from almost nothing in some deserts to hundreds of tonnes of dry matter per hectare in tropical forests, and the rates of dry matter production also differ greatly. Crops of well watered and fertilized C3 cereals in temperate zones have the potential to produce more than 25 t dry matter ha^{-1} in an 8-month growing season and in the tropics sugar cane (C4) forms over 80 t dry matter ha^{-1} $year^{-1}$. All organic matter is derived from photosynthesis and accumulation of inorganic matter in vegetation (usually < 10 per cent of dry matter) requires photosynthetic energy.

The process may be analysed by considering an agricultural crop, such as wheat, sown with the seed at uniform spacing as a monoculture in homogeneous soil. A mature crop effectively covers the soil so the leaf area index (LAI = area of leaf/area of soil) is large, greater than, say, 4, and intercepts 80−90 per cent of the incident radiation. Within the crop the light profile (Fig. 13.1) decreases exponentially from the crop surface to the soil surface, and the relation between absorbent and total energy available is given by

$$I' = I_0 \exp(-KL) \qquad \qquad [13.1]$$

where I' is the light flux at a point in the canopy below a LAI of L, I_0 is the flux above the canopy and K is the canopy extinction coefficient. K depends on crop architecture, e.g. narrow, erect leaves in a cereal crop with light penetration deep into the canopy, result in a smaller value than broad-leaf crops such as cotton for a given LAI. Also, K will change with growth stage, presence of ears or flowers, etc., but is usually very constant for a given crop and period. From a knowledge of K and the incident radiation (or from incident and transmitted radiation) an estimate of the intercepted radiation may be obtained. However, it includes the reflected component, so for a better estimate of the energy flux the net radiation above the crop is measured. Total absorbed energy can be partitioned into photosynthetically active radiation (PAR, 400−700 nm waveband); environmental physics texts such as Monteith and Unsworth (1990) consider the problems.

With an estimate of the energy intercepted by the crop and of the fraction absorbed in the range used for photosynthesis over a given time interval,

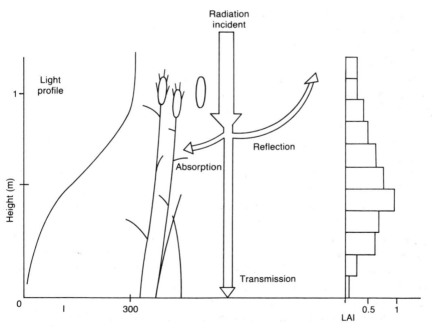

FIG. 13.1   Light interception by the leaf area under (LAI) in layers of a cereal crop and the use of incident photosynthetically active radiation.

the efficiency of energy use in dry matter production may be assessed if the crop production over the same period is known, e.g. by harvesting the plant. The overall efficiency depends not only on the relation of gross photosynthesis to light, CO_2 (but this is relatively constant from day to day and over the life of an annual crop) and temperature but the response of photo- and dark respiration of both leaves and other plant parts to conditions. Over a day/night cycle $P_g - R_l + R_d$) of all components, leaves, roots, stems and fruits, determines the daily net carbon balance and hence the increment of growth.

Total net photosynthesis of a crop over a growing season per unit ground area (m^2) may be considered a function of light energy absorbed per unit leaf area, the response of net photosynthesis to light, total leaf area (m^2) and the number of days of assimilation:

$$\text{Total net photosynthesis } m^{-2} \text{ season}^{-1} = \text{photosynthesis } m^{-2} \text{ leaf day}^{-1} \times$$
$$m^2 \text{ total leaf } m^{-2} \text{ ground} \times \text{day season}^{-1} \qquad \text{[13.2]}$$

Assimilation may be calculated (although difficult to achieve in practice) for an entire growing season by integration of the photosynthetic rate of individual leaves over light, temperature, CO_2 supply, water stress, etc., and then over the entire leaf area for each day. The daily assimilation may then be summed over the season to give total assimilation. Net photosynthesis is a function of light intensity (photon flux), CO_2 concentration in the atmosphere within the vegetation, temperature, water supply, nutrition and the physiological state of the plant, for example, leaf age and reproductive state. To combine knowledge of the partial processes in plant growth in such a way that the total productivity and efficiency of the vegetation can be calculated is still extremely difficult to achieve with very great accuracy. This is desirable for several reasons — the need to predict the production of agricultural crops and the effects of adverse environments on it and the urgent need to assess the likely effects of potential changes in global climate are but two. To achieve this, mathematical simulation modelling is employed. The response of a plant process to particular conditions in the environment is determined experimentally and the relation expressed as a mathematical function. A group of these functions expressing the responses of the most important plant processes are combined into a plant or crop (because most of this work has considered the simplified situation in homogeneous stands and has direct links to profitability of agriculture) or vegetation model. Then, using estimates of environmental factors the productivity may be calculated. An advantage of this approach is that the effects on plant production of combinations of unusual conditions may be assessed, even if experimentally very difficult to achieve, relatively rapidly. A brief example is given. Net photosynthesis of a crop may be calculated if the response of CO_2 assimilation P_n, of a leaf to light flux is known. This may be expressed in several ways; one of the simpler, frequently used functions is the rectangular hyperbola (eqn 13.3):

$$P_n = \alpha I \tau C /(\alpha I + \tau C) - R_d \qquad\qquad [13.3]$$

where α is light utilization efficiency (quantum efficiency) given by the initial slope of the response, τ is leaf conductance to the flux of CO_2, R_d is the dark respiration of the leaf and C is the CO_2 concentration. Values may be assigned to these parameters based on experiments. Assimilation by leaves may still increase at light intensity where the rectangular hyperbola is 'saturated' so in many studies non-rectangular hyperbolae are used; these include a 'curvature factor' which is selected to fit the observed data better.

From knowledge of the incident and transmitted radiation or of K and LAI the radiation incident on a leaf in the canopy can be determined. Using this as input to eqn 13.3 and with the parameters of the model established the net photosynthesis may be calculated for that leaf. The calculation may be repeated for different leaves. Also, responses to other factors may be included. By summing the instantaneous values of assimilation over time, a net assimilate supply is calculated. The total respiratory consumption of carbon may be determined from a temperature response function for the process and subtracted from the assimilate to give a net increment of dry matter over time. The product may then be apportioned (using the best information available) in relation to environmental conditions or physiological state of the crop, into organs, e.g. leaves or reproductive and vegetative organs bearing yield. In this way, basic biological information obtained on the small scale and short term may be used to solve large scale, long-term practical problems.

There is considerable interaction between environment and plant growth which controls P_n (net rate of photosynthesis) and total net photosynthesis; poor nutrition, for example, slows leaf growth and causes early leaf senescence. A smaller leaf area over the life of the crop, even if the rate of assimilation is little affected, decreases total assimilation and growth. Water stress for short periods may decrease stomatal conductance and inhibit P_n more than area, and thereby inhibit growth. Vegetation is subject to many fluctuations in environment which increase or decrease growth 'stresses' (p. 268) and each combination may influence different plant processes, with different quantitative and qualitative effects, although modifying the same plant mechanisms. For each plant species and community, net photosynthesis per unit leaf area is optimal under a particular range of conditions. Differences caused by environment and between species are explicable in terms of the interactions between environment, biochemistry and physiology. Understanding of environmental control of net photosynthesis is developing rapidly, particularly at the level of organization of the leaf, as discussed in Chapter 12.

Light is the driving force for production, and temperature, nutrition and water regulate production; they are also responsible for seasonal variations, with particular combinations of conditions important in different years. Conditions during development of the photosynthetic system modify the structure of light-harvesting, energy-transducing and enzyme systems and thereby alter the efficiency of assimilation. However, although understanding of the mechanisms is developing, quantitative models of the changes in biochemistry and physiology in relation to environment are not well developed, despite the potential importance in understanding the control of productivity in vegetation.

Maximum production of photosynthate depends on the photon flux incident upon the canopy and its absorption by leaves in different layers, and on the efficiency of conversion to assimilate. About 10–15 per cent of incident PAR (photosynthetically active radiation) is reflected and transmitted by leaves and only part of the energy not captured by upper leaves in the canopy is absorbed by other foliage, thus limiting overall efficiency of light capture to *c.* 85–90 per cent of incident PAR. As P_n is saturated at photon fluxes of half or less of full sunlight, particularly in C3 and shade plants, efficiency decreases markedly at intensity above saturation of P_n. Leaves adapted to bright light are more effective at using radiation at the top of vegetation canopies and shade leaves use dim light in the lower canopy. In dim light, assimilation is a linear function of intensity and is at its most efficient. Plants with larger photochemical efficiency in dim light will have greater growth rates for a given light absorption and may be more successful in competition, for example, in dense vegetation or other shady habitats. Natural vegetation canopies are stratified, with the success of different species in occupying niches mainly dependent on their ability to capture light. Plants of very dim light may be unable to adjust to bright light without photochemical damage and some have developed structural and other characteristics, such as the ability to change leaf orientation to light, for protection. Stratification of leaves of different efficiency within a canopy allows very effective light absorption and high productivity. C3 plants are more efficient in dim to intermediate light intensities, whereas C4 are more efficient in bright light. Canopy architecture influences the efficiency of light utilization; thus C3 crops suffering nutrient or water stress produce less leaf than unstressed crops and so intercept less light overall. However, the light intercepted is of higher intensity where assimilation may be higher but the conversion is less efficient. In contrast, C4 crops would absorb less total light with less leaf area but absorb it with higher efficiency. Vegetation does not receive bright radiation at all times. Many habitats are dimly lit for long periods, with clouds and twilight, when the sun's elevation is low. Efficiency in dim light is then of great importance and C3 species would have an advantage.

Oxygen concentration in the canopies of actively photosynthesizing crops is effectively constant. However, CO_2, which is at much lower partial pressure, may decrease substantially, for example, by 20 per cent from 34 to about 28 Pa within dense, vigorously assimilating vegetation, where turbulent transfer of gases is restricted. Internal CO_2 partial pressure will then decrease further, depending on stomatal conductance, C3 crops are more affected than C4 because of the inefficient Rubisco reaction. Stomatal conductance may also decrease due to water shortage, high temperature, etc., and further restrict CO_2 supply and decrease photosynthesis; again the effect is relatively greater in C3 plants.

Response of A to temperature is also very important in determining assimilation (see Ch. 12). Warm conditions favour vegetation with a high temperature optimum for assimilation, for example, C4 over C3 species. Often relatively small differences in temperature have large effects on productivity, particularly when plants approach the limits of their adaptation. Differences between seasons in crop production may be related to interactions between temperature and light; thus C3 crops may produce more in a cool, relatively dimly lit but long-growing season than in a warm, bright but short season. The effects of conditions, both singly and in combination, on growth become more pronounced as they depart from the 'broad optima' to which particular plant species are adjusted. Thus, the productivity and yield stability of maize decreases more the further into temperate zones that the plant is grown, because of the cooler and shorter growing seasons. However, the complex responses of growth and yield to environment involve many biochemical and physiological responses in addition to those of photosynthesis.

Respiration is the main process other than assimilation which determines dry matter production. Total net photosynthesis (eqn 13.2) if calculated from net photosynthesis during illumination, allows for both photo- and dark respiration of leaves in the light, but neglects respiration during the light period for other organs, and for the whole plant in darkness. Plant productivity is, therefore, dependent on respiratory losses throughout the period of crop growth (eqn 13.4):

$$\text{dry matter production} = \text{total net photosynthesis m}^{-2}\text{ ground day}^{-1}$$
$$\times \text{ number of days season}^{-1}$$
$$- \text{ total respiration m}^{-2}\text{ ground day}^{-1}$$
$$\times \text{ number of days season}^{-1} \qquad [\textbf{13.4}]$$

As with photosynthesis the rate of respiration per unit dry mass of different organs must be correctly integrated over total mass, age, and conditions throughout the life of the crop. Respiration is required to provide energy and substrates for all biochemical processes including turnover of cell structure and also formation of new growth, although part of the requirement may come directly from photosynthesis. Respiration is therefore often divided into 'maintenance' and 'growth' components corresponding to the demands of the existing system and to the requirements for synthesis of new cell components. However, although this distinction is a useful concept to analyse the interaction between respiration and plant functions, the type of respiration is identical; only the use to which the products are put differs. It is difficult to distinguish between the two functions of respiration.

Two mechanisms of dark respiration consuming carbohydrates have been identified, coupled in two ways with the electron transport chain in the mitochondria and ATP production. In one, the normal electron transport is coupled to 3 ATP synthesis and in the other, electrons are consumed by an alternative oxidase pathway. Electron flow in this gives only 1 ATP, and 'short circuits' the normal process and is less energetically efficient. It may provide a spill-over mechanism to regulate carbohydrate and reductant energy in plants. Respiration is important as it proceeds throughout each day and uses much assimilate, perhaps 50 per cent of total production. Small rates of respiration, as in shade adapted plants, minimize loss of energy and carbon and thus increase the efficiency of energy use under limited light. Plants adapted to bright light have relatively large respiration; presumably energy and carbon are not limiting factors in their growth. Plants appear to require rapid respiration for rapid growth, and in species adapted to adverse conditions metabolic activity is much lower. Perhaps large respiration is an inevitable consequence of maintaining a large assimilatory capacity. There is evidence that plants, e.g. rye grass (*Lolium*) with small rates of respiration may be selected by breeding, resulting in increased production; this appears easier than selection for enhanced photosynthetic capacity. Combination of low respiration with high capacity for photosynthesis gives the greatest dry matter production in a range of environments. Both processes may be optimized to different conditions so their role in determining total crop production may vary with environment. However, plants are integrated organisms so that the net carbon balance is related to growth processes ('sink demand') in the long term; plant breeding optimizes the most important processes (photosynthesis, respiration, organ growth) in relation to production, epecially yield, and to environment. Selection of single factors may not achieve balance between processes making the plant susceptible to changes in the environment. This might occur if specific processes are targeted, e.g. by molecular biological genetic engineering techniques.

Yield and light use efficiency

Photosynthetic rate is the major determinant of dry matter production; C4 and C3 crops, with average P_n in bright light of 30 and 13 μmol CO_2 m^{-2} leaf s^{-1}, respectively, have growth rates of 22 and 12 g m^{-2} ground day^{-1}. Yield of C4 crops is correspondingly greater than C3; average United States yields of the C3 plants wheat and soybean are 2.5 t ha^{-1} and of rice 5.8 t ha^{-1}, whilst those of the C4 species maize and sorghum are 7.8 and 4.5 t ha^{-1}, respectively. However, in experiments on many crops, P_n measured on individual leaves over short periods (minutes to hours) at high irradiance, has not correlated closely with dry matter production or yield. This apparent anomaly is readily understood as a consequence of the neglect of the factors which contribute to production. Measurements of CO_2

exchange over longer periods and of larger areas of crop have generally agreed better with production. Short-term estimates of P_n, particularly in bright light, give only the contribution of younger, active leaves to total assimilation and neglect the contribution of older, shaded leaves. Leaf area changes in response to conditions rather more than assimilation per unit area, and variation in leaf area may dominate production. Also, the amount of respiration per unit of dry matter and of the total crop is not usually measured. As both assimilation and respiration may change to different extents depending on conditions and are not included in short-term measurements the lack of correlation is perhaps not surprising. Whole crop measurements over longer periods are therefore expected to correlate better with production.

Returning to the concept that photosynthesis 'drives' biomass production, it is clear that it is the efficiency of total net assimilation interacting with the resources from the environment which determines the net productivity of vegetation. Efficiency depends on many genetically determined plant factors which cannot, currently, be determined from first principles. Therefore, the overall efficiency of production must be determined empirically by relating radiation interception to dry matter yield over the same interval. Crop production is closely related to light interception and therefore leaf area and the relation allows the efficiency to be determined. Efficiency of conversion of light to dry matter in cereals and sugar beet is close to 2 g dry matter MJ^{-1} of PAR absorbed (Fig. 13:2). Conversion of CO_2 to crop dry mass depends on the type of organic molecules synthesized and their relative proportion. One gram of CO_2 gives 0.4 g fats, 0.62 g starch and 0.5 g protein; an average value (allowing for the proportions of fats, proteins, etc., in dry matter) is 0.58 g dry matter. Plants producing much oil or protein produce less dry matter than those forming carbohydrates, but the energy yield may be similar. Each gram of CO_2 assimilated is equivalent to an energy content of 38 kJ g^{-1} in fats, 12.6 kJ g^{-1} in proteins and 17 kJ g^{-1} in starch, with an approximate value of $15-20$ kJ g^{-1} in dry matter from the leaves of a range of species. From the known total energy requirement for a given dry matter production and the energy in the crop, the efficiency of energy conversion of crops and natural vegetation can be calculated. For a C3 crop requiring 0.5 MJ g^{-1} dry matter and producing 25 tonnes dry matter per hectare the energy input is 1250 MJ m^{-2} and the energy content (taking the upper value of 20 kJ g^{-1}) is 50 MJ m^{-2}, an efficiency of about 4 per cent. Alternatively, assuming that the average energy of a mole of PAR photons is 0.2 MJ and that 1 g dry matter is equivalent to 1.7 g CO_2 (0.039 mol CO_2), then the quantum yield is less than 0.02 (mol CO_2/mol photons) an efficiency much smaller than the theoretical conversion efficiencies or those obtained under ideal conditions. The conversion efficiency of C4 crops is 50 per cent greater than C3 but still very inefficient in use of light energy.

Differences in total production between crops may reflect differences in

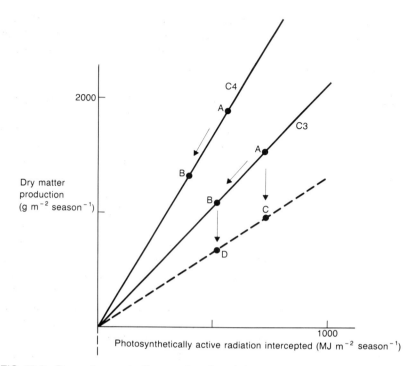

FIG. 13.2   Dry matter production as a function of photosynthetically active radiation for C4 and C3 crops with adequate water, nutrients and temperature (solid lines). Changes in leaf area decrease light interception and move production from A to B. Impaired metabolism caused by stress decreases production at a given light interception (broken line; A → C, illustrated for C3 only). Stresses which decrease both area and efficiency decrease production via A → B → D.

basic efficiency within the C3 and C4 groups, as well as variation in photosynthetic area; this determines the position on curve A (Fig. 13.2). Temperature usually determines the leaf area development of temperate crops and has little effect on efficiency so that productivity moves along the curves according to radiation absorbed. However, under nutrient or water deficiency the efficiency of conversion of light to dry matter decreases at a given light absorption (A→C). As metabolic processes seem to vary less than plant morphology, much variation comes from change in photosynthetic area.

Production and environmental stress

Water use of natural vegetation and crops depends on the atmospheric conditions, for example, humidity and wind speed, radiation load on the leaves and stomatal control. C3 plants have water-use efficiencies of very approximately $1-3$ mg dry matter g^{-1} water transpired, and C4 crops

3−8. These differences are related to the larger stomatal conductances and smaller assimilation rates, particularly under light saturating conditions, of C3 compared to C4 plants. In temperate climates the advantages of C4 plants are minimized but in bright but dry environments, their advantages increase.

Nutrition is of greatest importance to plant production. Nitrogen is essential for synthesis of many components, especially proteins and deficiency reduces the ability of plants to form organs. This applies particularly to leaves which have the largest proportion of the plants nitrogen because of the large content of Rubisco necessary for active CO_2 fixation. There is a strong relation between P_n, N and protein content of leaves. Much of the higher productivity of modern intensive farming comes from the use of industrially fixed N to supply plant needs. In N-deficient soils growth is decreased, with loss of production; constant removal of nutrients (e.g. by farming, grazing) requires that they are replaced otherwise plant production plummets. Natural vegetation has mechanisms for absorbing and retaining N in the plant and ecosystem, e.g. perennial plants absorb, store and recycle N from year to year. Plants adapted to such environments often have small N content, slow rates of assimilation and correspondingly slow biomass production, but accumulate large total biomass over long periods. Also, such vegetation may have enhanced ability to survive poor conditions than plants of N rich environments, e.g. many crops. Similar constraints on production come from deficiencies in all other nutrients if supply does not match demand, e.g. phosphorus and micronutrients are often particularly limiting in many tropical soils.

Responses of photosynthesis and plant growth to nutrients are not necessarily linearly related to the amount or rate of supply of a particular nutrient. In the case of N, photosynthetic rate generally increases as N content of the leaf rises (related to the increased amount of photosynthetic machinery) but at larger content the rate of increase slows; similar responses are seen for organ (e.g. leaf area) growth although growth seems capable of greater response than the composition of the leaf. Thus, the plant processes become 'saturated' and with abundant nutrition efficiency of conversion of light energy to dry matter may increase little and total light capture likewise. However, when N is deficient supplying it may lead to very large increase in both. The supply of N to the plant depends on nitrate reduction, and hence on the supply of reductant, carbon skeletons from photosynthesis and on the amount and turnover of enzymes as well as on NO_3^- supply. Thus, productivity is a complex function of environment and plant characteristics and is not directly related to maximum rates of P_n.

Photosynthesis and yield (i.e. a particular part of a crop required for human use) are indirectly related. Dry matter is distributed, 'partitioned', between harvestable organs (e.g. cereal grains and root tubers) and organs (e.g. cereal straw) which are not consumed. Distribution depends on the number of potential storage sites, their capacity for assimilate and supply of assimilate. In cereals, for example, yield is determined by the number of ears produced per unit ground area, the number of grains per ear and the mass per grain. Conditions which prevent grain or ear formation (nutrient or water stress) decrease yield irrespective of photosynthesis although there are generally some related effects of stress as the parts of the system are very closely integrated. Crop yield is then limited by the storage capacity for assimilate (often called the 'sink' capacity) under some conditions and the rate of production of assimilate ('source' of supply) in others. The relation of assimilation to yield and dry matter production depends on environment and plant characteristics and many processes, which occur at different times and respond in different ways to conditions, are involved. Despite many years of research into the 'source–sink' problem and the great social and economic importance of plant yield surprisingly little is understood of the factors which determine the growth of organs, particularly those bearing yield, so that prediction of the effects of environment on partitioning are largely based on extrapolations from observations and are not understood mechanistically.

Present crop yields, worldwide, are much smaller than potential because of poor nutrition, drought and poor varieties. Better husbandry would increase the food supply for a rapidly increasing world population without increasing intrinsic, or biochemically determined, photosynthetic efficiency. Plant breeding and selection have increased yields largely by improving harvest index (= yield of harvestable material/total dry matter produced). Also, selection of greater yielding crops appears to have decreased photosynthetic efficiency per unit leaf area. Thus, modern cereals have greater leaf area but smaller P_n than older varieties so that total productivity is similar under comparable nutrition, but harvest index is larger. Further improvements in yield may come from selection for larger harvest index but there are limits to the reductions that can be made in support and assimilatory organs. In the longer term, greater photosynthetic efficiency is needed. Genetic manipulation of enzymes (e.g. Rubisco) or of processes (e.g. photophosphorylation) which may limit assimilation may lead to higher efficiency and therefore to increased yield. Increased efficiency of production is, however, unlikely to result from changes in a single metabolic process in P_n and alterations to other, distantly related processes will be necessary.

References and Further Reading

Baker, **N.R.** and **Long, S.P.** (eds) (1986) *Topics in Photosynthesis*, Vol. 7, *Photosynthesis in Contrasting Environments*, Elsevier, Amsterdam.

Boote, **K.J.** and **Loomis, R.S.** (eds) (1991) *Modeling Crop Photosynthesis from Biochemistry to Canopy*, CSSA Special Publication Number 19, Crop Science Society of America, Inc., Madison.

Christy, A.L. and **Porter, C.A.** (1982) Canopy photosynthesis and yield in soybean, pp. 499–511 in Govindjee (ed.), *Photosynthesis*, Vol. II, *Development, Carbon Metabolism and Plant Productivity*, Academic Press, New York.

Coombs, J., Hall, D.O., Long, S.P. and **Scurlock, J.M.O.** (eds) (1985) *Techniques in Bioproduction and Photosynthesis*, (2nd edn), Pergamon Press, Oxford.

Evans, J.R. and **Farquhar, G.D.** (1991) Modeling canopy photosynthesis from biochemistry of the C3 chloroplast in Boote, K.J. and Loomis, R.S. (eds), *Modeling Crop Photosynthesis — from Biochemistry to Canopy*. CSSA Special Publication Number 19, Crop Science Society of America, Inc., Madison.

Gasser, C.S. and **Fraley, R.T.** (1989) Genetic engineering plants for crop improvement, *Science*, **244**, 1293–99.

Gifford, R.M. and **Evans, L.T.** (1981) Photosynthesis, carbon partitioning and yield, *A. Rev. Plant Physiol.*, **32**, 485–509.

Gifford, R.M. and **Jenkins, C.L.D.** (1982) Prospects of applying knowledge of photosynthesis toward improving crop production, pp. 419–57 in Govindjee (ed.), *Photosynthesis*, Vol. II, *Development, Carbon Metabolism and Plant Productivity*, Academic Press, New York.

Gifford, R.M., Thorne, J.H., Hitz, W.D. and **Giaquinta, R.T.** (1984) Crop productivity and photoassimilate partitioning, *Science*, **225**, 801–8.

Jones, H.G. (1983) *Plants and Microclimate*, Cambridge University Press, Cambridge.

Mahon, J.D. (1990) Photosynthesis and crop productivity, pp. 379–94 in Zelitch, I. (ed.), *Perspectives in Biochemical and Genetic Regulation of Photosynthesis*, Alan R. Liss, Inc., New York.

Monteith, J.L. (1977) Climate and the efficiency of crop production in Britain, *Phil. Trans. R. Soc., Lond. B*, **281**, 277–94.

Monteith, J.L. and **Unsworth, M.H.** (1990) *Principles of Environmental Physics*, 2nd edition. Edward Arnold, London.

Neyra, C.A. (ed.) (1985) *The Biochemical Basis of Plant Breeding*, Vol. 1, *Carbon Metabolism*, CRC Press, Boca Raton, Florida.

Nobel, P.S. (1990) *Physicochemical and Environmental Plant Physiology*, Academic Press, San Diego.

Pearcy, R.W. (1990) Sunflecks and photosynthesis in plant canopies, *A. Rev. Plant Physiol. Plant Mol. Biol.*, **41**, 421–53.

Pearcy, R.W., Ehleringer, J.R., Mooney, H.A. and **Rundel, P.W.** (eds) (1990) *Plant Physiological Ecology, Field Methods and Instrumentation*, Chapman and Hall, London.

Portis, A.R. Jr (1982) Introduction to photosynthesis: Carbon assimilation and plant productivity, pp. 1–12 in Govindjee (ed.), *Photosynthesis*, Vol. II, *Development, Carbon Metabolism and Plant Productivity*, Academic Press, New York.

Thornley, J.M. and **and Johnson, D.N.** (1990) *Plant and Crop Modelling*, Clarendon Press, Oxford.

Van Keulen, H. and **Wolf, J.** (eds) (1986) *Modelling of agricultural production*. Weather, Soils and Crops Centre for Agricultural Publishing and Documentation, Wageningen, The Netherlands.

Zelitch, I. (1982) The close relationship between net photosynthesis and crop yield, *Bioscience*, **32**, 796–802.

Index